PREFACE

Concrete is a global material that underwrites commercial well-being and social development. Notwithstanding concrete's uniqueness, it faces challenges from new materials, environmental concerns and economic factors, as well as ever more demanding design requirements. Indeed, the pressure for change and improvement of performance is relentless and necessary.

The Concrete Technology Unit (CTU) of the University of Dundee organised this Congress to address these issues, continuing its established series of events, namely, Creating with Concrete in 1999, Concrete in the Service of Mankind in 1996, Economic and Durable Concrete Construction Through Excellence in 1993 and Protection of Concrete in 1990.

The event was organised in collaboration with three of the world's most recognised institutions: the Institution of Civil Engineers, the American Concrete Institute and the Japan Society of Civil Engineers. Under the theme of Challenges of Concrete Construction, the Congress consisted of three Seminars: (i) Composite Materials in Concrete Construction, (ii) Concrete Floors and Slabs, (iii) Repair, Rejuvenation and Enhancement of Concrete, and three Conferences: (i) Innovations and Developments in Concrete Materials and Construction, (ii) Sustainable Concrete Construction, (iii) Concrete for Extreme Conditions. In all, a total of 350 papers were presented from 58 countries.

The Opening Addresses were given by Mr Jack McConnell MSP, First Minister of the Scottish Executive, Sir Alan Langlands, Principal and Vice-Chancellor of the University of Dundee, Mr John Letford, Lord Provost, City of Dundee, Professor Adrian Long, Senior Vice-President of the Institution of Civil Engineers, Dr Taketo Uomoto, Director of the Japan Society of Civil Engineers and Dr Terence Holland, President of the American Concrete Institute. The Congress had six Opening and six Closing Papers dealing with the main themes of the Seminars and Conferences. Opening Papers were presented by Professor Gerard Van Erp, University of Southern Queensland, Australia Dr Peter Seidler, Astradur Industrieboden, Germany and Professor Kyosti Tuttii, Skanska Teknik AB, Sweden, Professor Surendra Shah, Northwestern University, USA, Dr Philip Nixon, Building Research Establishment, UK and Mr Hans de Vries, Ministry of Transport, the Netherlands. Closing Papers were presented by Dr Gier Horrigmoe, NORUT Technology Ltd, Norway, Professor Andrew Beeby, University of Leeds, UK, Professor Peter Robery, FaberMaunsell, UK, Professor Heiki Kukko, VTT Building and Transport, Finland, Dr Mette Glavind, Danish Technological Institute, Denmark and Professor Yoshihiro Masuda, Utsunomiya University, Japan. The Congress was closed by Professor Peter Hewlett, Chief Executive of the British Board of Agrément, UK.

The support of 23 International Professional Institutions and 32 Sponsoring Organisations was a major contribution to the success of the Congress. An extensive Trade Fair formed an integral part of the event. The work of the Congress was an immense undertaking and all of those involved are gratefully acknowledged, in particular, the members of the Organising Committee for managing the event from start to finish; members of the International Advisory and National Technical Committees for advising on the selection and reviewing of papers; the Authors and the Chairmen of Technical Sessions for their invaluable contributions to the proceedings.

All of the proceedings have been prepared directly from the camera-ready manuscripts submitted by the authors and editing has been restricted to minor changes where it was considered absolutely necessary.

Dundee
September 2002

Ravindra K Dhir
Chairman, Congress Organising Committee

INTRODUCTION

Concrete floors and slabs provide the work surface for the vast majority of global industrial enterprises. Since the 1970's there have been a number of significant developments in design and construction methods which have been necessitated by economic, loading, surface tolerance and surface abrasion requirements.

Design of concrete floors and slabs has advanced greatly in the past decade with novel design methods now being used to overcome difficult ground conditions or to allow wider column spacing in commercial developments. The advent of new overtopping materials has also lead to changes in design procedures to consider the adhesion and moisture dynamics in floor and slab construction.

As design methods evolve, so must the methods of constructing floors and slabs. Innovative techniques such as large area pours have streamlined construction and the interface between contractors and designers remains a key element in the successful execution of these elements. Floors and slabs are arguably the most crucial structural elements that are dependent on a close synergy between the designer and contractor due to the specialist techniques, such as laser screeding, often employed. This is reflected in the increasing number of specialist flooring contractors that are now prevalent within the construction industry.

Concrete floors are now being utilised under increasingly harsher environments, both physically and chemically and must not only withstand these conditions but perform to the highest level. Specialist flooring is now more commonplace as the life cycle and economic requirements of industry become more significant and new materials are constantly being developed to produce flooring and slab systems which meet the needs of today's working environment.

The Proceedings of this International Seminar; *'Concrete Floors and Slabs'* dealt with all of these subject areas and the issues raised, under three clearly defined themes: (i) Design Methods and Considerations, (ii) Construction Techniques, (iii) Specialist Requirements. Each theme started with a Keynote Paper presented by the foremost exponents in their respective fields. There were a total of 30 papers presented during the seminar which are compiled into these Proceedings.

Dundee
September 2002

Ravindra K Dhir
Moray D Newlands
Thomas A Harrison

ORGANISING COMMITTEE

Concrete Technology Unit

Professor R K Dhir OBE (Chairman)

Dr M D Newlands (Secretary)

Professor P C Hewlett
British Board of Agrément

Professor T A Harrison
Quarry Products Association

Professor V K Rigopoulou
National Technical University of Athens, Greece

Dr S Y N Chan
Hong Kong Polytechnic University

Dr N Y Ho
L & M Structural Systems, Singapore

Dr M R Jones

Dr M J McCarthy

Dr T D Dyer

Dr K A Paine

Dr L J Csetenyi

Dr J E Halliday

Dr L Zheng

Dr S Caliskan

Dr A Iordanidis

Ms P I Hynes (Congress Assistant)

Mr S R Scott (Unit Assistant)

INTERNATIONAL ADVISORY COMMITTEE

INTERNATIONAL ADVISORY COMMITTEE (CONTINUED)

NATIONAL TECHNICAL COMMITTEE

Professor S A Austin
Professor of Structural Engineering, Loughborough University

Professor A W Beeby
Professor of Structural Design, The University of Leeds

Professor T Broyd
Research & Innovations Director, W S Atkins

Professor J H Bungey
Head of Department, The University of Liverpool

Mr N Clarke
Publications Manager, The Concrete Society

Mr G Cooper
Development Director, Fosroc International Ltd

Mr S J Crompton
Divisional Manager, RMC Readymix Ltd

Dr S B Desai OBE
Visiting Professor, University of Surrey

Professor R K Dhir OBE (Chairman)
Director, Concrete Technology Unit, University of Dundee

Dr K Elliott
Senior Lecturer, University of Nottingham

Dr S Garvin
Director (Scotland), Building Research Establishment

Professor F P Glasser
Professor, University of Aberdeen

Mr P G Goring
Technical Director, John Doyle Construction

Professor T A Harrison
BRMCA Consultant, Quarry Products Association

Professor P C Hewlett
Chief Executive, British Board of Agrément

Mr A Johnson
Divisional Director and Manager, Mott MacDonald Special Services

Dr M R Jones
Senior Lecturer, Concrete Technology Unit, University of Dundee

Professor R J Kettle
Head of Department, Aston University

Mr P Livesey
National Technical Services Manager, Castle Cement Limited

NATIONAL TECHNICAL COMMITTEE (CONTINUED)

COLLABORATING INSTITUTIONS

Institution of Civil Engineers, UK
American Concrete Institute
Japan Society of Civil Engineers

SPONSORING ORGANISATIONS WITH EXHIBITION

Aggregate Industries plc
Babtie Group
British Board of Agrement
British Cement Association
Building Research Establishment
Caledonian Slag Cement
Castle Cement Limited
CEMBUREAU, Belgium
Danish Technological Institute, Denmark
Degussa Construction Chemicals, Italy
Dundee City Council
Elkem Materials Ltd
FaberMaunsell
Fosroc International Ltd
Heidelberg Cement, Germany
Heracles General Cement Company, Greece
Institution of Civil Engineers
John Doyle Construction
Lafarge Cement UK
Makers UK Ltd
MBT Admixtures
Netzsch Instruments, Germany
North East Slag Cement
Palladian Publications
RMC Readymix Ltd
Rugby Cement
ScotAsh Ltd

SPONSORING ORGANISATIONS WITH EXHIBITION (CONTINUED)

Scottish Enterprise Tayside

Thomas Telford Publishing

UK Quality Ash Association

Waste Recycling Action Programme (WRAP)

Wexham Developments

Zwick Testing Machines Ltd

SUPPORTING INSTITUTIONS

American Society of Civil Engineers

Austrian Concrete Society

Belgische Betongroepering, Belgium

Concrete Institute of Australia

Concrete Society of Southern Africa, South Africa

Concrete Society, UK

Concrete Association of Finland

Czech Concrete Society

Entreprises Generales de France, France

European Concrete Societies Network

Fédération de I'Industrie du Béton, France

German Society for Concrete and Construction

Hong Kong Institution of Engineers

Indian Concrete Institute

Institute of Concrete Technology, UK

Instituto Brasileiro Do Concreto, Brazil

Irish Concrete Society

Japan Concrete Institute

Netherlands Concrete Society

New Zealand Concrete Society

Norwegian Concrete Association

Singapore Concrete Institute

Swedish Concrete Association

CONTENTS

THEME 3: SPECIALIST REQUIREMENTS

OPENING PAPER

HOW POLYMERS IMPROVE CONCRETE FLOORS - POSSIBILITIES TODAY AND PROSPECTS FOR THE FUTURE

P Seidler

Astradur Industrieboden AG

Germany

ABSTRACT. Industrial floors are required to resist extreme environmental stresses. The planning and the installation are therefore not easy tasks. Unnecessary damage is an all too common result. A properly dimensioned and placed concrete slab is not expensive, and is sufficient for many practical demands. There are possibilities to improve the slab – increasing workability and reducing cracks – by modifying the concrete by addition of powerful additives. If the strength of the surface needs to be further improved, and/or needs protection against chemical attack, a protective layer of a reaction polymer is essential: impregnation, coating, overlay, repair mortar with or without an additional topcoat [2]. In general, the application of a further component or layer is only worthwhile if it brings significant added benefits. The cost-benefit analysis indicates that a system consisting of fewer elements is better when the costs and properties are the same. A plain concrete from Portland cement is therefore preferable unless pronounced added benefits are involved. Reference is made to future developments under consideration by RILEM TC 184 "Industrial Floors".

Keywords: Abrasion, Adhesion, Additive, Chemical attack, Durability, Impact, Industrial floor, Polymer dispersion, Reaction polymer (Impregnation, Coating, Overlay, Topcoat), Repair mortar, Resistance, Strength, Thickness.

P Seidler, is co-owner of Astradur Industrieboden AG producing since 30 years high performance polymers for upgrading and repair of industrial floors: polymer impregnated concrete (PIC), coatings and polymer concrete (PC) overlays. He studied physical chemistry at Karlsruhe (Germany), Lausanne (Switzerland) and at the Sorbonne. Thesis about "Surface grafting of acrylic acid to PTFE" at CNRS in Paris. Study of management at INSEAD. He is editor of the books "Industrial Floors" and "Reaction Polymers on Site" and chairman of the five international colloquia "Industrial Floors". This event is to be repeated for the 5th time from 21st to 23rd of January, 2003. The proceedings are available on a CD-ROM. He is also chairman of RILEM TC "Industrial Floors", member of the board of ICPIC (International Congress on Polymers in Concrete) and a member of ACI, GDCh and ICRI. He is responsible for the international ePortal www.industrial-floors.com.

WE HAVE ALL THE MEANS FOR BETTER CONCRETE

Adam Neville, who is a part of the history of concrete, gave 10 years ago, a most notable lecture about "Concrete in the Year 2000" at a RILEM Conference 1992 in Athens. I like to quote his closing words, which he spoke as an "apologia", or as a "justification" for the scepticism expressed in his lecture:

There is no doubt that concrete is a most useful and versatile material, and it is highly unlikely that it will be supplanted by any other material in the foreseeable future. My pessimism stems from the fact that we have not improved the quality of ordinary concrete as much as we could by simply using existing knowledge and technology.

What I have tried to do is to ram home the message that we have in our possession all the means necessary to do better in the use of ordinary concrete. So there is a method in my madness: I am trying to incite those concerned with concrete construction into doing better, and then better still.

But if one looks at the field of floors in industry, one cannot help but conclude that concrete is actually nearly **useless** as an industrial floor, if only because of its joints, cracks and because of an always dusting surface. Because of this relative "uselessness", polymers in and/or on concrete are indispensable. Improvement of the situation is possible by adequate training of skilled workers. I shall return to this point later.

HISTORY OF MORTARS AND CONCRETES

During the history of mortars and concretes since the successfull experiments by John **Smeaton** and Robert **Stevenson** using hydraulically setting cements in the construction of the Eddystone lighthouse at Plymouth (UK) already in 1756, nearly 250 years ago, and Bell Rock lighthouse at the Firth of Forth in 1811, and the invention of Portland cement by Joseph **Aspdin** in Wakefield (UK) in 1824, there have been no great advances regarding its properties as an industrial floor.

Figure 1 Bell Rock lighthouse in Scotland (1811)

COSTS OF AN INDUSTRIAL CONCRETE FLOOR

Evidently the low price of concrete and much higher strength compared to clay, which was used in past days, have been the decisive factor for its use, see Table 1.

Table1 Market prices for a concrete slab as an industrial floor

WORK	DETAILS	PRICE	%
Polyethylene sheet, twice	**material** and labour	**0,80 €/m²**	**3%**
Concrete B 35, 180 mm	**material**	**13,30 €/m²**	**50%**
Concrete B 35, placing and levelling	labour	4,10 €/m²	15%
Super plasticizer 3 kg/m²	**material**	**0,60 €/m²**	**2%**
Helicopter trowelling	labour	1,80 €/m²	7%
Forming of running joints	labour	1,00 €/m²	4%
Dowelling of running joints	labour	3,60 €/m²	13%
Sawing of joints	labour	0,60 €/m²	2%
Curing with a polyethylene sheet	labour	1,00 €/m²	4%
Concrete slab on ground	**material and labour**	**26,80 €/m²**	**100%**

In comparison with these costs, improvement with polymers immediately makes the concrete very much more expensive, even if only a few percent of polymer in relation to the amount of cement are added:

Table 2 Cost of materials for industrial floors (Germany)

1. Aggregate	9 €/1.000 kg
2. Portland cement (silo)	90
3. Steel reinforcement	900
4. Polymer dispersion	2,700
5. Reaction polymer	9,000

THE MEANING OF "STRENGTH" AND "RESISTANCE"

An industrial floor not only has to be cheap. It must also stand up to numerous different demands. It must be strong.

Already at my inaugural lecture for the 2nd International Colloquium "Industrial Floors '91". I dealt with the multi-faceted word "resistance" which may be considered as largely synonymous with the word "strength". I repeat here what I said 10 years ago in Table 3.

Table 3 The multi-faceted term "resistance"

Fork-lift truck	fork-lift truck	**resistant**
Pallet truck	pallet truck	**resistant**
Small metal wheels	metal wheel	**resistant**
Containers with metal feet	metal foot	**resistant**
Loads up to 1,500 kg and more	bending tension	**resistant**
Rolling barrels	impact	**resistant**
Falling metal parts	damage	**resistant**
8 m stack height	pressure	**resistant**
Spark formation	spark	**resistant**
Explosive gas and powders	explosion	**resistant**
Aggressive acids	chemical	**resistant**
Sulphur compounds	sulphur	**resistant**
Groundwater protection	liquid	**resistant**
Surface	scratch	**resistant**
Food production	microbe	**resistant**
Mould	mould	**resistant**
Wet areas	water	**resistant**

SOME SHORTCOMINGS OF CONCRETE

Some important properties for an industrial floor are not provided by concrete as we currently know, or at least not to an adequate extent. Generally speaking, concrete is:

Table 4 Disadvantages of concrete as an industrial floor

- not free of joints
- not free of cracks

- not strong enough
- not homogeneous from the surface to the core

- not dust-free
- not abrasion-resistant enough
- not impact-resistant enough
- not sufficiently resistant to chemicals

- not a moisture barrier
- mostly available in grey, and not in all colours
- does not have homogeneous surface colour

- very thick, at a layer thickness of $\geq$ 180 mm
- heavy, at an installed weight of $\geq$ 400 kg/m^2
- difficult to smooth with a high degree of precision
- frequently cannot be placed without reinforcement

- **difficult to plan**
- **difficult to place, needing training**
- **difficult to repair**

Often these disadvantages are little known because people only look at the **price** and do not carry out any **benchmarking** or a **value analysis**, even with regard to **fitness for the purpose** and **durability.** **Field analyses** are only conducted sporadically, without the results obtained being systematically put into practice. The work of A. M. Vaysburd et al. is an exception. McDonald reported about this work at the 10th ICPIC in Honolulu (USA) in May 2001.

MIRACLES MADE OF CONCRETE, BUT NOT FOR FLOORS

In the research laboratories of the cement industry, concrete has not been given much attention as a material for industrial floors, even though elsewhere engineering miracles are possible with concrete, provided it is done properly. Numerous bridges and the "Troll" drilling platform in the North Sea have proved it over and over again.

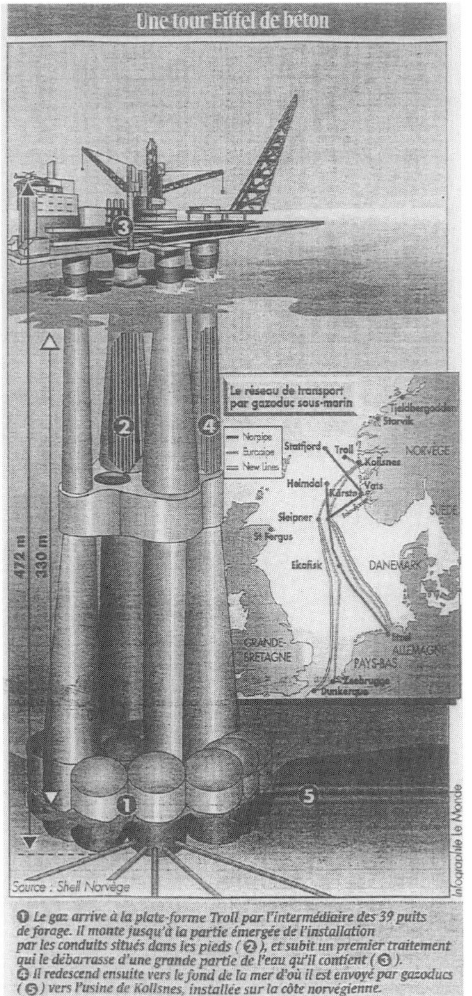

Figure 2 An Eiffel Tower of concrete, the "Troll" drilling platform, rising 472 m

SPECIAL CEMENTS FOR FLOORS WITHOUT SHRINKAGE

True, there is such a material as "shrinkage-compensating" concretes, which are placed without any joints and show no cracks. But they are not sufficiently "foolproof" in use. J. Holland emphasised these facts with urgency at the 4th International Colloquium "Industrial Floors '99".

This subject will again be one of the main topics when the experts in the field of industrial floors meet for the fifth time from **21 to 23 January 2003** at the Technical Academy in Esslingen (Germany) **for "Industrial Floors '03".** A special workshop:

"How to reduce shrinkage in Portland cement concrete?"

will be held within the frame of the colloquium to identify the possible solutions and the limits to what is possible. The cement industry from around the world is invited to make its contributions to the discussion.

PROBLEMS CAUSED BY CEMENTITIOUS OVERLAYS

As it is too difficult to get a concrete surface – and this surface layer is what really counts – so that it is immediately "strong" and "smooth enough to wallpaper", often a second layer is applied, namely a cementitious overlay with a thickness of ≥ 10 mm and a weight of ≥ 20 kg/m^2. It may cause additional problems at the interface with the concrete sub-base. Stresses resulting from different thermal expansion and shrinkage can cause the overlay to debond. There is an old and venerable work dealing with the many reasons why a cementitious overlay can fail. They are shown in the following tables.

Table 5 Types and causes of damages

Types of Damage

- insufficiently hardened
- is powdering, porous
- shows cracks and breaks
- is rough
- is contaminated with oil or chemicals
- is worn, has grooves

Causes and Responsibility

- Construction management
- Manufacturer
- Contractor
- External influences

Table 6 Responsibility of damages

Responsibility of Construction Management

- premature stress
- heating
- radiant heat
- draught

Responsibility of the Manufacturer

- bad grain-size distribution
- too many fines
- insufficient grain strength
- too many filterable constituents
- contamination
- humus acids
- swelling grains (such as marl or brown coal)
- frozen sand
- contaminated water
- too small proportion of binder
- too large proportion of binder
- wrong proportion of additive
- too much water
- insufficient mixing

Responsibility of Contractor

- too long storage before placement
- insufficient compaction
- too early smoothing
- too much smoothing
- too little thickness
- drying out not uniform
  - due to strong sun radiation
  - due to draught
- too large distances between joints

External Influences

- effect of frost during solidification and hardening
- improper maintenance
- improper use

WHAT REACTION POLYMERS CAN ACHIEVE, AT A COST

Many of these problems can be solved with polymers. However, it is important to distinguish between polymers **in** concrete and polymers **on** concrete. Problems that can be solved with polymers are marked grey in Table 7.

Table 7 Disadvantages of concrete which polymers can compensate for

- not **free of joints**
- not **free of cracks**

- not **strong** enough
- not **homogeneous** from the surface to the core

- not **dust-free**
- not **abrasion-resistant** enough
- not **impact-resistant** enough
- not sufficiently **resistant to chemicals**

- not a **moisture barrier**
- only available in grey, and not in **all colours**
- does not have **homogeneous surface colouring**

- **thin layer thickness** ≥ 0…5 mm
- **light installed weight** of ≥ 0,2…10,0 kg/m²
- **to smooth with a high degree of precision**
- mostly placed **without reinforcement**

- **difficult to plan**
- **difficult to place, needing training**
- **difficult to repair**

The difficulties for planning, placement and repair still remain even when polymers are used.

IMPROVEMENTS WITH POLYMER DISPERSIONS

Concretes and cement overlays are far more reliable in terms of freedom from joints and cracks, as well as abrasion and other properties, when suitable dispersions of polymers are added. There is an extensive literature on this subject and, by way of example, we merely refer to the numerous and wide-ranging works of Y. Ohama and to the recent contributions at the 10th ICPIC in Honolulu (USA) of R. Letsch, M. Maultzsch and J. E. McDonald. M. Puterman has reported critically about the improvements .

A VERY SIMPLE METHOD OF BENCHMARKING

At the ICPIC Workshop held in Bled (Slovenia) in 1996, the problem of the lack of transparency in the market for repair mortars was addressed. The discussion led to a "EUREKA" project in which A. Zajc and collaborators (IRMA/Slovenia), N. Swamy (University of Sheffield/UK), and astradur Industrieboden AG (Rodalben/Germany) were participating. First of all, a matrix was developed for comparing materials, which is now also being discussed by the RILEM Technical Committee TC 184. RILEM is the international union of major materials testing institutes. The matrix with 67 items, which is also suitable for use in benchmarking, is shown in Table 8.

Table 8 (Part1): Matrix of material properties and other important data of impregnations, coatings, overlays and top coats

1 Designation	**Product 1**	**Product 2**	**Product 3**
2 Main Characteristics			
Basis (composition)			
Solvent			
Use			
Components			
Fibres (% by weight)			
Thickness, min			
Thickness, max			
Density			
Price per kg (max package)			
Price per L (max package)			
Average (Ø) thickness			
Price per m2 (Ø thick: material)			
Price of application per m^2			
Setting time			
Layers, min			
Primer			
Waiting time 1, min (20°C)			
Waiting time 1, max (20°C)			
Levelling			
Waiting time 2, min (20°C)			
Waiting time 2, max (20°C)			
Topcoat (sealing)			
Abrasion			
Testing method			
Strength (4*4*16) N/mm^2			
Package, max			
3 Other Technical Data			
Quality system			
Flow behaviour			
Testing method			
Outdoor use			
Wet substrate			
Filling, min			
Filling, max			
Aggregate grain, min			
Aggregate grain, max			
Cleaning instructions			

Table 8 (Part2): Matrix of material properties and other important data of impregnations, coatings, overlays and top coats

Designation	Product 1	Product 2	Product 3
4 Safety Data Sheet			
Health hazards			
Symbol(s)			
R- and S-sentences			
Transport prescriptions			
5 Application			
Surface per hour and worker			
Mixing on site			
6 References			
Start of production			
Object reports			
7 Sources of Knowledge			
Last Date			
Technical data sheet			
Safety data sheet			
Label(s)			
Instructions for application (video)			
Instructions for cleaning			
Instructions for maintenance			
8 General Properties			
Skid proofness			
Package, min			
Temperature min			
Temperature max			
Characterisation (fingerprint)			
9 Samples			
Before setting (1 kg)			
After setting (4*4*16) max filled			

PROTECTION AGAINST IMPACT AND CHEMICAL ATTACK AS A FUNCTION OF STRENGTH AND LAYER THICKNESS

Recently we were discussing the feasibility to make nine different kinds of industrial floors from only one epoxy resin. The nine different kinds show different layer thickness, starting with an impregnation and ending with a porous, cheap repair mortar, not needing much binder. This would give a practical **modular system** of "one epoxy resin for nearly everything". These technical possibilities have advantages and disadvantages, which must be considered. In my view they create confusion on two levels:

1. They make decisions difficult for architects and owners, who do not know the signification of layer thickness.
2. They increase the amount of items stocked on site thus increasing complexity and the probability of applying the wrong material or the wrong composition.

Table 9 Layer thicknesses of a multipurpose-systems

	Layer	Thickness
1	Impregnation	0…0.1 mm
2	Levelling impregnation	0…1 mm
3	Self-levelling coating, thin	1…2 mm
4	Self-levelling coating, thick	2…5 mm
5	Self-levelling overlay	5…7 mm
6	Overlay, thin	5…7 mm
7	Repair mortar, liquid-proof	2…≥ 20 mm
8	Overlay, thick	≥ 7 mm
9	Repair mortar, liquid-porous	2…≥ 20 mm

The multi-purpose system is not suitable for high chemical resistance and crack-bridging. It is our practical experience that we do not need normally so many different kinds of floorings. The normal choice is for a new construction or a renovation is:

1. an impregnation
2. (and only for renovation) levelling, then a
3. coating, which may be completed by a
4. topcoat

THE FUTURE: CHEMISTRY AND ROBOTRONICS

To sum up, I would like to put forward some propositions for concrete slabs on ground and for reaction polymers improving the surface of concretes. These propositions, in my view, will influence future developments. I am aware that a great deal of discussion is still needed on the propositions about "The concrete slab as an industrial floor" and "Reaction polymers for industrial floors". I consider it as very important to keep these propositions in mind, and to remember that it is up to all of us who are responsible for the highly interesting field of "industrial floors" to ensure that further progress is made. In doing this, the words of Albert Einstein should be considered (quoted by A.M. Vaysburd et al.):

Pure logical thinking cannot yield us any knowledge of the empirical world: all knowledge of reality starts from experience and ends in it. Propositions arrived at by purely logical means are completely empty of reality.

Table 10 Propositions on concrete slabs as industrial floors

1. Laying under roof, therefore independence from the climatic situation outside.
2. No shrinkage of the cement, therefore no joints and Cracks.
3. No overlay, but monolithic.
4. No curing agent.
5. Less waiting time (< 28 days) for load-bearing strength .
   (watervapor diffusion-open overlays are a possibility).
6. No use of hard aggregate.
7. No dust formation.
8. Homogenous colour at a reasonable price.
9. Fibre reinforcement.
10. Laying by a robot, according to "Good Site Practice".

Table 11 Propositions on reaction polymers for industrial floors

Reaction Polymers for Industrial Floors

1. **Recognised trade** of "industrial floor applicator" with four special fields – computer assisted training:
    - concrete
    - overlays
    - reaction polymers
    - special floors
2. Products which are really **foolproof:**
    - no physiological risk
    - no risk of explosion
    - no volatile organic compounds (VOC)
    - excellent spreading, wetting and adhesion
    - wide tolerance for mixing ratios
    - no risk from moisture (blisters by osmosis)
3. Easy **measuring** of properties:
    - relevance of properties
    - constancy of composition by characterisation
    - transparency of competition
4. **Standardized** application:
    - AGI Worksheet A 80
    - robots for application
5. Better **training** for technical advisors and sales staff:
    - transfer of knowledge from engineers to foremen to workers
6. Value analysis and **expert system:**
    - for customers, designers and contractors
    - for easier decision making
    - directory of experts
    - directory of international prices
    - literature database on CD-ROM e.g. www.industrial-floors.com
7. Information and ordering via **Internet** (knowledge base)
8. No substrate preparation: a dream?
9. Good Site Practice as a standard:
    - Quality system according to ISO 9001-2000
    - Monitoring of products and processes

REFERENCES

1. ANN. “Problem: Joint spalls – spalls that occur adjacent to joints in floors, slabs, or pavements” Concrete Construction, p. 74, 11/00.

2. ARBEITSGEMEINSCHAFT INDUSTRIEBAU AGI (Germany). Work Sheet A 80 “Industrieböden aus Kunstharz” (1995) – English version to be published.

3. ICRI (USA). “Selecting and specifying concrete surface preparation for coatings, sealers, and polymer overlays” including nine surface profile chips (1998).

4. SNFORES (France). “Règles et Recommandations” without date.

5. S. CHANDRA, Y. OHAMA, “Polymers in Concrete”, Boca Raton 1994, 204p.

6. M. S. EGLINTON, “Concrete and its chemical behaviour” (1987), 136 p.

7. R. GEBHARDT, “Fließbeton und seine Fugenausbildung” in “Industrieböden”, 4. Ed. 2001, p. 59-100 – English version translated by Bill Depuy – to be published.

8. M.S. MCGOVERN, “Cracking Down on Repair Materials”, Concrete Construction 08.99, p. 50-54.

9. P.C. HEWLETT (ED.), “Lea’s Chemistry of Cement and Concrete” 4th ed., London 1998.

10. P.C. HEWLETT, P. SEIDLER, “Betonoberfläche und Adhäsion” in “Industrieböden”, 4. Ed. 2001, p. 216-232.

11. W. H. KUENNING (Ed.), ACI COMMITTEE 515, “Guide for the protection of concrete against chemical attack by means of coatings and other corrosion-resistant materials” (1966), p. 1305-1391.

12. J. C. MASO (Ed.), “Pore Structure and Construction Materials Properties”, Proc. 1st Int. RILEM Congress in Versailles, London 1987, 378 p.

13. J. E. MCDONALD, “Performance Criteria for Polymer-Modified Cement Repair Materials”, Proc. 10th ICPIC 2001 Honolulu (USA), CD-ROM.

14. W. S. PHELAN, “Guide for Concrete Floor and Slab Construction” ACI 302, Manual of Concrete Practice, p. 302.1 R-1 to 45 with 61 references.

15. M. PUTERMAN, Y. OHAMA (Ed.), RILEM TC 151 “Adhesion Technology in Concrete Engineering – Physical and Chemical aspects” in preparation.

16. M. PUTERMAN, “Parameters Affecting the Properties of Fresh and Hardened PCC Mortars”, Proc. 10th ICPIC 2001 Honolulu (USA), CD-ROM.

17. S.L. SAKAR, "Understanding Floor Covering Failure Mechanism (influence of humidity)" Concrete Repair Bulletin 03.99 p. 6-1 with 16 references.

18. H. SCHORN, "Shape and Distribution of Polymer Particles in PCC – Investigated by Environmental Scanning Electron Microscope (ESEM)", Proc. 10th ICPIC 2001 Honolulu (USA), CD-ROM.

19. P. SEIDLER (Ed.), CD-ROM "Industrial Floors" with contents of the Intern. Colloquia "Industrial Floors '87, '91, '95, '99" Fraunhofer IRB-Verlag 1999, mainly in German language.

20. P. SEIDLER (Ed.), "The Concrete Slab as an Industrial Floors and it's Upgrading by High Performance Polymers – A Guideline for Design and Practice" - to be published 2003.

21. P. SEIDLER, "Problems of Characterisation of Coatings", ICPIC-8 Oostende (Belgium) 1995, to be published.

22. P. SEIDLER, "Industrial Floors: Why Apply a Topcoat to Impregnation, Coating or Overlay?", Proc. 4th South African Conference on "Polymers in Concrete", 20-23 June 2000.

23. R. SIEVER, "Sand – an Archive of the History of Earth", New York 1988; German edition Heidelberg 1989, 254 p.

24. M. SPRINKEL, "25 Year Experience with Polymer Concrete Bridge Deck Overlays", Proc. 10th ICPIC 2001 Honolulu (USA), CD-ROM.

25. A.M. VAYSBURD, P.H. EMMONS, J.E. MCDONALD, R.W. POSTON, K.E. KESSNER, "Selecting Durable Repair Materials: Performance Criteria – Field Studies", Concrete International, December 2000, p. 39 – 45 with 4 references.

26. W.W. WALKER, J.A. HOLLAND, "Design, Materials, and Construction Considerations for Reducing Curling and Shrinkage", in Proc. 4th Int. Coll. "Industrial Floors '99" Ed. P. Seidler, TAE (Germany), Vol. I, p. 171-180.

www.industrial-floors.com
www.icpic.org
www.rilem.org
www.astradur.com
www.aci-int.org
www.icri.org

THEME ONE:

DESIGN METHODS AND CONSIDERATIONS

Keynote Paper

DESIGN OF TWO-WAY SLAB SYSTEMS FOR STRENGTH AND SERVICEABILITY

A Scanlon

Pennsylvania State University

United States of America

ABSTRACT. The current state-of-the-art for design of two-way slab systems is presented. Design for both strength and serviceability criteria are discussed. The history of the development of two-way slab design methods from the early twentieth century to the present time is presented. Design methods for flexure include elasticity-based and plasticity-based methods. Design for punching shear and transfer of moments between slab and column are shown to be critical to the design for strength. Design for serviceability is concerned primarily with deflection and crack control. Deflection control in particular is critical for satisfactory structural performance. The paper concludes with a discussion of expected future developments in two-way slab design.

Keywords: Two-way slabs, Concrete, Flexure, Shear, Deflection, Cracking.

A Scanlon is Professor of Civil Engineering in the Department of Civil and Environmental Engineering at The Pennsylvania State University. His research interests focus on safety and serviceability of concrete structures.

INTRODUCTION

Two-way slab systems are among the most common structural forms used in floor and roof construction in reinforced concrete buildings. In North America, column-supported slabs are categorized as follows:

Flat Plates: Slabs of uniform thickness over an entire panel, supported on columns, usually without column capitals; limited to relatively short spans and low levels of live load.

Flat Slabs: Slabs with drop panels adjacent to columns to provide enhanced punching shear capacity, sometimes with column capitals also for shear capacity; applicable to longer spans and heavier live load.

Waffle Slabs: Essentially a two-way joist system with solid full-depth slab adjacent to columns for shear capacity; spans and loadings similar to flat slabs.

Two-Way Slabs with Beams: Uniform depth slab supported on beams spanning along column lines in both directions; beams provide extra strength and stiffness to accommodate longer spans and higher live loads, however formwork is more complicated than for other systems.

Two-way slab systems are also used for mat foundations supporting columns. In this case the slab load is produced by soil pressure. Two-way edge-supported slabs are commonly used for floor systems and vertical walls of liquid-retaining tanks.

Two-way column-supported slabs have been in use since the early twentieth century when patented systems were introduced. Since no analytical tools were available at that time these systems were only accepted on the basis of load tests. Reinforcement often consisted of wires draped to provide negative moment capacity in the vicinity of columns and positive moment capacity at mid-span.

Reinforcement patterns were often fairly complicated with reinforcement spanning along orthogonal column lines and along diagonals between columns. Design methods gradually evolved until the 1950's when an extensive investigation was undertaken at the University of Illinois, resulting in the design methods that are currently embodied in the ACI Code [1]. An interesting history of the development of slab design and analysis in the first half of the twentieth century is given by Sozen and Siess [2]. Behavior and design of slab systems is described in detail by Park and Gamble [3].

This paper provides a brief summary of the current state-of-the-art and practice for design of two-way slab systems. Suggestions for likely future developments are also provided.

DESIGN FOR STRENGTH

Flexural Strength

Design for strength requires consideration of both flexural strength and shear strength. In the design for flexural strength the designer needs to determine an appropriate moment field to provide a basis for an orthogonal layout of flexural reinforcement. As a slab is loaded from zero load to ultimate load the actual moment field in the slab can change drastically as localized stiffness changes resulting from cracking initially and eventually yielding of the reinforcement take place. Because of the relatively low reinforcement ratios used in most slab systems, significant redistribution of moments can occur as a result of the high flexural ductility available and the high degree of statical indeterminacy associated with slab systems.

At early loading stages when the slab can be considered to be uncracked, small deflection elastic plate bending theory [4] can be used to determine moments in the slab. For column supported two-way slab systems, high moment gradients and high moment intensities occur in the vicinity of the columns leading to cracking that produces redistribution of moment to negative moment regions away from the column and to positive moment regions in the mid-span region. As loading continues, nonlinear analysis techniques can be used to trace the load-deformation response of the system.

Closed form solutions of the fourth order partial differential equation of plate bending theory are limited to simple cases. For practical designs, resort must be made to simplified analysis methods or numerical methods. Prior to the availability of finite element computer programs, the equivalent frame method [5] was developed to provide positive and negative moments in given spans. These moments are then distributed transversely using factors based on numerical and experimental studies. This approach as outlined in the ACI Code remains the most popular method in practice for two-way slab design and computer programs are available to perform the tedious calculations involved. The method does have limitations including difficulties in application to slabs with irregular column layouts. For regular slab layouts satisfying certain criteria outlined in the code, the Direct Design Method can be used in which design moments are determined for critical sections in column and middle strips simply as fractions of the total static moment.

With the advent of finite element computer programs much more flexibility became available for the analysis of slab systems of regular or irregular configurations [6]. General-purpose programs and special-purpose programs specifically developed for slab design and analysis can be used. Slab systems of arbitrary configuration and general loading can be analyzed and the resulting moments M_x and M_y used for design of reinforcement. Because the analysis is based on elastic plate pending theory some averaging of the moments must be done before determining reinforcement layouts for column and module strips.

Moment intensities are either computed at integration points within elements or at element node points depending on the type of element used in the program. Output moments will generally be in the form of M_x, M_y , and M_{xy} components in the coordinate system used and/or $M_1$and M_2 principal moments oriented in general at an angle to the global coordinate system. Since reinforcement is generally provided in the orthogonal x and y directions, some authors recommend that a twisting moment component be added to the M_x and M_y moments [7]. It is useful to check that the resulting positive and negative design moments within a panel correspond to the total static moment.

$$M_o = w_u l_2 l_1^2/8$$

Where M_o = total static moment
w_u = factored uniformly distributed load
l_2 = transverse span length
l_1 = clear span length

This will also serve as a check against any errors in computed moments due to the approximations inherent in the finite element formulation.

Slabs can also be designed according to plasticity theory using either lower bound or upper bound methods. While elasticity theory requires satisfaction of equilibrium and compatibility conditions, the compatibility requirement is relaxed in plasticity theory and it only remains to satisfy conditions of equilibrium of design moments. The use of plasticity theory for slab systems relies on the high degree of flexural ductility that is available to permit liberal redistribution of moments as the ultimate load is approached. Care is required however to ensure that the reinforcement patterns developed are sufficiently close to elastic distributions of moments to avoid serviceability problems related to deflection or crack control.

The lower bound simple strip method [8] provides a very convenient method for slab systems, particularly edge-supported slabs. In this method the designer has wide latitude in assigning the load path to transfer for loads to the supports. This method is attractive because it provides an intuitive physical understanding of the load transfer mechanism of the system and reduces the analysis from two-way analysis to beam analysis.

An example is shown in Figure 1 for a uniformly loaded edge supported rectangular slab. Loads are transferred in one direction only for strips AA and BB and these beam strips can be analyzed by standard beam analysis methods. This approach is also very suitable for rectangular tank walls subjected to a triangular load distribution due to fluid pressure. For application to column supported slabs the advanced strip method is available.

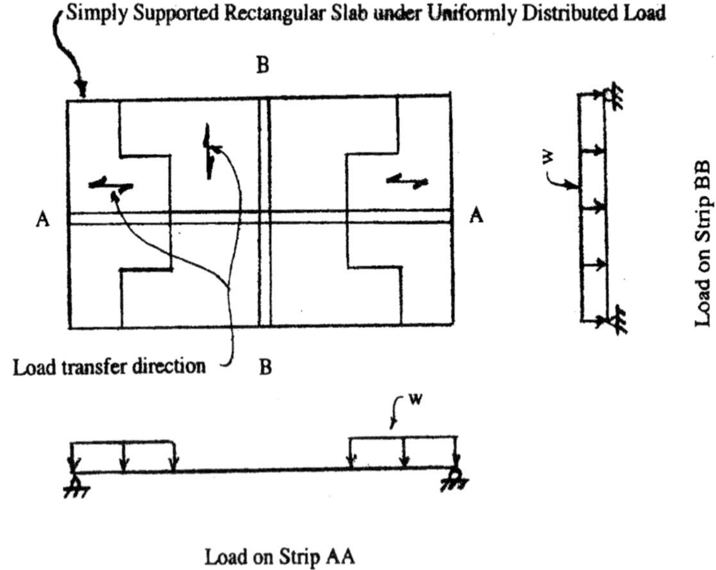

Figure 1 Application of simple strip method

The upper bound yield line method [9] is also convenient for determining moments at critical sections (along the yield lines). This method has not been widely used in North America perhaps because of concerns that many yield line patterns may be possible but all but one (usually) will provide an unconservative collapse load. The method appears however to be well accepted in Europe. Application of the method is demonstrated in Figure 2 for a column-supported two-way slab system. Three possible yield time patterns are demonstrated: a folding plate pattern in either span direction and a conical fan-shaped pattern around the columns. It can be shown [3] that the folding plate pattern will give a collapse load less than the conical pattern of the reinforcement pattern is uniformly distributed along the yield lines. However if the reinforcement is distributed between column and middle strips according to the usual rules, the conical fan failure will occur at a higher load than the folding plate mechanism so that the folding plate mechanism can be used as a basis for design.

Shear Strength

In general, slab systems are forgiving with respect to flexural failure due to the available ductility that allows redistribution of moments prior to failure, and because strain hardening in the steel can occur due to the typical low reinforcement ratios. Punching shear failure on the other hand is brittle, occurring with little warning. This mode of failure is responsible for most if not all slab collapses that have been reported. A detailed discussion of punching shear failure is given in ASCE-ACI 426 [10]. The ACI Code method for calculating punching shear strength requires computation of the critical perimeter b_o taking into account the effect of any openings that may be present. The shear capacity provided in the critical perimeter is then computed as

$$V_c = (2 + 4/\beta_c) \sqrt{f_c'}\, b_o\, d$$

Where β_c is the ratio of the long side to short side of a rectangular critical perimeter. This expression is based largely on test data that suggest that biaxial compressive stress conditions in the slab in the vicinity of the column enhance the shear capacity relative to beam shear capacity. As the column plan aspect ratio increases this effect decreases as reflected in the β_c term.

The effect of increased shear stresses resulting from transfer of moment between the slab and column must also be taken into account. The ACI Code method is based on a fictitious set of quasi-elastic shear stresses resulting from direct shear plus moment transfer stresses. More recently, attempts have been made to apply plastic truss, or strut-and-tie models to provide a more realistic treatment reflecting observed failure modes [11].

DESIGN FOR SERVICEABILITY

Deflection Control

While slabs are generally forgiving with respect to flexural strength they have a tendency to deflect more then anticipated. This can lend to unsatisfactory performance such as damage to non-structural elements and other serviceability problems. The simplest way to deal with deflection control is to select a slab thickness based on code-specified minimum thickness values expressed as a function of span length.

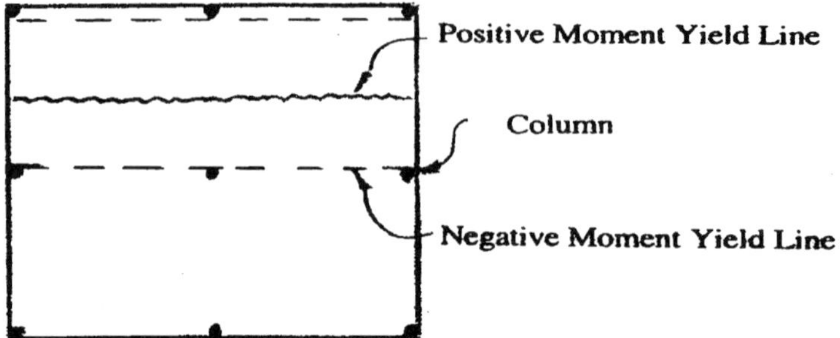

a) Folding Plate Yield Line Pattern

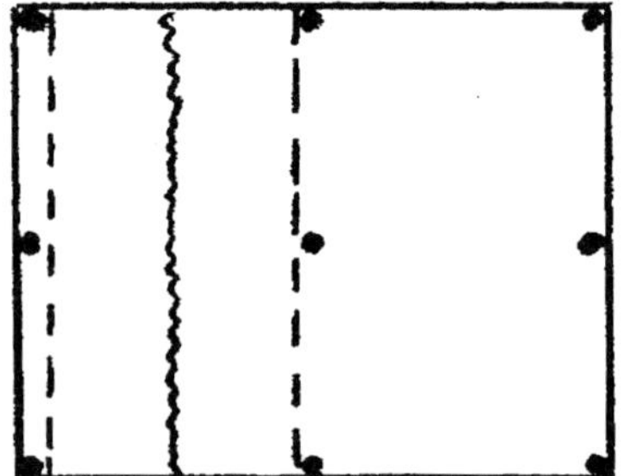

b) Folding Plate Yield Line Pattern

c) Conical Yield Line Pattern

Figure 2 Application of yield line method

The minimum thickness values in the ACI Code for example have given satisfactory performance for many years as long as care is taken during construction to avoid overloading the slab at early age. A slab thickness somewhat greater than the minimum thickness can be used to reduce the likelihood of deflection problems. However if it is decided to use a thickness less than the specified minimum thickness, the Code will permit this if deflections are calculated using criteria given in the Code and shown to be less than specified deflection limits.

Similar approaches are used in other codes although the complexity of criteria for deflection calculations varies considerably from code to code. The ACI Code, recognizing the uncertainties associated with deflection control provides relatively simple criteria for the deflection check. The effective movement of inertia concept is used to account for time-dependent effects of creep and shrinkage. A value of 2 is used for the time-dependent multiplier with a modifying coefficient to account for the presence of any reinforcement in the compressive zone. Given the high degree of variability associated with cracking, creep, and shrinkage as well as loading history, it should not be surprising that calculated deflections often do not correlate well with field-measured deflections. In terms of load and resistance factor design concepts one can think of the deflection check as being made with load and resistance factors set equal to unity.

Guidance on methods for calculating two-way slab deflections is given in the ACI 435 report on Deflection Control [12]. Calculation methods include:

- Analytical solutions of the plate equation
- Equivalent frame analysis
- Crossing beam analogy
- Finite element method

Of these, the crossing beam method based in design moments scaled to service load levels is the most readily understood by structural engineers because it provides a clear picture of the behavior of the slab and the deflected shape. Deflection can be calculated at the mid-span of column strips, mid-span of middle shapes and by superposition, at mid-panel as shown in Figure 3. These deflections can be calculated by standard beam analysis based on the moment diagrams.

Finite element analysis also can be used although care must be taken to assess the effect of cracking since most available software packages used in design offices assume linear material properties.

Because of high elastic negative moment intensities in the vicinity of columns, the presence of shrinkage restraint stresses, and high levels of loading at early age during construction, cracking is often under-predicted. These effects can be compensated for by using a reduced cracking moment, $M_{cr,}$ and estimating the maximum load expected during construction [12]. Two deflection limits are specified in the ACI Code, a limit on immediate deflection due to live load, and a limit on additional time-dependent deflection occurring after installation of non-structural elements. The second limit requires the engineer to assume a loading history and the time at which non-structural elements are installed. The limit also includes an instantaneous application of full live load at the end of the loading history. The incremental time-dependent deflection limit usually governs over the live load deflection.

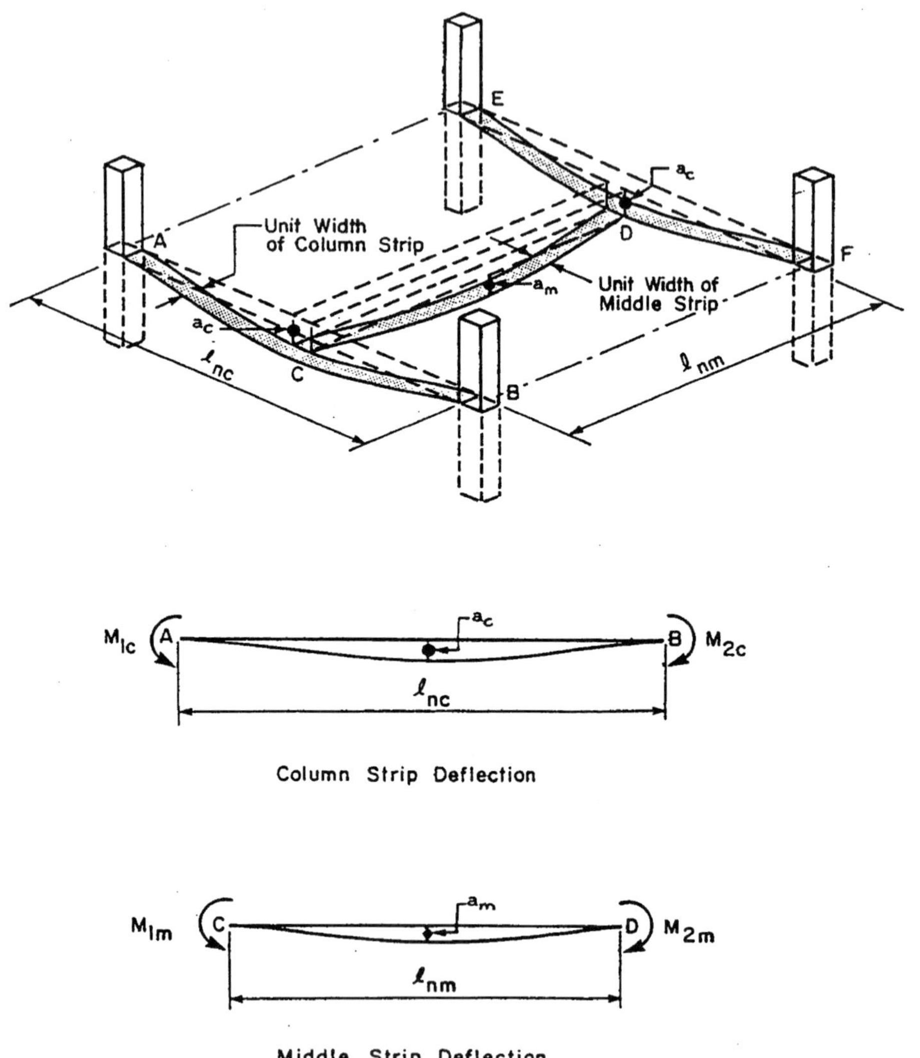

Figure 3 Crossing beam analogy for deflection calculation

Crack Control

Code provisions include requirements for crack control with the intent of limiting crack width under service loads. The ACI Code recently switched from a requirement derived from the Gergely Lutz crack width equation to a simple requirement to limit bar spacing. Nawy [13] has also provided an equation for crack width calculation.

OTHER CONSTRUCTION SYSTEMS

Post-tensioning can be used in two-way slab systems making use of prestressing to reduce the required slab thickness for a given span and load. This type of construction requires specialized expertise provided by post-tensioning contractors.

Lift slabs are sometimes used where the slabs are cast on the ground and then lifted into place using hydraulic jacks followed by attachment to support columns. Special techniques are required to ensure safety during the lifting operation.

FUTURE DEVELOPMENTS

Two-way slab systems can be expected to remain popular as floor framing systems in the future. However there will be opportunities both in design and construction for further improvements in two-way slab systems.

Improvements in strength and stiffness of concrete may allow some increase in practical span lengths without requiring excessive slab thickness. However as slab spans increase, deflection control will tend to dominate the design. Improvements in post-tensioning technology will also allow increased span lengths.

Increasing confidence in results of computer-based analyses using the finite element method will provide engineers with the tools needed to design slabs of arbitrary configuration including optimizing design parameters such as drop panel dimensions and thickness.

Developments in FRP technology may result in the replacement of steel reinforcement.

Improvements in construction technology and materials will allow more rapid construction of multi-story systems.

REFERENCES

1. ACI COMMITTEE 318. Building Code Requirements for Structural Concrete (318-99) and Commentary (318R-99), American Concrete Institute, 1999, 391 pp.

2. SOZEN, M A, and SIESS, C P. Investigation of Multiple-Panel Reinforced Concrete Floor Slabs: Design Methods – Their Evolution and Comparison, Journal of the American Concrete Institute, 1963, Vol 60, No 8, pp 999-1027.

3. PARK, R, and GAMBLE, W L. Reinforced Concrete Slabs, 2nd Ed John Wiley & Sons, Inc, New York, 2000, 716 pp.

4. TIMOSHENKO, S, and WOINOWSKY-KRIEGER, S. Theory of Plates and Shells. McGraw Hill, New York, 1959.

5. CORLEY, W G., and JIRSA, J O. Equivalent Frame Analysis for Slab Design. ACI Journal, V. 67, No. 11, 1970, pp 875-884.

6. HRABOK, M M, and HRUDEY, T M. Finite Element Analysis in Design of Floor Systems, Journal of Structural Engineering, ASCE, Vol 109, No 4, 1983, pp 909-925.

7. MACGREGOR, J G. Reinforced Concrete: Mechanics and Design, 2nd Ed. Prentice Hall, 1992, 848 pp.

8. HILLERBORG, A. Jamviktsteori for Armerade Betongplattor. Betong, 1956, Vol 41, No 4, pp 171-182.

9. JOHANSEN, K W. Yield-Line Formulae for Slabs, translated by Cement and Concrete Association, London, 1972, 106 pp.

10. ASCE-ACI COMMITTEE 426. The Shear Strength of Reinforced Concrete Members: Slabs. Journal of the Structural Division, ASCE, 1974, Vol 100, No ST8, pp 1543-1591.

11. ALEXANDER, S B, and SIMMONDS, S H. Ultimate Strength of Slab-Column Connections. ACI Structural Journal, 1987, Vol 84, No 3, pp 255-261.

12. ACI COMMITTEE 435. Control of Deflection in Concrete Structures (ACI 435R-95), American Concrete Institute, 1995, 77 pp.

13. NAWY, E. G. Crack Control through Reinforcement Distribution in Two-way Acting Slabs and Plates. ACI Journal, 1972, Vol. 69, No. 4, pp 217-219.

COMPUTATIONAL MODEL OF CONTINUOUS FLOORS BASED ON SUBSTITUTING BEAM SYSTEMS

A Plotnikov

Chuvash State University

Russia

ABSTRACT. In this paper the method of estimating the allocation of loads between coordinate directions supported on an outline of continuous floors, both grids systems and intense states is discussed based on a common beam-model. The calculations and finite outcome of the equations that take into account reinforcement in two directions through conditional coefficients anisotropy, and through quantitative metrics are reduced. The link beam-model and method of a limiting equilibrium for calculation of a carrying capacity is shown. The pattern of stress distribution distinguished from classical methods of the theory of elasticity and plasticity is observed in existing overlapped reinforcement. In different reinforcement schemes and comparable relations of the sides of floors, the corner between a crack on a bottom face plate and its side varies in limits from 30^0 up to 60^0, depending on the coefficient of rigidity (cut) of the reinforce (Asy/Asx from 0,4 up to 1,6).

Keywords: Reinforcement, Beam–model, Grids, Continuous floors, Rigidity, Limiting equilibrium, Stress distribution, Corrosion.

A N Plotnikov is Deputy of the Dean of Building Faculty the Chuvash State University on Scientific Activity (Russia), higher teacher of faculty of building constructions. His research focuses of the schemes of reinforcement on the intense state of continuous plates and grid systems, registration of this appearance in designing new constructions and on inspection of existing constructions, including maintenance in conditions of corrosion.

INTRODUCTION

In existing calculation practice of floors supported on an outline continuous monolithic and modular, and also grids, (especially wide-span reinforcement concrete floors), there is a necessity of the registration of influence of the reinforcement on their rigidity. In a commonly used method of the theory of elasticity and plasticity, a combination solution is often found by methods of finite differences in the last limiting state to determine stresses and deflections in reinforced concrete units. This is often by the common theory of plates however, it is impossible to consider floors as a plates, as they more often than not are orthotropic with systems of cylindrical rigidities. In a method of limiting equilibrium with solutions by static means to give a minimum value of a carrying capacity (lower bound) and a dynamic way, giving maximum value (upper bound) [2] the recommended optimum coefficients of an anisotropies depend on a relationship of the sizes of the element in question.

ANALYSIS OF THE REFERENCES

The influence of reinforcement on allocation of loads in the Russian literature for the first time was justified S M Krylov [3] in a part of the boundary conditions of increase of the moments in cuts of statically indefinable constructions, which contain parameters of the fixture. The estimate of carrying capacity and deflections of plates under the theory of yield-lines conterminous with a kinematic way of a method of a limiting equilibrium practices also. Among the authors of this theory A A Gvosdev [2, 3], C T Morley [12], J D Renton [13], M Save [14] etc.

Worldwide theories including the Russian theory of reinforced concrete operates with parameters of reinforcement on the boundary of yieldlines for definition of a limiting resistance supported on an outline of plates and estimate is stress - deflexions state. V M Bondarenko and S V Bondarenko [4] parse an existing technique of reallocation of stress and score the following: in accordance with increase of load the decrease turning and increase bending moments is observed; similarly statically to indefinable beams systems in plates there is a reallocation of stress between separate cuts; allocation of bending moments move from center in peripheral zones of a plate; the reallocation of stress from one coordinate direction on another, owing to distinction in rigidities because of different working heights is characteristic for plates. For supported on an outline of plates of a floor being the continual system, the registration of influence of the fixture becomes inevitable. On the registration of forces originating in the fixture, the theory of deflexion of reinforcement concrete with flaws N I Karpenko is constructed [2]. The important outcome of this operation is the prediction of the scheme of a break of plates, the theory of yieldlines A A Gvosdev- C T Morley is put in a basis of a method, the computational methods of deformations designed at the generalized intense state in view of the various schemes of cracks.

In operating normative documents the calculation of plates, supported on an outline, is yielded under the underformed scheme, without the registration of change of geometry and space operation. V S Syrianov [5] gives computational methods of strength supported on an outline and three sides of plates under the deformable scheme in view of space operation, and also principles rational reinforcement of such plates, which allow to reach essential lowering of expenditure of materials, expenditures of labor, money resources. B J Davranov [6] concludes, that the operation it is not enough reinforcement supported on an outline of plates in a stage of corrupting is characterized by reaching of limiting resistances of the reinforce at considerable reallocation of stress on cuts of a break, that is a substantiation of applicability for calculation of their strength of a method of a limiting equilibrium.

As has shown A M Kobeisy [7] orthotropi of reinforcement concrete plates stipulated by any other reinforcement in two directions, renders influence on the value of corners of a break. In square plates, supported on an outline, the influence of coefficient orthotropi is unessential, and the corners of a break are close to 45^0. In rectangular plates, supported on an outline, orthotropi has an effect more noticeably on the value of corners of a break, which themes deviate from 45^0, than more coefficient orthotropi more.

The reduced here references testify to the accumulated considerable base for creation of more simple engineering computational methods of reinforcement concrete floors. For an estimate of allocation of stress in plates (floors), especially, at the enlarged spans, it is important to know, what influence render the physically existing parameters reinforcement, what declination of a crack to the side of a plate, at the moment of its derivation and at increase of load.

OUTPUT OF CUMULATIVE DISTRIBUTION FUNCTIONS OF STRESS

To enter into the designed formulas of a method of a limiting equilibrium a varying corner of crack depending on parameters of reinforcement (cut of the reinforcement, of working height of cut of a construction) the beam - model allows substituting.

The interest represents a relation of bending stress between two directions of the schedule of a plate, that is defined by conditional rigidities of directions. The plate of an floor or grid is substituted at calculation by the system of two mutually perpendicular intersected beams iterating height of cut, spanning, amount of the fixture on unity of width and its binding, thus influence of a torque conditionally is not taken into account.

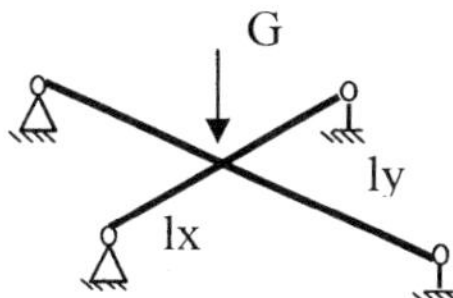

Figure 1 Substituting beam - model

The equilibrium in a cross point of beams is considered, in this case fixing of beams on the extremities of a role does not play, the allocation of stress between two directions is important only, we accept turning fixed.

$$\begin{cases} q_x l_x + q_y l_y = G \\ f_x = f_y \end{cases} \tag{1}$$

Under identical conditions fixed of beams X and Y and identical application of load on small length of a site we equate expressions of sags in a point:

$$\frac{1}{r_x} l_x^2 = \frac{1}{r_y} l_y^2 \tag{2}$$

The expressions of curvature are written with usage of the formulas of the Russian norms of calculation of reinforcement concrete constructions [1], being a basis of a method of limiting states: at the first stage is stress - deflexion state without the registration of cracks in concrete, at the second stage and last limiting state by a method sequential loads is iterative - depending on a relation Mxy and Mcrc, and, in one direction the system can work without cracks, and in the friend - with cracks in the stretched zone.

Expressions of curvature of a unit without cracks in the stretched zone

$$\left(\frac{1}{r}\right) = \frac{M\,\varphi_{b2}}{\varphi_{b1}\,E_b\,I_{red}} \tag{3}$$

M - moment from external load,
φ_{b1} - coefficient which is taking into account influence of a short-term creep of concrete,
φ_{b2} - coefficient of a long-lived creep of concrete on deflexion of a unit without cracks.

Expressions of curvature of a unit with cracks in the stretched zone is written

$$\left(\frac{1}{r}\right) = \frac{M}{h_0\,z}\left[\frac{\psi_s}{E_s\,A_s} + \frac{\psi_b}{\xi b h_0\,E_b v}\right] \tag{4}$$

z - distance from a barycentre of a sectional area of the reinforce S up to a point of the application of equal in effect stress in an oblate zone of cut above a crack,
ψ_s - coefficient of stretched concrete on a site with crack,
ψ_b - coefficient an extreme oblate filament of concrete on length of a site with crack,
ξ - relative height of an oblate zone of concrete at presence of cracks ,
v- coefficient describing an elasto-plastic state of concrete of an oblate zone,

From a stage of operation of reinforcement concrete without cracks up to the last limiting state the position of a zero line varies at decrease of height of an oblate zone in cut with a crack. At the first stage the barycentre of an oblate zone is defined, in opinion of the author of the given paper, as for a delta circuit, that corresponds of stress in an oblate zone, actual application of equal in effect force. Such approach is logical, as the position of a neutral line is defined from an equilibrium condition of the static moments of units of cut, at a level of microstructure the small particles of a material most remote from a neutral line have a major potential energy, as it is taken into account by a major shoulder equal in effect.

Both on first, and at the subsequent stage of operation of a unit, the expression of curvature is output from $\left(\frac{1}{r}\right) = \frac{\varepsilon_{bm} + \varepsilon_{sm}}{h_0}$, that is, ratio of the total of deformations oblate and stretched of zones to working height of cut. Thus the change of the value Z will happen more smoothly, especially, at transition to a stage after derivation of cracks in the stretched zone. The value ξ, as show experimental and calculation dates [11], varying in limits 20% unessentially has an effect for the value of curvature.

The allocation of stress between intersected beams, in main, depends on a tentative relation of an amount of the reinforce, that is, relative height of an oblate zone X of appropriate beams, further reallocation from change of this value in connection with advanced growth of height and width of a crack of one of directions.

Let us consider four cases of reallocation of stress, the expressions for which are obtained by record and equating of the equations of curvature of two directions in a cross point of beams:

1. In beams X and Y the cracks are not present.

$$q_x = \frac{G \cdot l_y^3 \cdot J_{red,x}}{l_x \cdot (l_y^3 \cdot J_{red,x} + l_x^3 \cdot J_{red,y})} \qquad q_y = \frac{G \cdot l_x^3 \cdot J_{red,y}}{l_y \cdot (l_y^3 \cdot J_{red,x} + l_x^3 \cdot J_{red,y})} \tag{5}$$

2. In a beam X the cracks are not present, in a beam Y of a cracks is.

$$q_x = \frac{G \cdot l_y^3 \cdot \varphi_{b1} \cdot E_b \cdot J_{red,x} \cdot [K_y + F_y]}{l_x (l_y^3 \cdot \varphi_{b1} \cdot E_b \cdot J_{red,x} \cdot [K_y + F_y] + l_x^3 \cdot \varphi_{b2} \cdot h_{oy} \cdot z_y)}, \qquad q_y = \frac{G \cdot l_x^3 \cdot \varphi_{b2} \cdot h_{oy} \cdot z_y}{l_y (l_y^3 \cdot \varphi_{b1} \cdot E_b \cdot J_{red,x} \cdot [K \ + F_y] + l_x^3 \cdot \varphi_{b2} \cdot h_{oy} \cdot z_y)}, \tag{6}$$

3. In beams X and Y of a cracks is $q_x = \frac{G}{C \cdot l_y + l_x}$ и $q_y = C \cdot \frac{G}{C \cdot l_y + l_x}$ (7)

where $K = \frac{\psi_s}{E_s \cdot A_s}$, $F = \frac{\psi_b}{\xi \cdot b \cdot h_0 \cdot E_b \cdot \nu}$, $C = \frac{l_x^4}{l_y^4} \cdot \frac{h_{oy} \cdot z_y}{h_{ox} \cdot z_x} \cdot \frac{[K_x + F_x]}{[K_y + F_y]}$ (8)

At calculation in a stage of maintenance as a first approximation influence of the external moment on relative height of an oblate zone (5, 6) is not taken into account and ξ starts on a stage of operation without cracks.

The numerical analysis was carried out with several variable parameters: change of a relation Lx/Ly - (0,6 - 0,78 - 1,0 - 1,29 - 1,67) - traditional parameter for supported on an outline of floors, change of a relation of squares of the reinforce of two directions Asx/Asy - (0 - 0,67) - new parameter, change of working height of cut of two directions h_{0x}/h_{0y} - 1, change of a level of load - (20, 60, 100 кN), specifying the analysis of reallocation of stress by a method of a sequential loading.

THE ANALYSIS OF OUTCOMES OF NUMERICAL SOLUTION

As show outcomes of calculations on the computer on designed algorithm, the units transit stages of operation elastic, at small loads and inelastic, with cracks of width up to 0.35 mm at boosted loads, thus the difference of sets of the graphs on steps (20,60,100) unsignificantly and happens smoothly. At increase of load and development of the process crack-derivations this relation varies in the side of increase on the long side Y. At decrease of an amount of the fixture in a direction X - the loads in this direction decrease 1:0.53 (approximately in 1.8 -1.9 times).

A graphics, from the point of view of change of the parameter reinforcement, carry asymptotic logarithmic character at a single relation of lengths of the sides and coming nearer to a linear relation for the narrow plates, the escalating of the reinforce thus goes on the short side.

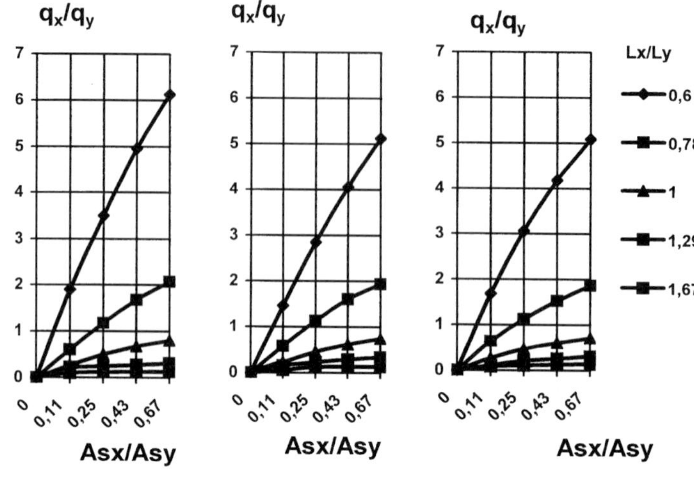

G=20 кN , h_{ox}/h_{oy}=1 G=60 кN , h_{ox}/h_{oy}=1 G=100 кN, h_{ox}/h_{oy}=1

Figure 2 Graphics of allocation of loads between directions of plates on beam-analogy depending on a relation of an amount of the reinforce, relation of the sides, working height of cuts

RELATION OF RIGIDITY ON REINFORCEMENT IN A METHOD OF LIMITING EQUILIBRIUM

The plates, supported on an outline, calculate by a kinematic way of a method of a limiting equilibrium more often, in which start, that the cracks on a bottom face of a plate are directed on bisectrixes of corners. In a reality the slope angle of cracks can be various. It depends on a rigidity of a plate in different coordinate directions (reinforcement). It is very important to spot a real pattern crack-derivations of reinforcement concrete plates, supported on an outline, for calculation of their carrying capacity and construction of the reinforce.

Conditionally we substitute a plate by the system of two mutually perpendicular intersected beams. The equation of the moment in a beam is written as (see fig. 3): $M=\frac{q*x^2}{2}$ (9)

here q - load intensity come on one of coordinate directions, particular on beam-analogy depending on a relation reinforcement , X - current plane coordinates of a plate.

Moment of crack-derivations in first beam shall write as: $M_{crc1}=\frac{q_{crc1}*x^2}{2}$, and moment of crack-derivations in 2 beam : $M_{crc2}=\frac{q_{crc2}*x^2}{2}$, during derivation of crack in the second beam value of the moment in the first beam: $M_1=\frac{q_1*x^2}{2}$ (10)

At this time start to be derivate angular cracks. Therefore, distance Y between ordinates of two graphs (Figure 3) at the moment of derivation of cracks in 1-st beam M_{crc1} also is half of length of a crack in one direction, ie passing from beam-system to a plate, distance for y'=2y is a maximum length of a longitudinal crack up to a place of appearance of angular cracks, and it in turn enables to spot a corner of their declination.

$$y_1 = \frac{q_{crc1} * \left(\frac{l_1}{2} - x\right)^2}{2} + a \;;\; y_2 = \frac{q_1 * \left(\frac{l_1}{2} - x\right)^2}{2} \tag{11}$$

$$y_1\Big|_{X=0} = \frac{q_{crc1} * l_1^2}{8} + a \;;\; y_2\Big|_{X=0} = \frac{q_1 * l_1^2}{8} \;;\; a = \left[(q_1 - q_{crc1}) * l_1^2\right]/8 \tag{12}$$

Having substituted in an input equation (12) and considering Y1 = Y2, we shall receive:

$$(q_1 - q_{crc1}) * l_1^2 = 4 * q_1 * \left(\frac{l_1}{2} - x\right)^2 \tag{13}$$

$$x^2 - l_1 * x + \frac{l_1^2}{4} * \left(1 - \frac{q_1 - q_{crc1}}{q_1}\right) = 0 \tag{14}$$

Solving this quadratic equation we shall receive:

$$X = \frac{l_1}{2} * \left(1 + \sqrt{1 - \frac{q_{crc1}}{q_1}}\right) \;;\; \text{and } y = x - \frac{l_1}{2}, \quad y = \frac{l_1}{2} * \sqrt{1 - \frac{q_{crc1}}{q_1}} \tag{15}$$

$$\mathrm{tg}\,\alpha = \frac{l_1}{2} * \frac{2}{l_2 - y'} = \frac{l_1}{l_2 - y'} \;;\; y' = 2y = l_1 * \sqrt{1 - \frac{q_{crc1}}{q_1}} \tag{16}$$

$$\text{Therefore: } \mathrm{tg}\,\alpha = \frac{l_1}{l_2 - l_1 * \sqrt{1 - \frac{q_{crc1}}{q_1}}}, \tag{17}$$

where l_1 and l_2 - spanungs (to smaller and greater beams); q_{crc1} - distributed load in a smaller beam at the moment of derivation of cracks in it; q_1 - distributed load in a smaller beam at the moment of derivation of cracks in a major beam.

The carrying capacity of a floor can be spotted under the transformed known formula:

$$\frac{gl_1^2}{12}(3l_2 - l_1 ctg\alpha) = (2M_1 + M_I + M_I')l_2 + (2M_2 + M_{II} + M_{II}')l_1 \tag{18}$$

EXPERIMENTAL DATA

For check of the output formula (17) the pattern of corrupting of an experimental plate was used. Were compared theoretical length of a crack and experimental. Two models continuous reinforcement concrete plates (P1 and P2), square in the schedule were made, with various reinforcement [9, 10].

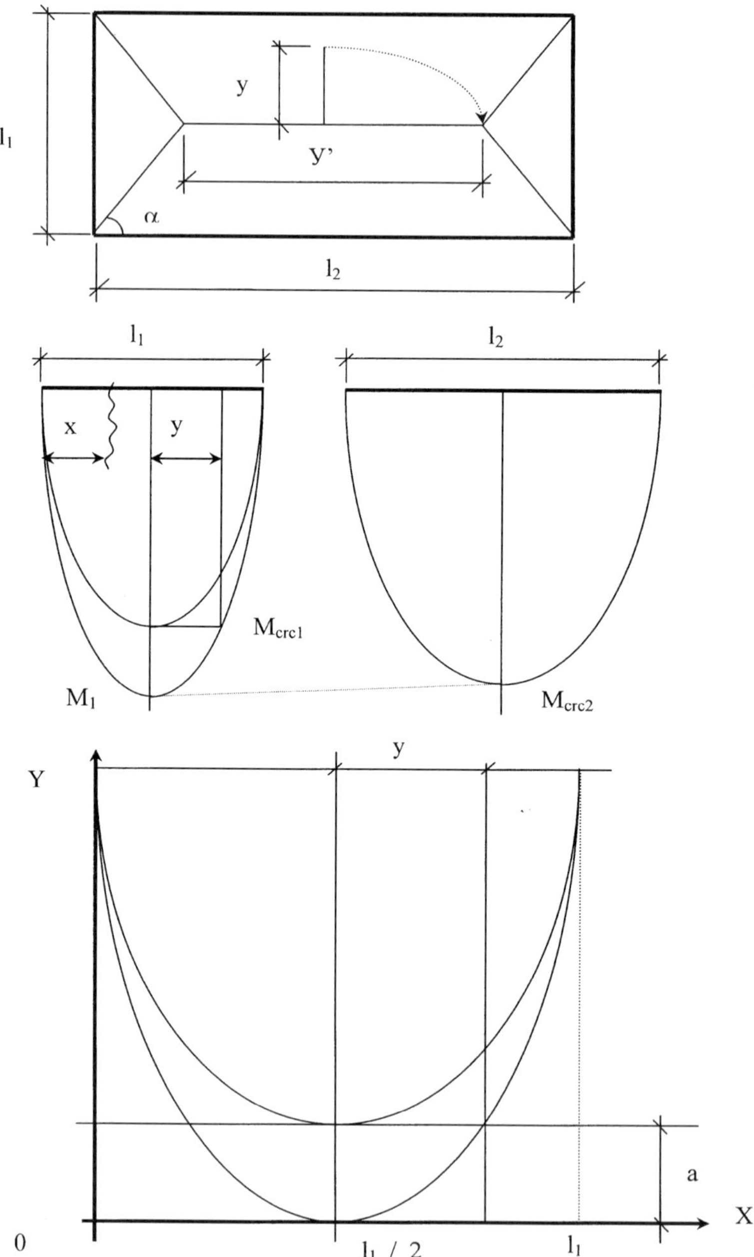

Figure 3 The scheme to definition of a pattern of layout of cracks on a bottom face of an plate in a method of a limiting equilibrium

A size of plates: 620 x 620 mm, width of 40 mm, concrete of the class B10, reinforce - with a low contents of carbon a wire by a diameter of 4 mm. The plate P1 had identical on two mutually perpendicular directions reinforce - in each direction allocated till 7 rods. The plate P2 had various reinforce - in one 5 rods, and in the friend - 9 rods allocated. The test bed consist of a basic outline tested plate, press and system traverse. The outline was made by 8 dynamometers working by a principle of a strain measurement, located on corners and middle of the sides of tested plates. To plates by press affixed the concentrated stress converted into uniformly distributed load by means of the system traverse. The pattern crack-derivations had the three-disk system in two directions, that hampered definition of a slope angle of cracks. But, by reviewing derivates in center of a plate of a rectangular site, restricted breaks, it is visible, that length of this site rather precisely corresponds to length of a central crack calculated theoretically.

Lattice supported on an outline of a floors have a common nature of allocation of stress between coordinate directions with continuous plates. It is visually shown due to a method of grids [8, 9, 13]. The projection of cracks on a bottom face of lattice floors on a horizontal plane, as show experimental dates, gives a pattern similar to continuous plates [9].

The produced calculation of a number of plates, identical on sizes, but having various coefficients of reinforcement, and also distinguishing by layout of the reinforce, confirms influence of the schemes of reinforcement supported on an outline of plates on character of allocation of cracks on their surface and, accordingly, on allocation of stress in a plate. The obtained slope angles of angular cracks differ from a classical corner ($\alpha=45^{o}$), limiting equilibrium, accepted now in a method.

CONCLUSIONS

1. The coefficients of relationships between rigidity and parameters of reinforcement (coefficients orthotropi) influence primary crack-derivations on floors, and therefore, influence allocation of stress between two directions of the schedule.

2. For output of the expression of the slope of angle of a crack, it is possible to use beam-analogy, considering the directions of a plate as independent beams having levels of the moment of crack derivations.

3. The allocation of stress between directions depends on the level of load. At the operation load $P = (0.4 – 0.6)\ P_{max}$, depending on coefficients of a relation reinforcement (0.4 – 1.6), the stress differs by 50 % and the limitation superimposes a range of change of declination of a corner (from 30^{0} up to 60^{0}).

4. The given operation allows estimation of the stress-deflection state of existing reinforced concrete floors. The method also determines that where damage is occurring eg non-uniform carbonation of concrete and corrosion of the reinforcement in separate areas, there is a reallocation of stress, which can be estimated only by means of the measured parameters.

REFERENCES

1. SNiP 2.03.01-84. Concrete and reinforcement concrete constr. Rus. Moscow 1989.

2. KARPENCO N I, The theory of deformation of reinforced concrete with cracks. Rus Moscow 1976.

3. KRYLOV S M, Reallocation of stress in statically indeterminable reinforced concrete constructions. Rus Moscow 1964.

4. BONDARENKO V M, BONDARENKO S V, Engineering methods of the nonlinear theory of reinforced concrete. Rus Moscow 1982.

5. SYRIANOV V S, Space operation of reinforced concrete plates, supported on a outline. / a Thesis on Ph D. Moscow 1990.

6. DAVRANOV V J, Features of operation it is not enough reinforced supported on a outline of plates of floors of buildings / a Thesis on Ms D. Rus Moscow 1992.

7. KOBEISI A M Influence orthotropi of a reinforced on the shape of collapse of reinforced concrete plates / A Thesis on Ms D. Rus Moscow 1992.

8. PLOTNIKOV A N, AIVASOV R L, Influence of a relation of rigidities of beams and boundary conditions on a response of a supported outline cross - ridge floors/Building constructions. Cheboksary 1993.

9. PLOTNIKOV A N, Location and reallocation of stress in supported on a outline reinforced concrete grid-beams composite overlapping / Materials of the All-Russia scientifically practical conference of the young scientists Moscow 2000.

10. IVANOV A A, SALMIN E A, AMOSOV A G, PLOTNIKOV A N, Experimental researches of allocation of intensity of basic pressure of a supported outline of anisotropic reinforced concrete plates / Materials of scientific student's conference. Cheboksary 1998 - 2000.

11. CALCULATION OF REINFORCED CONCRETE CONSTRUCTIONS ON STRENGTH, RESISTANCE OF CRACKS AND DEFLEXIONS / A S Salesov, E N Kodysh, L L Lemysh, I K Nikitin – Moscow 1988.

12. C T MORLEY, On the yield criterion on an orthogonally reinforced concrete slab element. Journal Mech Phys Solids 1966, Vol 14.

13. J D RENTON, On the gridwork analogy for plates. Journal Mech Phys Solids 1965 Vol 13.

14. M SAVE, A consistent limit-analysis theory for reinforced concrete slabs. Magazine of concrete research, Vol 19 N 58 1967.

DESIGN ISSUES IN THE ALIGNMENT OF STEEL FIBRES IN CONCRETE SLABS USING A MAGNETIC FIN

R P West

Trinity College Dublin

Ireland

ABSTRACT. The inclusion of steel fibres in concrete slabs for shrinkage crack resistance is inherently inefficient due to their random orientation through the slab depth. This paper describes the development of a prototype device which has been used successfully to align the fibres onto a horizontal plane at a chosen depth below the surface, using compact permanent magnets. A fin, housing the magnets, is attached to a vibrating screed and is dragged through the fresh concrete. The various conflicting design parameters are discussed in detail, including the choice of magnet, fin and concrete workability requirements. The effectiveness of using this device is evaluated and perceived potential benefits are identified.

Keywords: Steel fibre, Shrinkage cracking, Slabs on grade, Alignment, Magnets.

R P West is a Senior Lecturer and former Director of the Structural Laboratory at Trinity College Dublin. His research interests in concrete lie in durability, rheology and new materials. He is a former Vice-Chairman of the Council of the Irish Concrete Society. He is currently the Irish liaison Engineer for Eurocode 2.

BACKGROUND: REINFORCING FLOOR SLABS ON GRADE

The conventional means of restricting the formation of long-term shrinkage cracks in concrete slabs on grade is to reinforce the slab with a light steel mesh, placing it, preferably, at 5cm from the top of the slab. Although this method is relatively cheap in material terms, it is prone to be ineffective due to difficulties in locating the mesh properly during the casting operations. Achieving an acceptable level of workmanship such that the reinforcement is fully effective, despite being a primitive technique, is often not achieved on site.

The increased use of steel fibres as an alternative to mesh in these circumstances reflects the technical advantages that are perceived by specifiers and clients. Millions of short thin fibres are added to the concrete at the mixing stage and disperse themselves relatively uniformly, but randomly in orientation, throughout the mix. This allows any shrinkage cracks to be distributed over the surface as micro cracks, rather than as a number of unacceptably large discrete cracks. While the issue of depth of placement is no longer relevant, material costs are inevitably higher as the steel fibres are recognised as being an expensive additive to concrete.

In concrete mix design, fibre related parameters include the shape and size of the fibre, the fibre's aspect ratio (length to effective diameter) and dosage rates. All of these parameters can be optimised to improve the slab's performance. The tensile strength of the fibres is rarely an issue as it is the anchorage/bond of the fibres to the cement matrix which usually determines the pull-out characteristics. In particular, the shape, diameter and length of the fibres can be chosen to ensure that this is the case.

As well as improving the shrinkage resistance [1], fibres also improve the concrete's post-cracking ductility (toughness) and its impact and abrasion resistance. Improvements in the flexural and shear strength are also of importance [2], although, despite popular belief, there is only a negligible increase in the compressive strength of concrete that contains fibres. On a practical level, the increases in joint spacing, with subsequent lower maintenance costs, are a major advantage in industrial superflat floors.

However, in addition to its high cost, there are several other disadvantages to fibre reinforced concrete. Firstly, the expensive fibres are distributed throughout the depth of the slab, not just at the surface where they are most needed. Further, the fibres are randomly orientated, instead of being aligned in the direction of maximum tensile stress (that is, horizontally). For example, many fibres will have a vertical component in their orientation, a direction in which there is no significant tendency for the concrete to shrink. Both of these facets mean that fibres, overall, are inherently inefficient and, hence, are costly and wasteful, which is clearly undesirable. Secondly, there are workability problems (such as slump loss and balling of the fibres) that require changes to the mix proportions.

Furthermore, unless the concrete is properly designed and finished, it is possible that the steel fibres will protrude out through the concrete surface, rendering the slab unsafe and unusable. While labour costs are arguably less, skilled workers, as well as careful mix design, are required to achieve an acceptable level of finish.

ALIGNMENT OF STEEL FIBRES USING PERMANENT MAGNETS

In terms of costs, it was believed [3, 4] that there could be significant advantages if it were possible to show that an aligned fibre concrete slab could maintain its primary functions while reducing the fibre dosages and slab depth. In terms of performance, it was anticipated that, *inter alia*, aligning the fibres would improve the shrinkage characteristics and flexural strength, while also reducing the risk of fibre protruding the top surface.

The concept of aligning steel fibres *per se* is not a new one – successful alignment has been achieved in the pre-cast industry using electromagnets. This was done without aligning them on a single plane in the concrete; they were simply rotated into a horizontal direction without moving them vertically through the slab depth. However, advances in magnet technology have made possible the use of small, but targeted, strong, permanent magnets in many industrial applications in recent years.

In this context, a Swedish inventor, Bjorn Svedberg, took out a patent on the concept of attracting and aligning steel fibres onto a horizontal plane at any chosen level below the surface of the concrete in a controlled fashion using compact magnets. His idea (Figure 1) was to place a magnet inside a fin and to drag this fin through fresh concrete, attracting near-by fibres to the magnet and, somehow, aligning the fibres on a single plane in the wake of the fin. At this point, the author was approached to develop a practical alignment device such that proof of its technical viability could be established.

It was decided to attach the fin to a vibrating screed to take advantage of the rheological fluidity of the concrete during vibration, thereby increasing the mobility of the fibres under the influence of a given magnetic field. The fin tail could be placed at a depth such that the final fibre layer would indeed be, say, 5 cm below the surface. The magnet inside the fin would have to be designed to attract far-field fibres and to drag them around the leading circumference of the fin, releasing and depositing them in the wake of the tail of the fin. Hence, the magnet would have to be rotated at a particular angular velocity as the screed passes with a corresponding horizontal velocity along the guide rails of the screed.

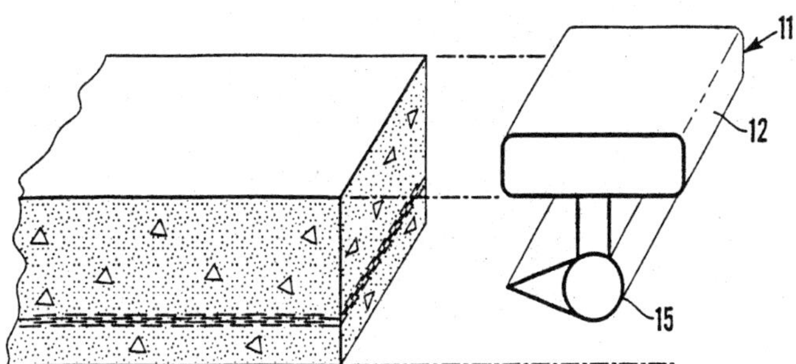

Figure 1 Svedberg's original patent sketch of the concept of using a fin to align fibres. (Note, in the patent sketch, 11, 12 = Vibrating screed, 15 = magnetic fin)

Alignment Device Design

The following parameters were considered by the author in the design of a prototype alignment device:

- magnetic field strength: strong enough to pull fibres from as large a catchment area as possible,
- magnetic field shape: a dipolar shape (Figure 2), not uniform, makes it easier to drag fibres around the leading edge of the fin and ensures that fibres attracted to the fin will experience a period without the influence of a magnetic field, making the release of fibres out the back of the fin easier,
- angular velocity of magnet: a slow enough rotation of the magnetic field to expose the fibres for a long enough period to enable them to become attached to the fin, yet not so slow that the release of fibres out the back is staggered due to lack of co-ordination between angular velocity and screed horizontal speed,
- horizontal velocity of screed/fin: set at a practical concrete placing and finishing rate,
- concrete rheology: appropriate to magnet and fibre type, the concrete's workability needs to be fluid enough to promote fibre mobility,
- drag on the fin: minimise the diameter of the fin and optimise its shape so that the magnitude of the force required to drag the fin though the concrete is achievable in practice,
- damage to the concrete in the wake of the fin: the flow of concrete around the fin may be disruptive to the homogeneity of the concrete, thereby reducing density and strength [5],
- robustness of prototype: the alignment device should be robust and reliable such that it can repeatably align fibres for a wide range of fibre types and workabilities.

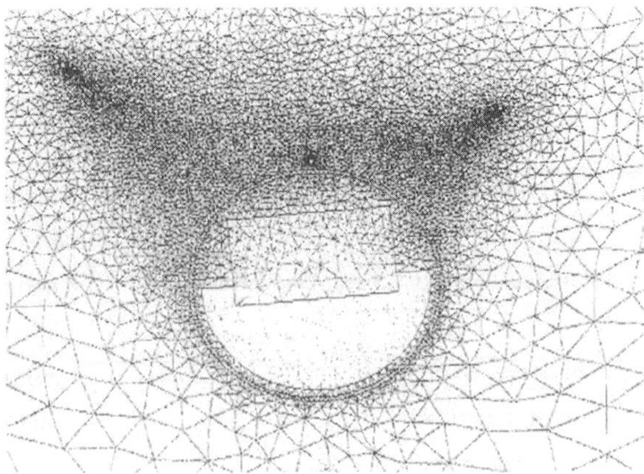

Figure 2 Targeted di-polar magnetic field which facilitates the dual role of far-field attraction and release of fibres from the fin

ENGINEERING ISSUES

The question arose as to whether these competing design requirements could be reconciled to successfully align the fibres. Consideration of how the alignment device would function suggested that the following design objectives needed to be addressed (Figure 3):

- fibres should be attracted from as far afield as possible, including all fibres from the top surface of the slab. One would need a strong magnetic field to achieve this, but not so strong that the fibres wouldn't be released out the back of the fin. If this release were not to occur, the fibres would accumulate on the fin, thereby blocking the magnetic field (which would now pass only through the attached fibres, freezing the attraction process),
- non-viscous concrete is required to allow the maximum quantity of fibres to be attracted to the fin. However, the concrete must also be sufficiently viscous to remove the fibres off the tail of the fin by a shearing/dragging type action,
- a suitable field shape must be chosen to drag the fibres around the leading circumference of the fin and yet release them unto the tail for removal,
- magnetic field shielding (by provision of a steel plate on the top surface of the tail of the fin) will be required to enhance the release of the fibres as the magnetic field rotates. If this were not provided, it was believed likely that the magnet strength and rotational velocity would be such as to drag the fibres back (against the direction of flow over the fin) towards the front of the fin,
- a suitable fin shape and size must be chosen to minimise the drag arising from the fin passing through the concrete - is it physically possible to manhandle the fin through the concrete?

Magnet Design

In selecting the magnet, field shape, strength and magnet size needed to be designed. In particular, the following were undertaken:

- the tripartite relationship between magnetic field strength, radius of influence (the movement of a single fibre with time) and viscosity of concrete needed to be established. This is also affected by fibre size as larger fibres experience a greater force, but have a larger natural resistance to movement,
- wallpaper pastes of varying powder to water ratios, with similar viscosities to cement
- pastes, were used to observe the movement of fibres under the influence of different magnets (Figure 4),
- the percentages of fibres attracted to 10 different magnets which were stationary or rotated in the paste over various time periods were established. It was possible to observe far field fibres rotating, initially slowly, as they were pulled end-on towards the magnet.
- the time of exposure to the magnetic field as it rotates is one parameter which determined the velocity of magnet rotation and screed movement.
- the effect of adding aggregates to the paste was established, such that the larger particles provided an impediment to the passage of fibres towards the fin.

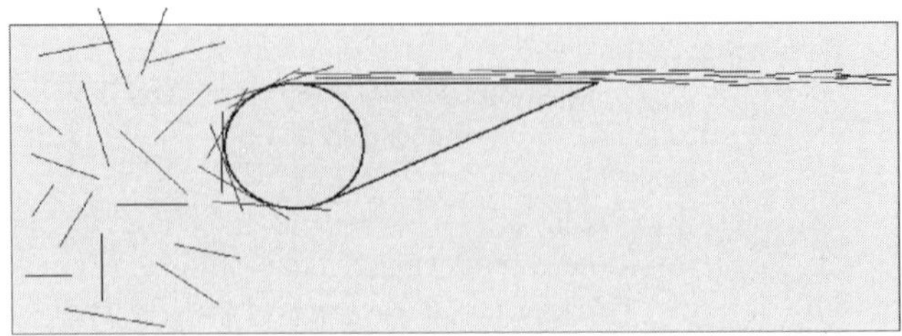

Figure 3 Concept of attracting far-field and protruding fibres onto a common horizontal plane

Following these fundamental tests, a closed perspex channel was constructed (Figure 5) which allowed the movement of fibres in a flowing viscous medium under the influence of static and rotating magnets to be observed. Flow velocities and magnet rotational velocities were varied for a number of magnets and each test was recorded on video for study [6].

Subsequently, the most promising magnets were selected for a reduced study using a variety of concrete mixes. Fibercon 25 mm crimped and Dramix 60 mm hooked fibres, at dosage rates of 40 kg/m^3, were incorporated into different mixes to evaluate the percentage of fibres attracted to the static or rotating magnets when the concrete was unvibrated or vibrated.

The static results gave an indication of the effectiveness of the field shape and strength while the rotating magnets allowed the maximum number of fibres to be attracted, taken from a complete cylindrical catchment around the magnet (see Figure 4(a) and (b). The radius of this cylinder determined the maximum quantity of fibres which could be attracted, assuming a uniform dispersion of fibres in the original mix, at the selected dosage rate.

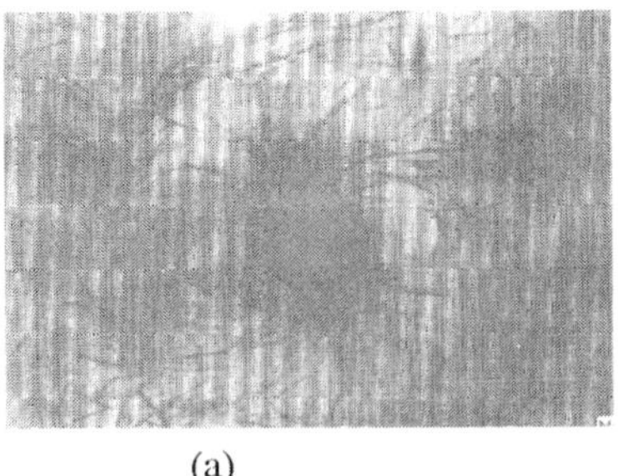

(a)

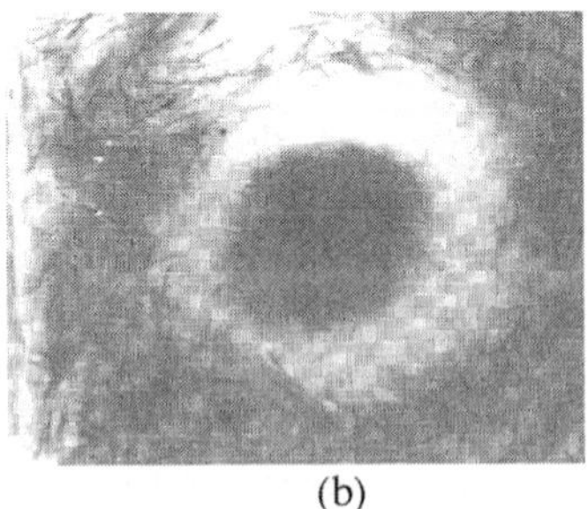

(b)

Figure 4 Plan photograph of fibres attracted to a typical magnet in a viscous medium: *a) static and b) rotated*

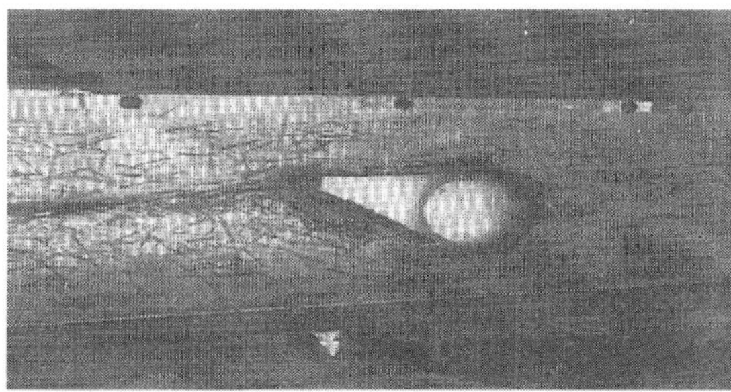

Figure 5 Perspex channel trials with steel fibres added to wallpaper paste. The influence of the magnet (not shown) and fin may be seen in aligning the fibres in the wake of the tail

It was concluded that a bi-directional magnet was selected (it being better at dragging the fibres around the leading edge than a uniform field), with a specific strength depending on the fibre type. Further, a range of rotational velocities could be recommended for particular screed velocities in order to avoid fibre layer staggering. These results were transmitted to a Swedish company, HYAB Magnater AB, which specialises in magnet design and manufacture. They conducted their own adaptive meshing Finite Element Analysis of the proposed magnets to confirm field spread, shape and strength (Figure 2) and to recommend the thickness of tail shield (placed at the top of the trailing edge of the fin) required to nullify the field in order to enhance the release of fibres.

Prototype Design and Testing

The construction of a prototype was undertaken (Figure 6), attaching the stainless steel fin to a truncated conventional screed. The fin enclosed a rotating cylinder that housed the magnets that had been manufactured by HYAB. An additional motor was added to drive the cylindrical magnet shaft at varying rotational velocities. Adjustments were also provided to allow variation in the location of the fin in front of the screed, the fin height below the slab surface and the fin angle.

An extensive series of small-scale tests were conducted on 3.2 x 1.2 x 0.2 m concrete slabs. For example, a Grade 35 concrete (maximum aggregate size 20 mm) with normal Portland cement and a slump of 150 mm was mixed with Fibercon's 25 mm crimped fibres at a dosage rate of 40 kg/m^3. The velocity of the screed was fixed at 0.6 m/min and the rotational speed of the magnet was varied from 15-25 rev/min.

The force required to drag the fin and screed along the screed rail was measured using a winch with a calibrated strain gauge, resulting in a range of forces from 200 to 400N [3] being required to pull a 1.2 m wide fin through the concrete.

The fibres remaining on the fin when it was removed from the concrete at the end of its run were weighed. The conclusion from this was that 97% of the total quantity of fibres in the path of the fin had been deposited in its wake. To verify the success or otherwise of the test, a wall-saw was used to cut a 20 mm thick slice along the centre of the main longitudinal axis of the slab, following an adequate curing period. These were then inspected using specialised X-ray photography from which it was clearly evident that the fibres had been aligned on a chosen horizontal plane (Figure 7). The fibres do indeed form a dense matrix of well-bonded planar reinforcement. There was no evidence of staggered alignment.

Proof-of-concept of aligning fibres had thus been established and it was, therefore, possible, with refinement, to ensure that no fibres would protrude out the top of a concrete slab. It was also possible to pre-determine the depth at which the fibres could be aligned by adjusting the fin height. It was noted that if fibres did accumulate on the fin there was a risk that magnetic blockage would occur, thereby freezing the alignment process. Nonetheless, there did appear to be scope for reducing fibre dosages and slab depths, a matter that merits further investigation.

Figure 6 Prototype alignment device in place on shutter, prior to pouring concrete

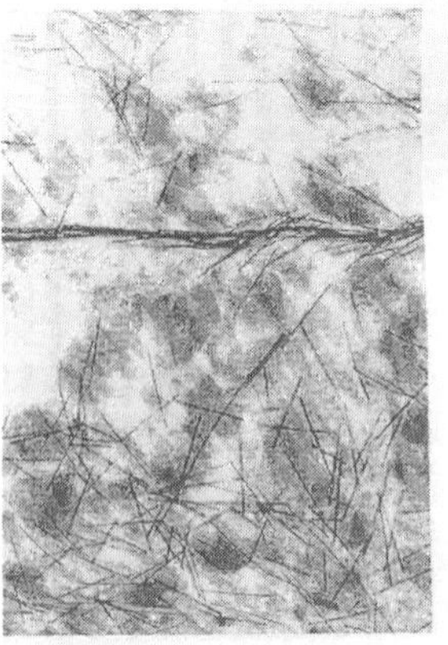

Figure 7 X-ray of 20 mm thick concrete strip, 200 mm deep, with aligned Fibercon fibres

Aligned Fibre Slab Performance

Regardless of any secondary advantages of aligning the fibres, there must be strong evidence that one of the primary functions of the fibres, namely shrinkage cracking control, is not impaired by the alignment process. With a view to monitoring the long term shrinkage performance of plain concrete, randomly distributed fibre concrete, aligned fibre concrete (of varying dosages and fibre type) and conventional mesh, a series of slabs were tested [1]. Indications were that there is no significant difference in the long-term shrinkage strain results as a consequence of aligning the fibres. However, aligned fibres are more effective than randomly distributed fibres, so some scope for reduced dosages may still exist.

Taking advantage of the post-cracking ductility of fibres *per se* is further exploited by aligning the fibres at a chosen depth. While it is recognised that neither fibres nor mesh are used alone to provide significant flexural resistance, the aligned fibres do enhance the performance of the concrete. In particular, there appears to be a discernable and statistically significant benefit in aligning the fibres in terms of reduced deflections and increased ultimate strengths over their non-aligned counterparts [2]. This relies on individual fibres being fully bonded to the matrix, bridging the tensile cracks as they occur. This fact would allow a reduction in the fibre dosages for similar performance levels to unaligned fibres. Quantifying the extent of this reduction is currently under investigation.

It appears, therefore, that it is possible to align the fibres such that they form a continuous matrix and there appears to be potential for passing a low voltage electrical current through the fibres with a view to heating up the slab. This has exciting possibilities in the areas of early age strength development and accelerated drying, aiding in cold weather concreting and de-icing of pavements in exposed trafficked/public areas.

CONCLUSIONS

The relevant competing design parameters, which influence the feasibility of aligning steel fibres onto a chosen horizontal plane within a concrete slab have been discussed. Factors affecting the decisions to be made regarding magnet and fin design and workability requirements have been described. It is concluded that the concept of aligning fibres in concrete slabs using permanent magnets is proven to be technically viable. Perceived benefits include elimination of the problem of protruding fibres, reduced cost due to reduced fibre dosages for similar performance and numerous potential applications due to heating a slab containing a continuous matrix of interconnected aligned fibres. Prototype refinements are being undertaken to optimise the effectiveness of the alignment, improve its scope and increase its robustness and manufacturability. Full-scale trials are planned and a market survey has been undertaken to establish the industry's response to the possibility of a new market product.

ACKNOWLEDGEMENTS

The Author wishes to acknowledge with thanks the financial support of Readymix (Ireland) plc for the current phase of the experimental work. Thanks are also due to Mr Bjorn Svedberg for his technical advice.

REFERENCES

1. HANNANT, D., Fibres in industrial ground floor slabs, Concrete Society Technical Report 28, Slough, 1994.

2. PULL, M., WEST, R.P., Flexural strength of aligned steel fibre reinforced concrete, Confidential Technical Report, Trinity College Dublin, 2000, pp 198.

3. WEST, R.P., IJOMAH, H.N., The alignment of steel fibres in concrete slabs using permanent magnets, 2001, submitted for publication.

4. IJOMAH, H.N., WEST, R.P., The heating of concrete with aligned steel fibres, 2001, submitted for publication.

5. WEST, R.P., CUOMO, G., Strength loss in the wake of a cylinder dragged through a concrete slab, Colloquium on Concrete Research in Ireland 2001, NUIG, 2001, pp 87-91.

6. WEST, R.P., HINDS, J.W.G., Assessment of magnets for use in the alignment of steel fibres in concrete slabs, Confidential Technical Report, Trinity College Dublin, 1999, pp 65.

SIMULATION OF PHYSICAL AND MECHANICAL PROCESSES IN CONCRETE FLOORS AND SLABS

E Schlangen

T Lemmens

T van Beek

INTRON

Netherlands

ABSTRACT. Cracks in concrete floors can be caused by many mechanisms. Thermal effects, internal shrinkage, drying, mechanical loading are all aspects that can lead to cracking. These cracks can cause failure. More often the cracks will be a problem for the durability aspects, water tightness and aesthetics. Cracks can be prevented and controlled by proper design of the concrete mixture, reinforcement and geometry. To check the design of a structure on the risk of cracks due to physical and mechanical phenomena described above, the stresses occurring in concrete should be made available to the engineer. The development of user-friendly software is an excellent way to provide this knowledge to the engineer. FEMMASSE developed a simulation tool called the module HEAT. With this module the engineer can simulate the physical and mechanical in young and old concrete. This model combines the effects of drying shrinkage with autogenous shrinkage, thermal dilation and external loads by using the state parameter approach. In this paper the theory used in the model will be explained. An outline will be given of the computer program. In an example the effect of the design of a structure on the risk of cracks will be presented.

Keywords: Finite element method, Temperature, Moisture, Floors, Slabs, Drying, Stresses.

Erik Schlangen is a Senior Consultant at INTRON. He also works as a Research and Development specialist for FEMMASSE. He received his MSc at Eindhoven Technical University, the Netherlands and his PhD. at Delft University of Technology, the Netherlands. He has published over 60 papers in the field of material engineering and fracture mechanics.

Ton Lemmens is a Senior Consultant at INTRON. He has many years experience in the design, fabrication and damage assessment of concrete floors.

Ton van Beek is a Consultant in Civil Engineering at INTRON. He received his MSc and PhD at Delft University of Technology, the Netherlands. His topics of interest include non-destructive test methods for civil structures and material properties of young concrete.

INTRODUCTION

Shrinkage due to hydration, thermal effects and drying is regarded to be a cause for damage in concrete structures for many years. Not only the cracks in the concrete can be a problem. Due to the deformation of the floor or slab overlays may be damaged (see Figure 1).

During hydration the volume of cement paste changes, which results in autogenous shrinkage of the concrete. If concrete is exposed to a dry environment, water will evaporate from the cement paste, which will lead to drying shrinkage. In reality, however, damage mostly occurs if there is a combination of both autogenous and drying shrinkage. Research has therefore been focussed on both of these phenomena. To translate this knowledge of shrinkage to the engineer in deformations and stresses the use of models has been widely accepted.

HEAT is developed specially to simulate the effects of the environment, like temperature and humidity, on the material state and properties of concrete. The complex calculation procedure has been incorporated in user-friendly software. In a simple manner a two dimensional drawing of the structure in view can be made. By applying external action on the boundaries of the structure, like humidity, temperature, radiation etc. the effect of the environment on the material state and properties can be simulated.

Young concrete can shrink during the hydration process due to autogenous shrinkage and due to the drying process. Both processes are highly influenced by the mix proportions of the concrete. The cement type, water cement ratio and the amount of aggregate determine the shrinkage. Shrinkage only causes cracks if the stresses exceed the tensile strength. The level of the stresses depends strongly on the restrained conditions the structure is in. The stresses are therefore depending on the geometry of the structure.

Another important issue when dealing with young concrete, especially in massive concrete structures, is the dilation and stresses due to the heat of hydration. Although it is incorporated in the model and in the simulations presented in this paper, the thermal effects will not be the focus of this paper. For thermal effects in concrete the authors refer to the paper COMPUTER SIMULATIONS AS A DESIGN TOOL FOR DURABLE CONCRETE also presented in Dundee.

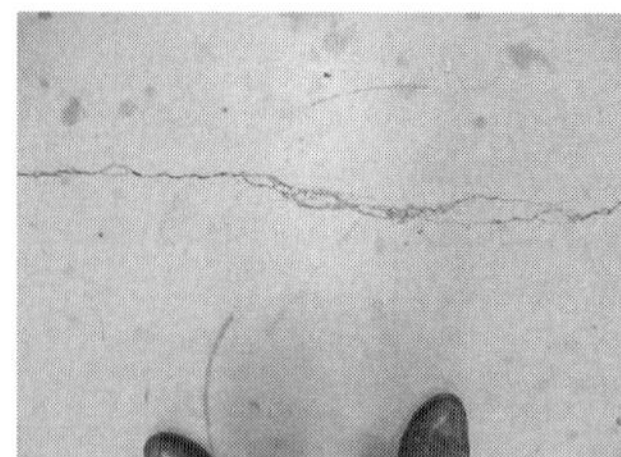

Cracks in concrete floors

Loss of adhesion in overlays

Cracked tiles

Figure 1 Examples of damaged floors in practice

THEORY OF SHRINKAGE IN CONCRETE

Shrinkage of (young) concrete is a complex phenomenon. Several non-linear processes, like formation of microstructure, development of physical and mechanical properties and transport of moisture are involved. All these processes can be simulated, both separately or combined, with the module HEAT of FEMMASSE. The module is based on the finite element method. The non-linear equations are solved using time integration. The formulation of all the equations describing the processes are not described in this paper. The reader is referred to [1] for details.

The microstructure of the concrete is formed during the hydration process of the cement. The degree of hydration (or the related property maturity) determines the transport properties of the material as well as the mechanical properties. In the module HEAT the state parameter concept is used. In this concept all the material properties are a function of the state of the material. This state (or the state parameter) can be the maturity, degree of hydration (DOH), temperature, moisture potential (also known as relative humidity) or moisture content, which can be chosen by the user, depending on the problem to be tackled.

In the simulation of the drying experiments, the material properties depend on the state parameters in the following way:

- Maturity and DOH of the material is calculated as function of the age, temperature and the moisture content.
- The diffusion coefficient for temperature and the heat capacity is a function of the moisture content.
- The diffusion coefficient for moisture is obtained from experimental data [3,4,5,6]. The diffusion coefficient for moisture is a function of the moisture potential. [7].
- The moisture content (capacity) in the concrete is a function of the moisture potential [8], the DOH and the temperature.
- The mechanical properties strength and stiffness are a function of the maturity; The viscoelasticity (relaxation of the stresses in the material) is also a function of temperature and moisture potential. For the viscoelasticity a Maxwell Chain model is used.
- The shrinkage of the concrete depends linearly on the change in moisture potential as found experimentally by Jonasson et al [9].

The temperature, moisture potential, stresses and displacements are calculated as follows:

- The temperature change in a time step is caused by the heat development due to hydration, the heat diffusion governed by the environment and the change of thermal capacity.
- The change of the moisture potential in a time step is caused by the moisture diffusion governed by the environment and the change of moisture capacity (due to self-dessication, change in temperature or DOH).
- The change of stresses and displacements are analysed in each step as a result of the change in temperature, moisture potential and due to the viscoelasticity (and external loads).

The calculation procedure and theory will be explained in more detail in the following section. In this section the procedure of setting up a simulation with the module HEAT is explained.

HEAT MODULE

The theory behind the simulations presented in this paper has been incorporated in user-friendly software [1, 10]. The exact formulation of the equations can be found in [1]. In this software the engineer has to follow a step by step procedure as shown in Figure 2. By following this procedure the engineer can gain insight into the problem of shrinkage in his structure. He can see the effects on the stresses of the proposed improvements.

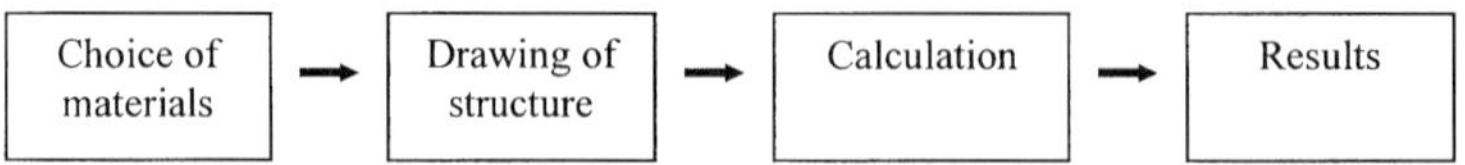

Figure 2 The HEAT procedure

Choice of Materials

The first step in the simulation procedure is the choice of materials. Different concrete mixtures have different material properties like autogenous shrinkage, modulus of elasticity, tensile strength and drying shrinkage. In young concrete these properties will not have a fixed value but will develop in time. The mix proportions determine the rate of the development.

The HEAT module has a huge database in which a lot of experimental data of many concrete mixtures are stored. For new mixtures, however, these properties will be different and new experiments will be conducted to gain the data needed for the calculations. In Figure 3 an example of the database in which the mix design controls the desorption isotherm is shown.

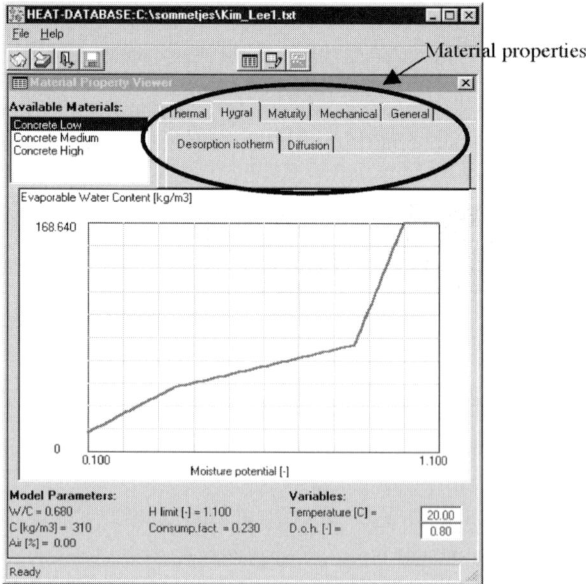

Figure 3 The material database

Design of Structure

When the materials (more than one is possible) of the structure are chosen the structure can be drawn in a design panel. Mechanical restrains have to be attached to the geometry. Next external loads have to be placed on the structure.

The external loads can be:

- Temperature
- Solar radiation
- Humidity
- Water pressure
- Forces

After dividing the structure in elements the calculation can start.

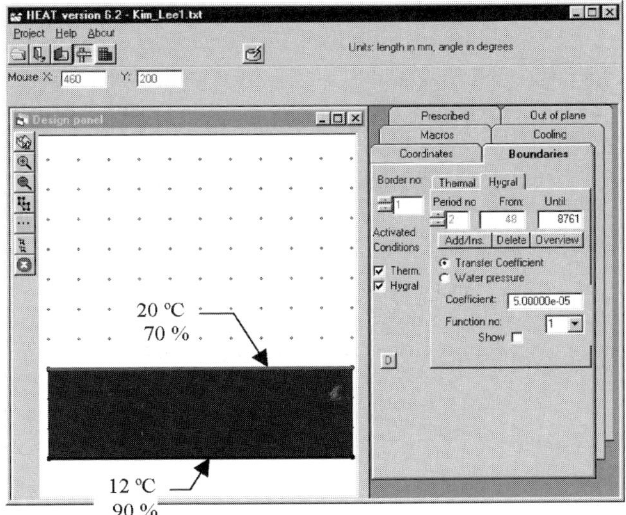

Figure 4 Drawing the structure in a two dimensional plane

Viewing Results

The results of the calculations can be shown in the two-dimensional plane, in line cuts and in graphs showing evolutions in time. By comparing the results of different simulations in which the concrete proportions have been changed, the geometry has been altered or other measures have been taken. The engineer can make the best choice for an optimal design.

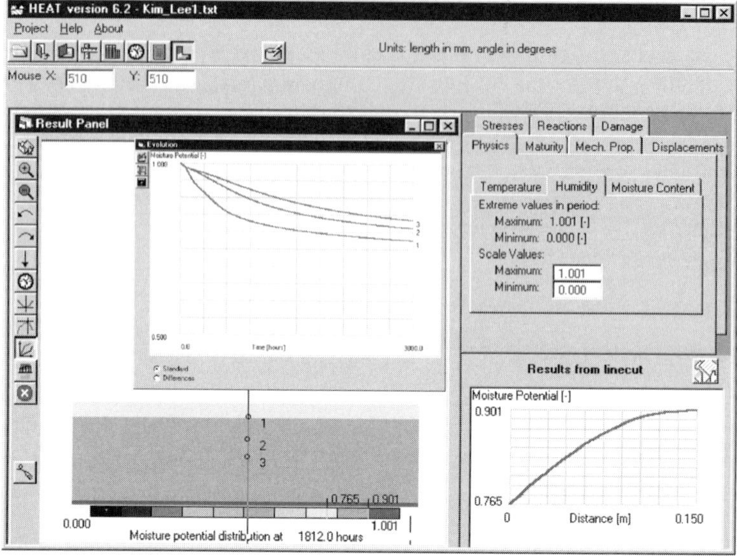

Figure 5 Results of the calculation

EXAMPLE OF A CONCRETE FLOOR

The Problem

In this example the effect of the mix design on the stresses due to shrinkage is shown. The structure is a new floor cast in a room with a relative humidity of 70%. The floor will be connected to the walls, which causes a restrained condition for the floor. After casting, the floor will be covered with plastic foil for two days to prevent drying. Figure 6 presents the layout of the structure.

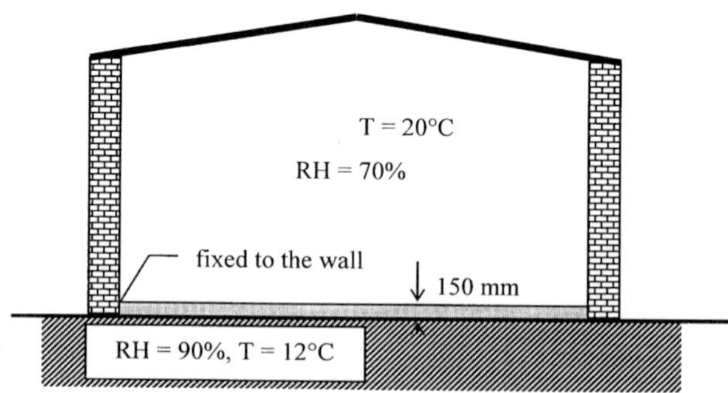

Figure 6 Structure of the floor and its environment

The Model

The floor has been modelled as a slab as could be seen in Figure 4. At the top of the slab the boundary conditions were the indoor environment (temperature 20° C and relative humidity 70%). The boundary conditions at the bottom were equal the temperature and relative humidity of the foundation. The deformations of the slab were kept completely restrained. The conditions were because of the connection of the floor with the wall. The material properties (see Table 1) used in this model were obtained from experiments from Kim, J-K and Lee, C-S [2]. The implementation of the results of the experiments into the HEAT model is explained in reference [11].

Table 1 Mix proportions of concrete

MIXTURE	W/C (%)	CEMENT CONTENT (kg/m^3)	fc (MPa)
A (low strength)	68	310	22
B (medium strength)	40	423	53
C (high strength)	28	541	76

Stresses in the Structure

The following simulation has been focussed on the stresses, which can occur due to shrinkage. Temperature effects due to hydration and the environment have been incorporated. When the concrete cracks the stresses will be lowered. The cracks have not been simulated in this example, but it is possible in the HEAT program to show the effect of cracks on the stresses by incorporating fracture mechanics. In Figure 7 it is shown that in the top layer of the floor the stresses will exceed the tensile strength. These stresses cause cracks in the top layer of the floor.

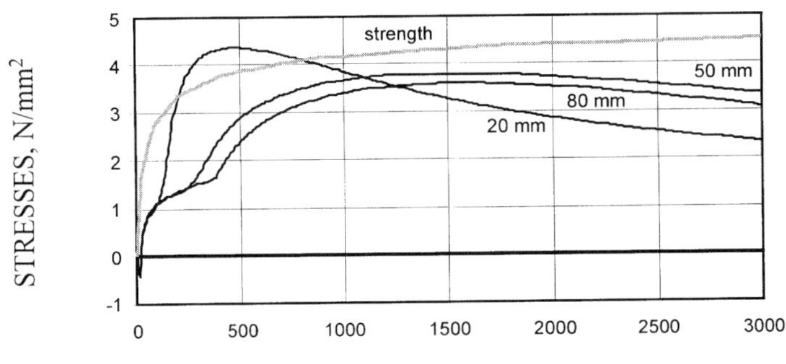

Figure 7 The stresses and strength development at 20 , 50 and 80 mm depth for mixture B

To prevent cracks to occur the mix design could be changed. With the knowledge that the stresses exceed the tensile strength it seems reasonable to use a mixture with a higher tensile strength, like mixture C (the high strength concrete). However another effect of using high strength concrete is that it has more autogenous shrinkage. From Figure 8 it can be seen that a structure with this mixture will have cracks too. By using mixture A with the lowest strength and the lowest autogenous shrinkage the stresses are so low that they do not exceed the tensile strength.

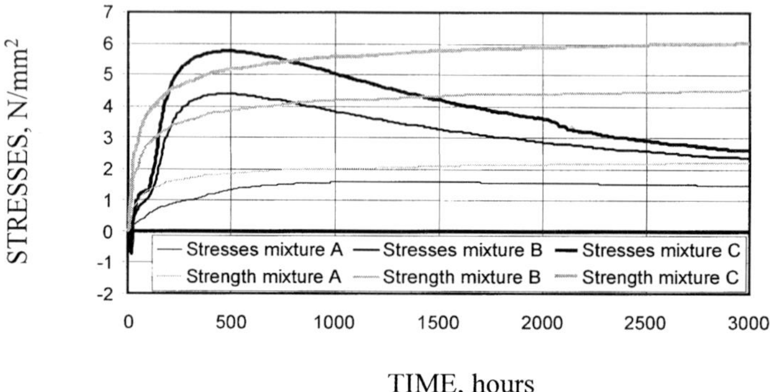

Figure 8 The stresses and tensile strength for the three mixtures at 20 mm from the surface

Another way to prevent cracks to occur is by preventing moisture evaporation in the early stage of hydration. By placing the foil for 7 days instead of 2 days the stresses will be lowered to a value below the tensile strength. The risk of cracks will be reduced significantly.

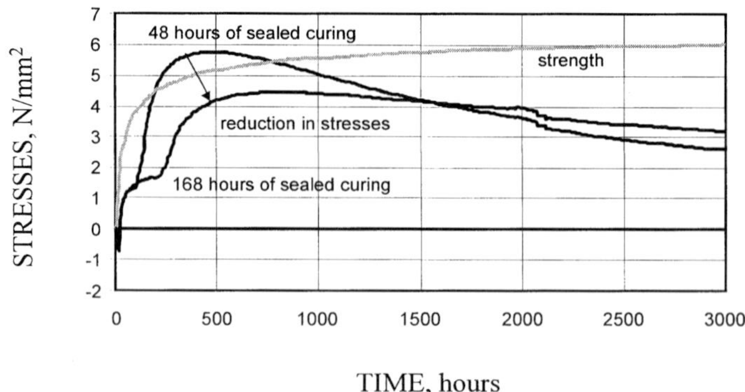

Figure 9 Reduction of stresses in mixture C by leaving the foil on the surface for 7 days

In this example it is shown how you can optimise the mix proportions to prevent cracks in a floor. The example also shows that taking adequate measures to prevent moisture exchange can reduce the stresses. In Table 2 the results are shown of the simulations. Also the effect of a very low relative humidity in the environment is presented, which causes cracks with all mixtures.

Table 2 Reduction of maximum stresses by adequate measures

RELATIVE HUMIDITY	MIXTURE A			MIXTURE B			MIXTURE C		
	σ	f_t	σ/f_t	σ	f_t	σ/f_t	σ	f_t	σ/f_t
50 %	2.8	1.8	1.6	5.9	3.5	1.7	7.4	4.8	1.5
70 %	1.6	2.0	0.8	4.3	3.8	1.1	5.8	3.1	1.9
70 % + 7 days foil							4.4	5.5	0.8

σ tensile stress (N/mm^2)
f_t tensile strength (N/mm^2)

FUTURE DEVELOPMENTS

The module HEAT is constant under development. At the moment various options are being incorporated to make it suitable for even more applications. One of the developments that is now being tested is the use of reinforcement in the concrete. The reinforcement can be used together with the already explained hydration, temperature and moisture effects. In Figure 10 two examples are given of the use of reinforcement.

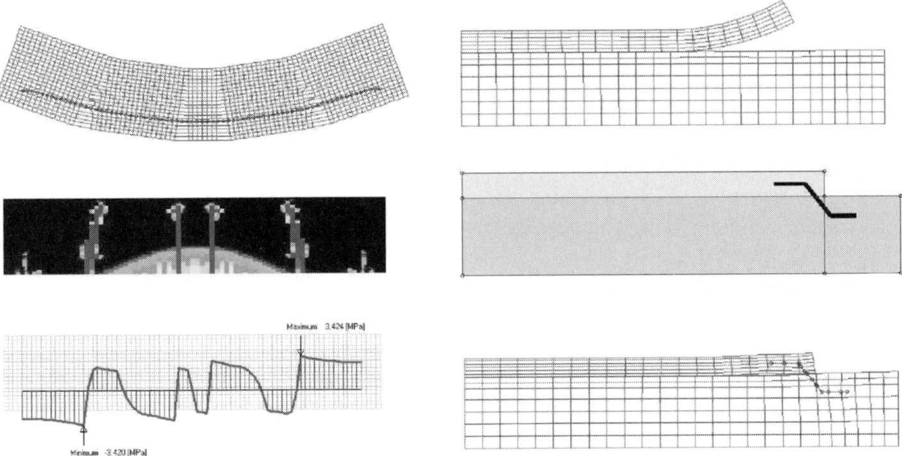

Figure 10 Cracking in a reinforced beam(a); deformed mesh (top), localized cracks (middle), shear stress at steel-concrete interface (bottom)

In Figure 10a, a simple 3-point-bending beam is shown with a reinforcement bar at the bottom of the beam. In the top figure the deformed mesh is given. In the middle figure the localized cracks are shown and in the figure at the bottom the shear stress in the interface between steel and concrete is given. In Figure 10b an application of another new development is given. The use of an interface element. This element is forming in this case the mechanical interface between a concrete slab and a concrete overlay. In the top figure it is clearly shown that debonding and curling of the overlay takes place. In the middle figure a steel reinforcement anchor is placed forming a fixation of the overlay to the slab. In the bottom figure the deformed mesh with the reinforcement is shown.

CONCLUSIONS

In this paper it is shown that numerical tools can be helpful for the engineer to predict the stresses in young and old concrete due to physical and mechanical processes, like hydration, temperature effects and drying shrinkage. By using these types of tools the design of the structure can be optimised and the effect of measures that are needed during construction can be visualised.

In practice the prediction of cracks can be difficult by the lack of information of the concrete mixture. The database available in the HEAT program can give the engineer the opportunity to show the effect of the choice of mixture. At the moment the model is often used to explain why cracks are found in the structure, while most of these cracks could be prevented in a simple manner. Especially also the new developments will give the engineer a tool to optimize the design of floor systems.

ACKNOWLEDGEMENTS

The authors are grateful to Peter Roelfstra for the fruitful discussion on the subject and for the collaboration in developing the module HEAT of FEMMASSE.

REFERENCES

1. FEMMASSE B V, Online Help/Manual module HEAT of FEMMASSE, 1990-2000, available on www.femmasse.com.
2. KIM, J-K and LEE, C-S, Moisture diffusion of concrete considering self-desiccation at early ages, in Cement and Concrete Research, Vol 29, 1999, pp 1921-1927.
3. HUNDT, J, AND KANTELBERG, H, Sorptionsuntersuchungen an Zementstein, Zementmortel und Beton, in Deutscher Ausschuss fur Stahlbeton, Vol 297, 1978.
4. ROELFSTRA, P E, A Numerical Approach to Investigate the Properties of Concrete, PhD-thesis EPFL – Lausanne, 1989, pp 199.
5. NILSSON, L-O, Hygroscopic moisture in concrete – drying, measurements and related properties, Lund Institute of Technology, Report TVBM-1003, 1980.

6. JONASSON, J-E, Modelling of Temperature, Moisture and Stresses in Young Concrete, PhD-thesis Lulea University of Technology, 1994, pp. 225.

7. BAZANT, Z P, Creep and shrinkage of concrete, mathematical modelling, in Proceedings 4th Rilem International Symposium, Evanstone, Illinois, USA, (1986).

8. NEVILLE, A M, Properties of Concrete, Longman Group Limited, Essex, UK, 1995, pp 844.

9. JONASSON, J-E, GROTH, P and HEDLUND H, Modelling of temperature and moisture field in concrete to study early age movements as a basis for stress analysis, in Thermal Cracking in Concrte at Early Ages, Rilem Proc. 25, ed. R. Springenschmid, E & F N Spon, London, 1994, pp 45-52.

10. ROELFSTRA, P E, SALET, T A M and KUIKS, J E, Defining and application of stress-analysis-based temperature difference limits to prevent early-age cracking in concrete structures, in Thermal Cracking in Concrete at Early Ages, Rilem Proc 25, ed R Springenschmid, E & F N Spon, London, 1994, pp 273-280.

11. BEEK A VAN, SCHLANGEN E, Simulating the effect of shrinkage on concrete structures, Shrinkage of Concrete - 'Shrinkage 2000', Rilem Pro. 17, Paris, France, 16-17 October 2000, Edited by V Baroghel-Bouny and P C Aïtcin.

PRELIMINARY RESULTS FOR LONG TERM TENSION STIFFENING EFFECTS IN RC MEMBERS

A W Beeby   A Antonopoulos

University of Leeds

R H Scott

University of Durham

United Kingdom

ABSTRACT This paper gives a brief description of a major collaborative project, financed by EPSRC and Industry, to investigate the behaviour over time of cracked tension zones in reinforced concrete members. The experimental work has been carried out at the Universities of Durham and Leeds and has consisted of long term tests on 44 prisms subjected to pure tension and 8 large slab slabs subjected to bending. The results so far show that tension stiffening reduces more rapidly with time than is normally assumed. It is hoped that a parallel theoretical study will result in the development of a significantly improved method for the calculation of the deformations of cracked reinforced concrete.

Keywords: Reinforced concrete, Deformations, Long term, Tension tests, Slab tests.

A W Beeby is Professor of Structural Design in the School of Civil Engineering at the University of Leeds. Prior to 1991 he had worked for the British Cement Association, formerly the Cement and Concrete Association, for 27 years. His main interests have always been the study of the behaviour of reinforced concrete and the introduction of research into design.

Dr R H Scott is Reader in Structural Engineering in the School of Engineering at the University of Durham, which he joined in 1978. Prior to this he had worked in industry for ten years. His main interest is in the behaviour of reinforced concrete structural elements, particularly the measurement of reinforcement strain and bond stress distributions.

A Antonopoulos is a Research Assistant in the School of Civil Engineering at the University of Leeds.

INTRODUCTION

Many attempts have been made to develop software which will accurately predict the behaviour of reinforced concrete under load and which will take into account time and load history. A number of these, commonly based on Finite Element systems, are available commercially though few, if any, can be considered to be fully reliable over the full range of practical problems [1]. Much research has been done or is in progress to provide improved models of behaviour that can be introduced into these programmes. Recent work on the development of computer systems for the prediction of the behaviour of slab systems indicated that there were problems with the prediction of the changes in deflection with time. It was believed that the problem did not arise from errors in the prediction of creep, which has been extensively studied and was considered to be predicted with adequate reliability, but with the understanding of the behaviour of the cracked tension zone. Though some studies have been carried out on this (for example, Stevens [2]), the data available covering the changes in deformation with time was found to be very limited indeed. The project described here aims to rectify this gap in our knowledge. The research has been carried out at the Universities of Durham and Leeds. Further work on incorporating the findings of the research into computer systems will be carried out by Ove Arup and Partners. The finance for the project comes from separate EPSRC grants to Durham and Leeds and industrial support from Arups, Cadogan Tietz, Giffords and the Concrete Society. The Concrete Society's contribution will be mainly in disseminating the results of the project. At the time of writing, the experimental work is close to completion, though the theoretical work and the full evaluation of the results still has some way to go. The conclusions drawn will, therefore necessarily be preliminary.

It may help if, at this stage, the nature of the problem being addressed and its practical significance is outlined.

There are a number of elements which make up the total deflection of a cracked reinforced concrete member. These are:

(1) The short term deflection. This is the deflection which occurs at the time that the load is applied and, for slabs, this is likely to make up about 50% of the total deflection.

(2) Deflection due to creep. Creep is a deformation of the concrete under constant stress. At normal service levels of stress this deformation is roughly proportional to the applied stress. In cracked reinforced concrete members, creep is dominantly a phenomenon affecting the concrete in the compression zone since the stresses in the concrete in compression are many times higher than the residual stresses in the concrete in the cracked tension zone.

(3) Deflection due to shrinkage. Shrinkage is a volume reduction in concrete due to loss of moisture from the pore system. This can occur either by the use of the water in the hydration of the cement (autogenous shrinkage) or by evaporation of water (drying shrinkage). If the shrinkage is restrained in a non-uniform manner over the depth of a section then the section will tend to warp, leading to a deflection. The major factor causing such restraint in a flexural member is the reinforcement. In a singly reinforced beam, therefore, the tension face is restrained while the compression face is not, leading to a shortening of the compressive face relative to the tension face and hence a deflection. In normal circumstances the contribution of shrinkage to the total deflection is small compared with other effects.

(4) An increase in deflection due to a reduction in tension stiffening with time. Since this is the subject of the project, this will be explained in greater detail below.

The phenomenon of tension stiffening can best be explained by reference to the behaviour of a very simple reinforced member; a concrete prism reinforced axially by a single bar. If such an element is loaded, the load-axial extension behaviour is as shown schematically in Figure 1.

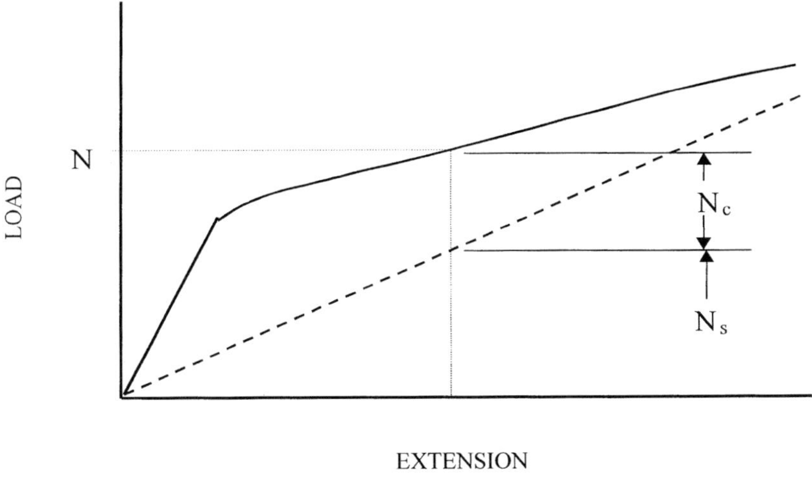

Figure 1 Schematic load – extension response of a member subjected to tension

It will be seen that, at low loads, the specimen behaves elastically (ie. extension is directly proportional to load). This phase continues until the stress reaches the tensile strength of the concrete and cracks start to form. As each crack forms, the stress in the concrete at the crack must reduce to zero and the whole of the tensile force applied to the specimen is carried by the reinforcement. This transfer of stress is accompanied by an increase in extension of the steel in the region of the crack. This, in turn, increases the overall extension and hence reduces the stiffness of the member. An increase in the number of cracks and possibly also reduction in bond results in a continuous decrease in stiffness with increase in load. If all the tension was carried by the reinforcement at all points along the member, the extension can easily be calculated from a knowledge of the size of the bar and its stress-strain response. This response is shown in Figure 1 as a broken line. In fact, the concrete between the cracks usually continues to carry some tension and the real extension remains smaller than the extension for the bar alone. Now consider the conditions in the member under a load, N. This results in an average strain ε_N over the length of the member. It should be clear from Figure 1 that, under this average strain, the average force carried by the reinforcement must be $N_s = A_sE_s\varepsilon_N$. By equilibrium, the force $N_c = N - N_s$ must be carried by the concrete in tension. This force can clearly be translated into an average stress in the cracked concrete. This stress, and its effect on the overall extension of the member is commonly known as "Tension Stiffening". It is well known that the tension stiffening reduces with time under constant load to about half its initial value. What does not seem to have been investigated to any major extent is the rate at which this reduction occurs.

Unlike creep, which is believed to result to a major degree from moisture movements within the concrete, loss of tension stiffening seems likely to be a cumulative damage process and there is no clear reason to expect the rate of loss of tension stiffening to be similar to the rate of creep.

The practical demand for a more reliable definition of the rate of loss of tension stiffening with time mainly arises from requirements to be able to predict the behaviour of slab systems reliably. Slabs tend to have relatively low percentages of tension reinforcement. Where the reinforcement percentage is in the region of 0.4 to 0.6%, it is found that tension stiffening has a major influence on the deflection and that the loss of about half this stress with time contributes about half to two thirds of the long term increase in deflection with time. Clearly, therefore, if computer programs are to be developed which will accurately predict the development of the deflection of slabs with time, an accurate knowledge of the rate of loss of stress is essential.

A more significant practical issue has come to light during the discussions which led to the formulation of the test programme and during the testing. This may be understood by considering the deflection calculation method given in BS8110 Part 2. In this document the assumption is made that the stress in the concrete at the level of the tension reinforcement after cracking is 1 N/mm^2 in the short term and 0.55 N/mm^2 in the long term. A critical design criterion is the increase in deflection that will occur after the installation of finishes and partitions. This is commonly limited to span/500. It is normally assumed that, at the time the finishes and partitions are installed, the short term properties apply and hence a tension stress of 1N/mm^2 is used to obtain the initial deflection. The final deflection is then calculated using the lower value of 0.55 N/mm^2 and the increment obtained by taking the difference between these two determinations. However, investigation of some previous results during the project formulation stage suggested that the tension stiffening might be lost at a much earlier stage than has normally been assumed. If, for example, the tensile stress in the concrete could have dropped to 0.55 N/mm^2 before any finishes or partitions are installed then 0.55 N/mm^2 could be used both for the calculation of the initial and final deflection. The effect of this would be to significantly reduce the calculated increment in deflection and to make the limit to the deflection increment much less critical. The calculated total deflection will not necessarily be changed since the deflection prior to installation of finishes and partitions will be increased. As will be seen below, the experimental work has confirmed this speculation.

DETAILS OF EXPERIMENTAL PROGRAMME

It was decided to concentrate on tests on members subjected to pure tension. All the tension tests have been carried out on 120 mm square prisms axially reinforced with a single bar of 12 mm, 16 mm or 20 mm diameter. The prisms were 1.2 m long with strain readings taken over the central 1 m. Three concrete strengths were used; nominally 30, 70 and 100 N/mm^2. To obtain the maximum useful information, two test programmes were carried out.

At Leeds, 9 simple test rigs were built in which the load was applied manually by tightening a nut on the stressing arrangement with a spanner. This meant that the load had to be adjusted manually at fairly frequent intervals during the early part of the testing. The frequency could be reduced after a few days. Sets of three identical specimens were tested under constant load in these rigs and the average surface strains monitored over time using Demec gauges. A total of 30 tests have been carried out.

In addition, in order to investigate the applicability of the results of the tension tests to flexural members, eight 3 m long slab specimens are being tested. Identical pairs of slabs were loaded back-to-back for periods of up to three months. Details of the slabs are shown in Figure 2. The reinforcement arrangement was chosen so that the individual bars were surrounded by an area of concrete as similar as possible to that used in the tension tests. The slabs were loaded to a level which gave the same steel stress, calculated on the basis of a cracked section, as was applied in the comparable tension tests.

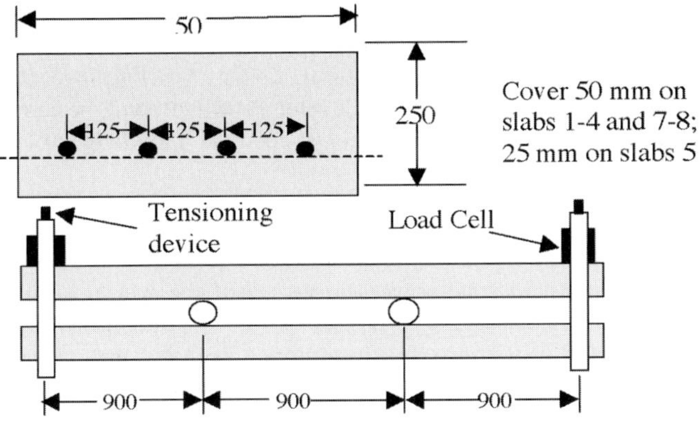

Figure 2 Details of slab tests

Figure 3 The test rigs at Durham with tension specimens in place

The results from these tests were processed as follows. Strains were measured at three levels down the side of the slabs so that the average distribution of the strain over the central constant moment zone could be established. This enabled the average position of the neutral axis to be established and hence, assuming that the concrete in the compression zone remains elastic, the lever arm of the internal forces. Knowing the applied moment, the tension and compression forces could be calculated. The strains measured at the level of the tension reinforcement enabled the average force carried by the reinforcement to be established and, by deducting this from the calculated total tension, the tension force carried by the concrete could be established.

This operation introduced a number of assumptions; notably that the concrete in compression behaved elastically and that the centroid of the tension force carried by the concrete coincided with the tension reinforcement. Establishment of the tension force in the concrete is therefore inherently less accurate than it is from the tension tests.

A procedure for fixing strain gauges at very close spacings along the length of reinforcing bars has been developed at Durham and has been used over a period of some 20 years in a large number of projects where detailed measurements of the variations of strain or stress along the reinforcement is necessary. The procedure for installing the gauges is to machine two bars in half and then mill a groove in the centre of each half bar. Electrical resistance strain gauges are fixed alternately in the grooves at the required spacing and the thin leads taken out to the end of the bar along the groove.

When the gauging is completed, the half bars are glued together to produce a bar which, from the outside, looks exactly like a normal bar [3]. Inevitably, this process is expensive and so the number of tests that can be carried out using such bars in any one project is limited. Durham's part in the experimental programme was to repeat a limited number of the tests in the Leeds programme using instrumented bars to provide a detailed picture of the change of the strain distribution along the bars with time. Two test rigs were built which used a dead weight and leverage system to apply a constant load to the specimens.

Figure 3 shows specimens set up in the two rigs. Identical pairs of tension specimens were tested, one in each rig. Generally, the loading procedure on the two specimens was slightly different with one being loaded directly up to the required maximum load and then maintained at this load for up to three months while the other was loaded in three stages with the load being held for some time at each load level. This gave information on two different load histories for each specimen.

PRELIMINARY RESULTS

Figures 4 (a) and (b) show the variation of the average tension stress in the concrete as a function of time from loading. Figure 4(a) is for a single specimen tested at Durham while Figure 4 (b) is the average of three results from identical specimens tested at Leeds. In both cases it will be seen that the stress reduces to a constant level within quite a short time. This can be seen better if a log scale is used for time. This has been done for the specimen in Figure 4(a) in Figure 5. The steady state stress was reached in 19 days from loading. In very many cases the time taken to reach the steady state condition was considerably shorter than this.

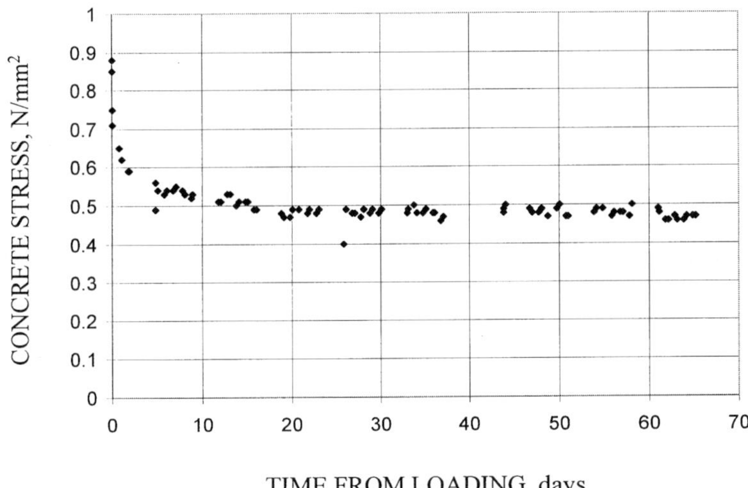

Figure 4(a) Result from tension test at Durham

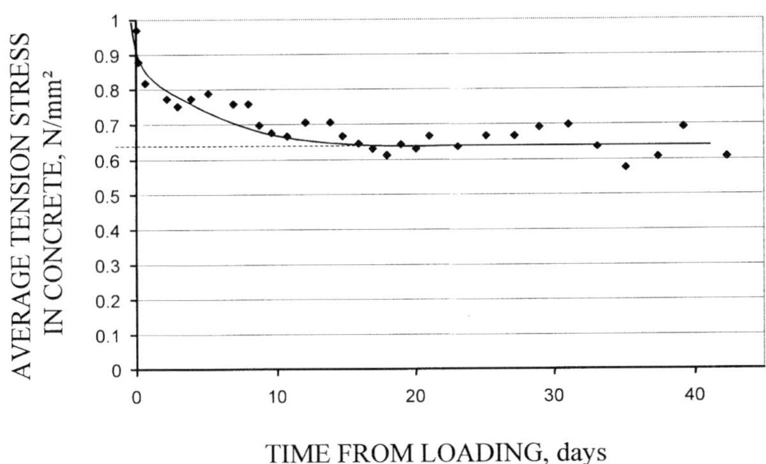

Figure 4(b) Average of 3 tension tests from Leeds

Figure 6 shows the average stress as a function of time for the average of two identical slab tests carried out at Leeds. Again, the stresses have reached a steady state in a relatively short time showing that the result is not a peculiarity of the tension tests. It is of interest to try to establish the mechanism causing the reduction in average stress with time. The detailed strain readings obtained at Durham permit this to be done since the stress in the concrete can be calculated at each gauge position.

Figure 7 shows the variation of the concrete stress between two cracks measured at different ages. It will be seen, firstly, that the increase in stress is close to linear up to a maximum at about mid-way between the cracks. This linear increase in stress implies a more-or-less constant bond stress over most of the length between the cracks.

There is also evidence of a length close to the cracks where the bond stress is close to zero. Initially, this unbonded length would appear to be about 20 mm on either side of the crack. The effect of time appears to be dominantly to increase this unbonded length. There is some marginal decrease in the bond stress but this is of much less significance than the increase in the debonding. The stress of close to –1 N/mm^2 (compressive) in Figure 7 seems to occur quite commonly in the region of the cracks and would seem to be the result of some cracks not forming completely round the whole circumference of the prism. This results in local bending, a higher strain in the reinforcement than the average and some compression in parts of the concrete section.

A more detailed treatment of the results will appear in a later paper.

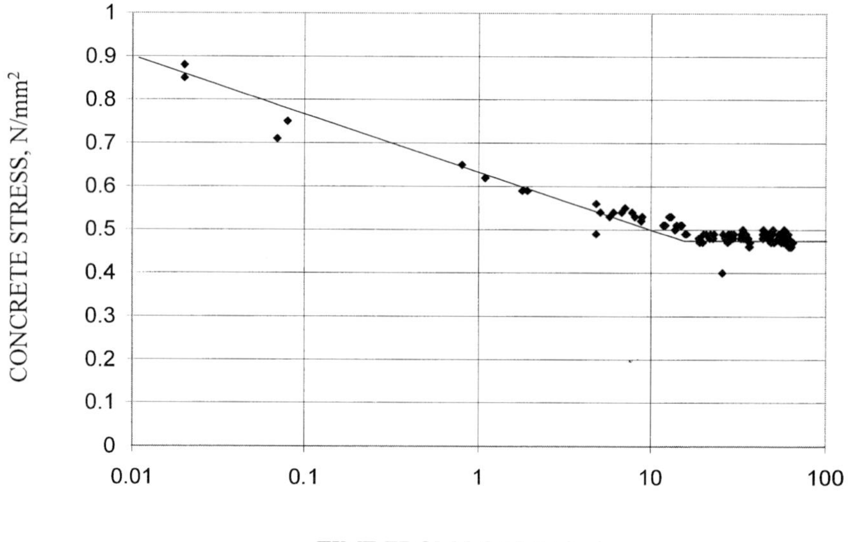

Figure 5 Figure 4(a) replotted on log scale

THEORETICAL DEVELOPMENTS.

It is possibly surprising that methods for calculating deflections of reinforced concrete members are almost all basically empirical while the necessary building blocks for a theoretical approach, based on more fundamental principles, exists. Current approaches are based on the following basic concept.

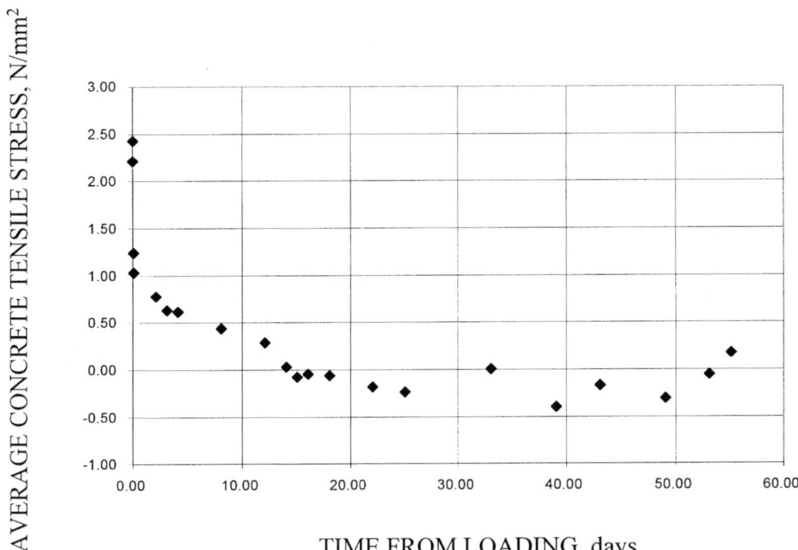

Figure 6 Average result from two slab

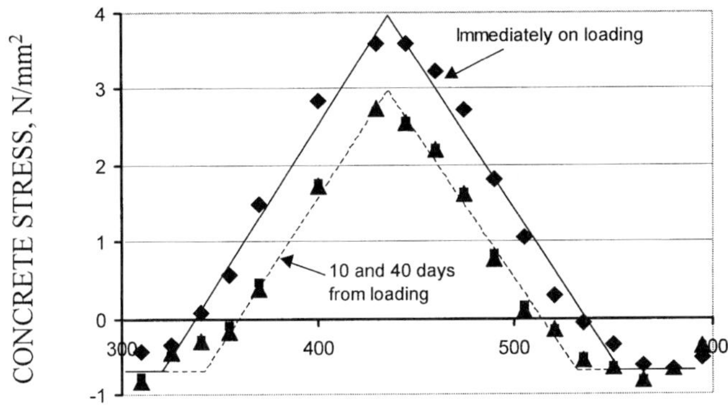

DISTANCE FROM LH END, mm

Figure 7 Effects of time on tension stresses between two cracks

Prior to cracking, the deformation of a reinforced concrete section subjected to bending (combined or not with axial load) is predictable using the theory of elasticity (combined, for long-term deformations, with creep and shrinkage). This deformation, a_I, is the uncracked deformation. After cracking, the deformation will increase as the stiffness is reduced by the development of cracking. A limit to the deformation after cracking is set by the assumption that concrete in tension is completely ineffective after cracking and carries zero tensile bending stresses.

This deformation, a_2, the fully cracked deformation, can also be calculated using the theory of elasticity. After cracking, and up till loads which will cause either steel or concrete to behave in a significantly non-linear manner, we thus have two bounds to the deformation such that:

$$a_1 \leq a \leq a_2$$

Current calculation methods provide an empirical means of deciding where between a_1 and a_2 the actual deformation lies. What is being considered in this section is a means of calculating the deformation on the basis of more fundamental principles. These principles have been available for years but do not seem to have been transferred from the more developed field of crack prediction to that of deformation calculation. The approach envisaged will be very briefly outlined below for the calculation of the deformation of prisms subjected to axial tension.

The conditions in a tension specimen may be written as:

$$\varepsilon_{av} = (f_{s2} - f_{tav}/\rho)/E_s \qquad (1)$$

where: ε_{av} = overall average strain
f_{tav} = average stress in concrete
ρ = reinforcement ratio, A_s/A_c
A_s = area of reinforcement
A_c = area of concrete
E_s = modulus of elasticity of reinforcement

The average stress in the concrete between two cracks varies as a function of the crack spacing and could be simplified from the form of distribution measured in the tests and shown in Figure 7 to, for example, that shown schematically in Figure 8.

Thus the overall average stress, f_{tav}.can be calculated either from sketches such as those in Figure 8 and a knowledge of the frequency distribution of each spacing or empirically from test data. Both approaches are being tried.

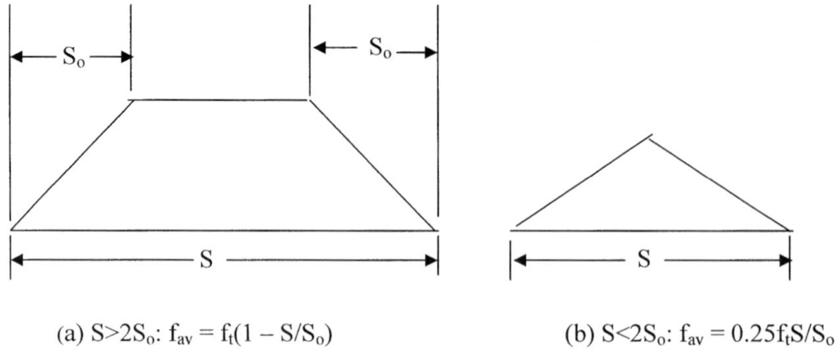

(a) $S>2S_o$: $f_{av} = f_t(1 - S/S_o)$ (b) $S<2S_o$: $f_{av} = 0.25f_tS/S_o$

Figure 8 Schematic variation of stress in concrete between cracks

The other essential step is the prediction of the crack spacing at any stage in the loading. Most current crack theories have concerned themselves only with the final average crack spacing, which occurs when all possible cracks have formed. There is little information on the prediction of the development of the crack spacing during the crack formation stage. However, work on cracking in the 1960s [4] showed that there was generally a good linear relationship between crack spacing and ($1/\varepsilon_{av}$). This relationship has now been confirmed from extensive data from tension tests, mostly carried out at the Ecole Polytechnique Federale de Lausanne [5]. The problem is to define the coefficients in the relation $S_{av} = a/\varepsilon_{av} + b$. Studies are in hand to establish these.

CONCLUSIONS

1. A preliminary study of the results from the project shows that the loss of tension stiffening, which it is well known occurs with time, occurs within a much shorter time than has generally been assumed. From the results so far, it appears that a steady state has been reached within less than 20 days from loading.

2. The reduction in tension stiffening appears to be largely due to extension of an effectively unbonded length which occurs very close to the cracks.

3. It appears a fruitful line of development to produce formulae for the prediction of the deformation of cracked tension zones from consideration of the crack spacing and its effect on the average stress in the concrete.

ACKNOWLEDGEMENTS

The financial support of EPSRC and the industrial partners, Ove Arup and Partners, Cadogan Tietz, Giffords and the Concrete Society, is gratefully acknowledged. Also gratefully acknowledged is the constructive guidance and support provided throughout the project by the Tension Stiffening Working Group set up by the Concrete Society and having as members representatives of the industrial partners.

REFERENCES.

1. VECCHIO, F J, Nonlinear Finite Element Analysis of Reinforced Concrete at the Crossroads? Structural Concrete Vol 2, No 4, Dec 2001.

2. STEVENS, R F, Deflections of reinforced concrete beams. Proceedings of the Institution of Civil Engineers, Part 2, Vol 53, September 1972.

3. SCOTT, R H and GILL P A T, Short term distributions of strain and bond stress along tension reinforcement, The Structural Engineer, Vol 65B, No 2, June 1987.

4. BEEBY, A W, An investigation of cracking in reinforced concrete slabs spanning one way – static loading. CIRIA Report 21, March 1970.

5. FARRA, B, Influence de la resistance du beton et de son adherence avec l'armature sur la fissuration. These No 1359 (1995), Department de Genie Civil, Ecole Polytechnique Federale de Lausanne, 1995.

THE IMPACT OF 500MPa REINFORCEMENT ON THE DUCTILITY OF REINFORCED CONCRETE SLABS: THE AUSTRALIAN EXPERIENCE

R I Gilbert

University of New South Wales

Australia

ABSTRACT. Steel reinforcement with a minimum characteristic yield stress of 500 MPa has been introduced recently in Australia. This 500 MPa reinforcement is classified according to its ductility, or strain capacity, as either Class L (low ductility) or Class N (normal ductility). Class L steel includes welded wire mesh, commonly used in slabs, and has a minimum specified uniform elongation of only 1.5%. Class N steel includes hot rolled deformed bars and has a minimum specified uniform elongation of 5%. The relatively low uniform elongation of some 500 MPa steel reinforcement has significant implications in the design and behaviour of reinforced concrete structures, particularly with regard to ductility. Fracture of the tensile steel will be the common failure mechanism in many under-reinforced beams and slabs and the usual assumptions made in ultimate strength design and analysis are in question. The ability of structures to redistribute internal actions at the ultimate limit state will also be compromised. These problems are not unique to Australia, as low ductility steels are often used in Europe and elsewhere. This paper addresses some of these design problems and their impact on current practice and design methodologies are illustrated using several examples.

Keywords: Ductility, High strength steel reinforcement, Moment redistribution, Reinforced concrete, Slabs, Ultimate strength design, Uniform elongation.

Ian Gilbert is Professor of Civil Engineering and Head of the School of Civil and Environmental Engineering at the University of New South Wales, Sydney, Australia. His main research interests are in the area of serviceability and the time-dependent behaviour of concrete structures. His publications include four books and over one hundred refereed papers in the area of reinforced and prestressed concrete structures. He has served on Standards Australia's Concrete Structures Code Committees since 1981 and was actively involved in the development of the Australian Standard for Concrete Structures AS3600-2001.

INTRODUCTION

Steel reinforcement with characteristic yield stress of 500MPa is now in the Australian market place and will soon be the only conventional reinforcement available. This steel has higher strength, but significantly lower ductility, than the 400 MPa tempcore bars previously used in concrete structures in Australia. A new Australian Standard for reinforcing steel, AS/NZS4671 [1], was introduced in 2001 with the specification of 500 MPa reinforcement covered for the first time. It replaced the existing Standards AS1302, AS1303 and AS1304. Also issued in 2001 was a revision of the Australian Standard for Concrete Structures AS3600 [2] to accommodate the use of 500 MPa steel in the design of concrete structures.

The proposed new Standards classify 500 MPa reinforcement according to its ductility, Class L (low ductility) and Class N (normal ductility). Class L steel includes cold worked wires and welded wire mesh. Class N steel includes hot rolled deformed bar. The mechanical properties of the new steels, specified by characteristic values, are summarised in Table 1. f_{sy} is the yield stress, with $f_{syk.L}$ and $f_{syk.U}$ being the lower and upper characteristic values; f_{su} is the tensile strength; and ε_{su} is the uniform elongation or the strain corresponding to maximum stress (just prior to the onset of necking).

Table 1 Characteristic mechanical properties of Australian 500 MPa reinforcement.

PROPERTY		500 L	500N
Nominal diameters (mm)		5 - 16	10 - 40
Characteristic Yield stress (MPa):	$f_{syk.L}$	500	500
	$f_{syk.U}$	750	650
Tensile to Yield stress ratio:	f_{su}/f_{sy}	1.03	1.08
Uniform Elongation:	ε_{su}	0.015	0.05

AS1302, AS1303 and AS1304 did not specify any limits on ε_{su}. For 400 MPa bars, with ε_{su} in excess of 0.10, this did not impose any difficulties. However, for some of the brittle welded wire meshes used in structures, with ε_{su} an order of magnitude less than this, the lack of attention to the ductility of reinforcing steel was a major concern. The alarmingly small values of ε_{su} of some commonly used meshes are now formally recognised and endorsed as Class L reinforcement. In the author's view, the formal endorsement of brittle steel (500L) for use in concrete structures, albeit with certain limitations, is a backward step. The relatively low value of ε_{su} of the new steel has very significant implications in the analysis and design of concrete structures, particularly with regard to ductility. The failure mechanism of under-reinforced flexural members may change, with ultimate curvatures often being governed by tensile steel fracture, rather than failure of the concrete in the compressive zone. This applies not only when Class L steel is used, but also in some cases for Class N steel (when ε_{su} is 0.05). The amount of moment redistribution permitted in AS3600 may no longer be appropriate and plastic design techniques may no longer apply. Indeed, even elastic analysis techniques may not be applicable at the ultimate limit states, when using Class L steels. Concrete structures are non-linear and inelastic, and must possess some ductility if the actual distribution of internal actions is to redistribute towards the elastic distribution assumed in design. This minimum ductility may not be available for members containing Class L steel.

Some of these design implications have been addressed in the recent revision of the Australian Standard AS3600, with varying degrees of success (as assessed earlier [3]), but designers need to be fully aware of the potential problems associated with the use of the new reinforcement. This paper outlines some of these problems and their impact on current practice and design methodologies. Several illustrative examples are also included.

DUCTILITY OF UNDER-REINFORCED SECTIONS IN BENDING

In the analysis and design of concrete structures at the ultimate limit states, many of the universally accepted theories and procedures are founded on the important assumption that individual cross-sections possess some degree of ductility. An idealised elastic-plastic stress-strain relationship for the reinforcement is commonly assumed, with the steel stress after yielding being maintained at f_{sy}, irrespective of how large the steel strain becomes. AS3600 and most other concrete codes are based on this assumption. When using hot-rolled bars, with ε_{su} in excess of 10%, this assumption is quite reasonable. However, for 500 MPa steel, especially Class L, this is not the case. Fracture of the steel becomes a common failure mode, particularly in under-reinforced members. The failure mode will be brittle (sudden) and at curvatures far less than predicted by current ultimate strength design theory.

Consider the flexural behaviour of the under-reinforced T-sections of Figure 1. Three different areas of tensile steel (A_{st}) are considered. The material properties are also given in Figure 1, where f'_c and f'_{cf} are the characteristic compressive strength and flexural tensile strength of the concrete, respectively, and E_c and E_s are the elastic moduli of the concrete and the steel, respectively. The moment-curvature response of each section is plotted in Figure 2. It is assumed that for the steel f_{sy} = 500 MPa and ε_{su} = 0.15. On each curve, the point corresponding to failure of the concrete in the compressive zone is shown. This is the point according to AS3600 (and ultimate strength theory) at which the ultimate moment M_u and the ultimate curvature κ_u are reached. As expected, M_u is almost directly proportional to A_{st} and κ_u (and hence ductility) decreases as A_{st} increases. The steel strain at every point on each curve is much less than ε_{su}.

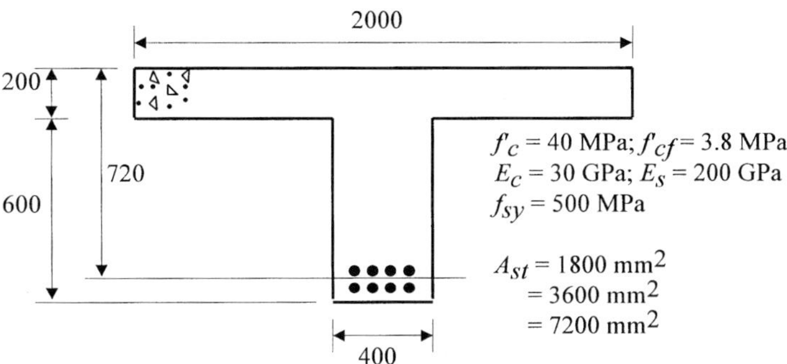

Figure 1 Typical T-Section in pure bending

Also shown on each curve are the points corresponding to tensile steel strains of 0.015 and 0.05 (ε_{su} for Class L and Class N steel, respectively). If steel with $\varepsilon_{su} = 0.015$ or 0.05 is used, the curves terminate at or near these points when fracture of the steel occurs. The peak or ultimate moments is not affected appreciably, but the ultimate curvature (the usual indicator of ductility) is reduced dramatically. When using Class L steel, the curvature at failure of each section is governed by fracture of the steel and is almost independent of A_{st} (curvature at failure actually increasing slightly as A_{st} increases).

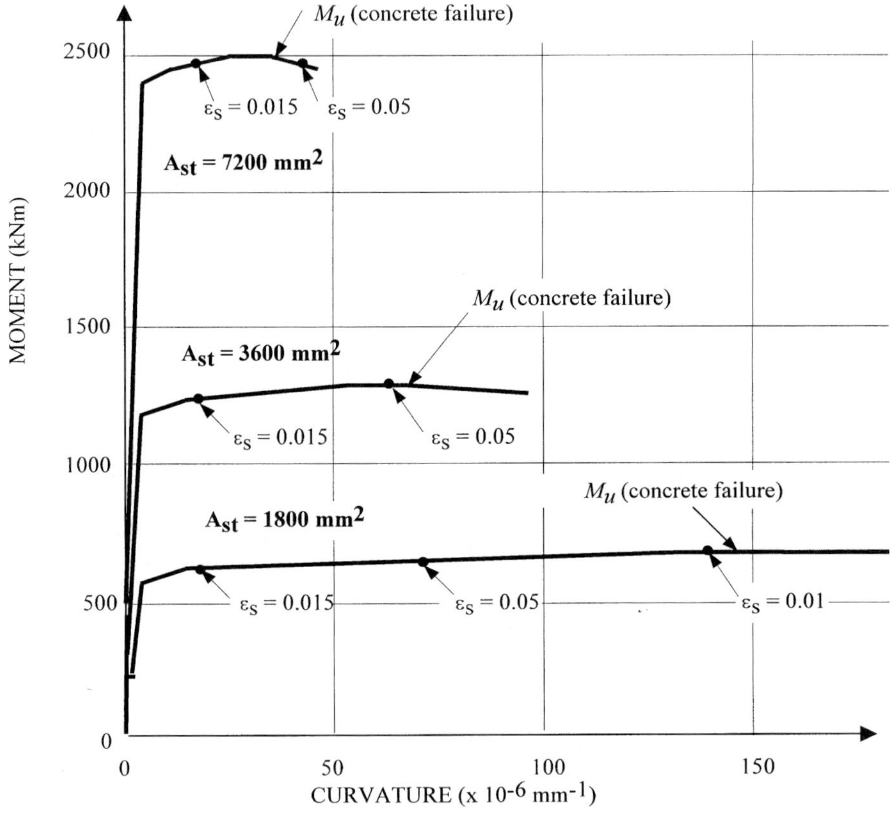

Figure 2 Moment vs Curvature for T-Sections shown in Figure 1

Graphs of ultimate curvature (κ_u) versus A_{st} for the cross-section of Figure 1 are shown in Figure 3. The solid curve depicts κ_u when the extreme fibre concrete compressive strain governs failure, while the dashed curves show the situation when Class L ($\varepsilon_{su} = 0.015$) and Class N ($\varepsilon_{su} = 0.05$) steels are employed. Where the dashed curves deviate from the solid curve, failure of the tensile steel occurs prior to compressive failure of the concrete. The concept of under-reinforced sections necessarily providing ductility is no longer valid. Tensile steel fracture will almost always be the failure mode when using Class L steel and will often be the failure mode when using Class N steel (if ε_{su} is at or near 0.05).

It is difficult to agree with claims that the acceptance and endorsement of this relatively brittle steel is a progressive advancement for the industry. Benefits resulting from the increased strength are outweighed by the problems created by the reduced ductility.

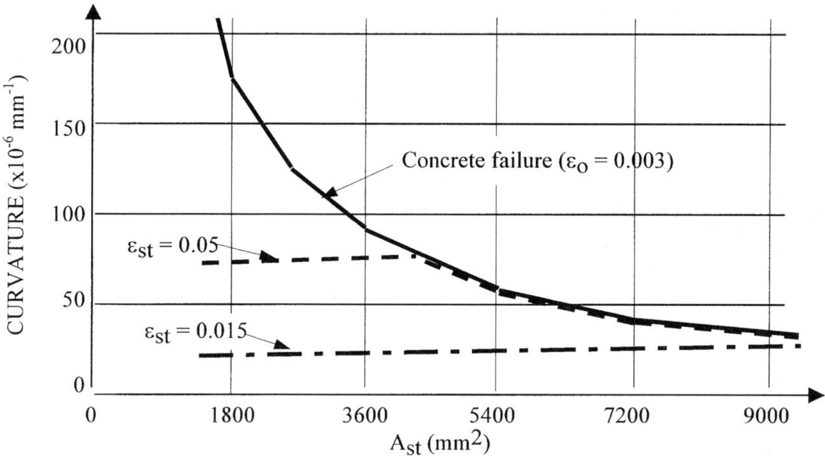

Figure 3 Ultimate Curvature versus tensile steel area for section in Figure 1

DUCTILITY - A FUNDAMENTAL DESIGN REQUIREMENT

Reinforced concrete structures are non-linear and inelastic. Stiffness reduces considerably when cracking occurs and, after cracking, stiffness depends on the quantity and location of the reinforcement. The stiffness of a reinforced concrete element varies from location to location depending on the extent of cracking and the reinforcement layout. In addition, the stiffness of a particular cross-section or region is time-dependent. The relative stiffness between cracked and uncracked regions changes with time due to creep, even when the steel quantities are the same. In addition, restraint to shrinkage and shrinkage warping can induce quite significant internal actions in reinforced concrete structures, as can other imposed deformations such as support settlements and temperature changes and gradients. All these factors cause the actual distribution of internal actions in an indeterminate structure to deviate from that assumed in an elastic analysis.

Despite these difficulties, AS3600 and other codes permit design using an elastic analysis based on gross section properties. This is reasonable provided the critical regions possess sufficient ductility to enable the actual distribution of actions to redistribute towards the calculated elastic distribution assumed in design. It is not at all certain, that some critical regions reinforced with Class L steel will have the necessary ductility.

When designing continuous members, designers often take advantage of moment redistribution, assuming an increase or decrease in the elastically determined peak negative moments, provided equilibrium is maintained and the peak negative moment regions are sufficiently ductile to allow the necessary rotation to occur. Simplified rules are specified in AS3600 when Class N steel is used, where the peak negative moments may be varied by up to 30% depending on the ultimate curvature at the negative moment regions.

Of course, the actual redistribution required as the ultimate load is approached may be larger (or smaller) than this, because the elastic peak moment and the actual value are unlikely to be the same.

The Standard also permits the use of plastic methods, such as yield line design, for slabs if Class N reinforcement is used. In yield line design, the fundamental assumption is that the yield lines possess enough rotational capacity to allow a collapse mechanism to develop. The designer is more or less free to decide the ratio of negative to positive yield line capacity and, frequently, it is advantageous to select a relatively low ratio to reduce the required quantity of top steel. Of course, this is only possible if the critical sections are very ductile.

One of the great advantages when designing indeterminate reinforced concrete structures for strength is the lack of need to accurately determine the distribution of internal actions, provided individual cross-sections are ductile and equilibrium is satisfied. With adequate ductility, the designer can be sure that the distribution of actions at the ultimate limit state will be the same as that assumed in design. With the introduction of 500 MPa steel (particularly Class L with $\varepsilon_{su} = 0.015$), this is no longer the case. When fracture of the steel becomes the failure mode of a critical section, the ability of the structure to find the desired design load path may be lost.

Consider the following examples.

Example 1

Consider a one-way slab simply-supported over a span of 5m. The slab is 200 mm thick and is subjected to a uniformly distributed dead load, *w*. The material properties are $f'_c = 40$ MPa; $f'_{cf} = 3.8$ MPa; $E_c = 25000$ MPa; $E_s = 200000$ MPa; $f_{sy} = 500$ MPa. The main tensile reinforcement in the bottom of the slab throughout the span is 400 mm^2/m at an effective depth of 170 mm. The critical section at midspan just complies with the minimum strength requirement of AS3600 (ie. the ultimate moment capacity M_u exceeds 1.2 times the cracking moment, M_{cr}). For this slab, the moment required to produce an extreme fibre concrete tensile stress of 3.8 MPa on the uncracked section is $M_{cr} = 26.1$ kNm/m and $M_u = 33.1$ kNm/m. The moment - curvature plot for the section at midspan is shown in Figure 4.

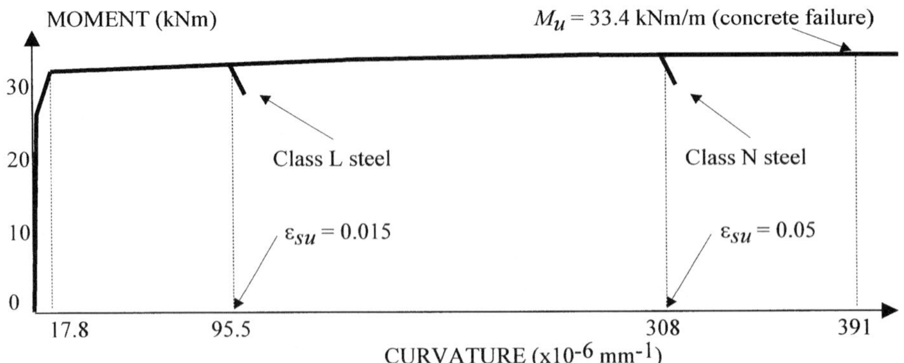

Figure 4 Moment -curvature relationship at midspan of simply-supported one-way slab

If Class L steel is used, the plot in Figure 4 would terminate at a curvature of 95.5 x 10^{-6} mm^{-1} (when the strain in the tensile steel reaches $\varepsilon_{su} = 0.015$). If Class N steel is used (with $\varepsilon_{su} = 0.05$), the plot would terminate at a curvature of 308 x 10^{-6} mm^{-1}. When the uniform elongation of the steel is greater than about 0.07, the ultimate curvature is controlled by failure of the compressive concrete and equals 391 x 10^{-6} mm^{-1}. The section is very ductile, provided the tensile steel does not fracture and can continue to carry f_{sy} under increasing deformation until the concrete in the top fibres of the section crushes - a consideration that, up until recently, a designer did not have to make.

The deflected shape of the slab just prior to failure when ductile reinforcement is used and when failure is governed by crushing of the concrete in the compression zone at midspan is drawn to scale in Figure 5a. The maximum deflection exceeds 225 mm (span/22), the failure is *gradual* and there is ample warning prior of failure. However, when Class L reinforcement is used, failure occurs when the tensile steel fractures at midspan at a maximum deflection of just 37mm (span/135). The deflected shape of the slab just prior to failure is shown in Figure 5b (also drawn to scale). There is little warning of failure, deflection is not alarming prior to failure, and the failure is sudden and catastrophic.

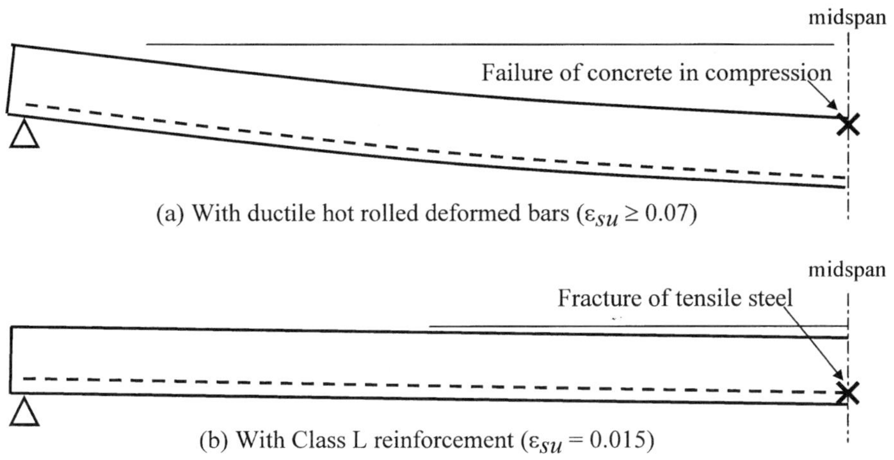

Figure 5 Deflected shape of one-way slab at the ultimate limit state

Example 2

Consider an internal, 6 m, fixed-ended span of a lightly loaded continuous one-way slab of overall depth 200 mm. The slab is subjected to a uniformly distributed dead load, *w*. The material properties are the same as for Example 1. The main tensile reinforcement in the bottom of the slab throughout the span is 400 mm^2/m at an effective depth of 170 mm and the main tensile reinforcement at the continuous supports is 400 mm^2/m at a depth of 30 mm below the top surface of the slab. As for Example 1, the critical sections (in this case, at each support and at midspan) just comply with the minimum strength requirement of AS3600 (ie. the ultimate moment capacity M_u exceeds 1.2 times the cracking moment, M_{cr}).

As in Example 1, at each critical section, the cracking moment is M_{cr} = 26.1 kNm/m. If the load on the slab is gradually applied, the negative moment at each support before cracking is $wL^2/12$ and the positive moment at midspan is $wL^2/24$. Cracking will first occur at the supports when the moment reaches -26.1 kNm/m and, soon after, the moment and the curvature corresponding to first yield of the tensile steel at the supports (M_{sy} = -32.0 kNm/m and κ_{sy} = -17.8 x 10^{-6} mm^{-1}) will be reached. At this load level, the section at midspan is uncracked and the bending moment diagram is shown as the dashed curve in Figure 6. Further increases of load will result in a very slight increase in moment at the support accompanied by large rotation and an increase in moment at midspan. Cracking occurs at midspan when the moment reaches M_{cr} and, eventually, the bottom steel at midspan will yield when the moment reaches M_{sy}. The moment diagram at this load level is shown as the solid curve in Figure 6. This load level will only be reached if the section at the support has sufficient rotational capacity to allow the necessary change of curvature at midspan. In this example, assuming a hinge length at each support equal to d = 170mm, the change of curvature necessary at each support is $\Delta\kappa_s$ = 199 x 10^{-6} mm^{-1}.

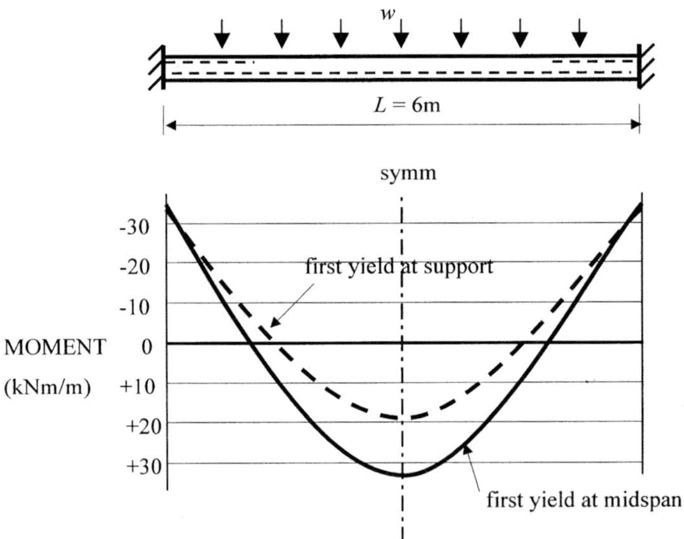

Figure 6 Bending moment diagrams for fixed ended slab

The moment-curvature relationship for the critical sections is plotted in Figure 7, and is identical to the relationship plotted in Figure 4 for the identical slab section of Example 1. It is evident from Figure 7, that when using Class L steels, the rotation at the supports necessary to develop the yield moment at midspan cannot occur. To develop the required curvature at the supports ($\kappa_{sy} + \Delta\kappa_s$ = 216 x 10^{-6} mm^{-1}), the necessary strain capacity of the tensile steel is 3.4%. When Class L steel is used, fracture will occur when the steel strain reaches 1.5%, at less than half the required curvature. Fracture of the steel at the supports will initiate a sudden failure of the span, probably when cracking at midspan first occurs. When Class N steel is used sufficient rotational capacity is available to develop M_{sy} at midspan, but insufficient rotational capacity is available to develop the ultimate moment M_u at midspan. The above analyses are simplistic and have considered only average strains and curvatures.

A detailed study of the ductility of reinforced concrete members must include strain localisation and the influence of crack spacing and bond breakdown in the peak moment regions - factors which undoubtedly affect ductility. However, even though strain localisation has not been considered in these simple examples, the results indicate that our traditionally held views on ductility of reinforced concrete are not applicable, if low-ductility steel reinforcement is used.

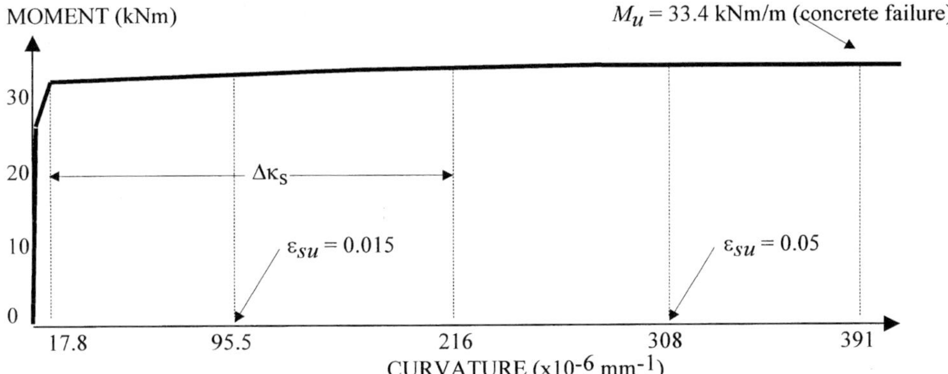

Figure 7 Moment-curvature relationship for critical sections of fixed ended slab

Example 3

Consider the two-span, one-way slab of Figure 8a subjected to a uniform load and supported by girders at A, B and C, as shown. If an elastic analysis (based on gross section dimensions) is undertaken assuming rigid supports (ie. girders A, B and C do not deflect), the bending moment diagram for the slab is shown as the solid line in Figure 8b, where $M_B = -wL^2/8$. The slab load resisted by the girder at B is 3.33 times that resisted by the girders at A (and C).

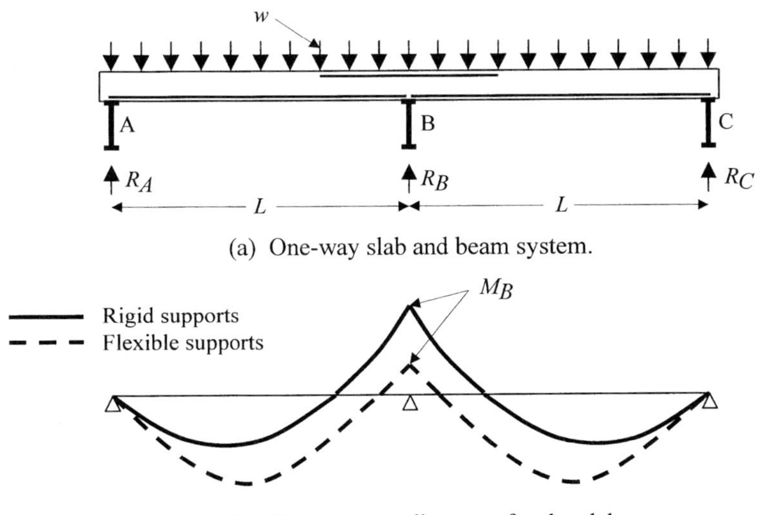

Figure 8 Effect of support settlement on bending moment distribution

If the deflection of the girder at B is v_B greater than that at A and C, then the negative moment at B decreases by $3EIv_B/L^2$ (where EI is the flexural rigidity of the slab) and the bending moment diagram for the slab is shown as the dashed curve in Figure 8b. Depending on the stiffness of the slab and the stiffness of the supporting girders, the value of the moment at B could well be positive and the positive span moment of the slab may exceed $wL^2/8$. For example, if L = 3000 mm, w = 20 kN/m^2, and the slab is 150 mm thick (with E = 28000 MPa), the moment at B assuming v_B = 0 (and assuming gross cross-sectional properties) is $M_B = -wL^2/8$ = -22.5 kNm/m. If v_B = 10 mm, then the change in moment at B resulting from differential support deflection is $3EIv_B/L^2$ = 26.25 kNm/m. The moment at B is therefore - 22.5 + 26.25 = +3.75kNm/m.

From the point of view of adequate strength, if the slab cross-sections are adequately ductile, it is quite reasonable to ignore the differential deflection of the supporting girders and to design the slab for the solid bending moment diagram of Figure 8a. This is often what is done in practice. When the cross-section is non-ductile, such an approach is not safe. For the example considered above, an enormous redistribution of moments is necessary if the solid elastic bending moment is used for the design of the slab. It may not even be safely possible to adopt such an approach when using class N steel with a uniform elongation as low as 0.05. The days of designers blissfully designing structures to satisfy equilibrium, and relying on ductility, may be at an end, and should be at an end, if brittle 500 MPa reinforcement is used.

CONCLUSIONS

The relatively low uniform elongation of 500 MPa steel reinforcement (particularly Class L reinforcement) has significant implications in the design of reinforced concrete structures, particularly with regard to ductility. The necessary ability of structures to redistributed moments as the ultimate load is approached may be significantly compromised and the failure mechanisms of elements may change significantly. It has been shown that for the design of flexural members, it is unwise to use Class L reinforcement in any circumstances. Structures containing Class L steel may not even have the ductility to justify the use of elastic analysis.

The introduction of 500MPa steel, with the low minimum uniform elongations currently specified, has been forced on the industry in Australia and is a backward step. If the minimum uniform elongation of Class L and Class N steels were 0.03 and 0.10, respectively, (rather than 0.015 and 0.05), then the concerns related to ductility would essentially disappear. The disadvantages associated with the introduction of 500 MPa steel, with the low ductility limits currently specified, outweigh the benefits.

REFERENCES

1. AS3600-2001, Australian Standard for Concrete Structures, Standards Australia, Sydney, (AS3600-1994, incorporating Amendment 1 (issued in 1996) and Amendment 2 (issued for public comment in 1999)(2001).

2. AS/NZS4671 - 2001, Australian/New Zealand Standard - Steel Reinforcing Materials, Standards Australia and Standards New Zealand, 2001.

3. GILBERT, R.I. (2001), "The Impact of 500 MPa Steel Reinforcement on the Ductility of Concrete Structures - Revision of AS3600", Proceedings, Concrete 2001, The Biennial Conf of the Concrete Institute of Australia, Perth, September, pp 597-603.

EFFECT OF LARGE OPENINGS AND COLUMN RECTANGULARITY ON PUNCHING SHEAR STRENGTH OF SLABS

S Teng **H K Cheong**

Nanyang Technological University

Singapore

K L Kuang

T.Y. Lin (South East Asia) Pte Ltd

Malaysia

ABSTRACT. This paper summarises one part of the research program on flat-plate structures conducted jointly at the Nanyang Technological University (NTU) and the Building and Construction Authority (BCA), Singapore. The paper focuses on punching shear design of slabs with openings and supported on rectangular columns. Twenty slab specimens had been tested in the programme and it was found as expected that the stresses in the slabs are concentrated mostly around the shorter ends of the rectangular columns. Openings reduce punching strength considerably and the best place for an opening is along the longer side of the column. A simple method to consider the effect of opening and column rectangularity is proposed and its application to the BS8110 Code is shown to be accurate. Comparison of the experimental results with the ACI Code and the Eurocode-2 predictions is also presented.

Keywords: Punching shear, Opening, Rectangular column, Shear strength, Building codes, Flat plate slab.

Professor S Teng, is an Associate Professor in the School of Civil and Environmental Engineering, Nanyang Technological University, Singapore. His current research interest is in the performance of flat plate slabs, deep beams, and finite element modelling.

Mr K L Kuang, is a Structural Engineer with T.Y. Lin (South East Asia) Pte. Ltd., Kuala Lumpur, Selangor, Malaysia. He received his M.Eng degree from Nanyang Technological University, Singapore.

Professor H K Cheong, is a Professor and Dean of the School of Civil and Environmental Engineering, Nanyang Technological University, Singapore. His current research interests include flat plate slabs, repair and retrofitting.

INTRODUCTION

In recent years, the flat-plate floor system has been increasingly adopted for high-rise residential buildings in Singapore. A typical residential building would have irregular column layout and very elongated (rectangular) columns due to architectural demand. A large opening may also be present next to a column. Unfortunately, the available design procedures in various codes of practice do not adequately cover this type of flat plate floors.

For this reason, research has been initiated at the Nanyang Technological University (NTU) with the support of the Building and Construction Authority (BCA), Singapore with the purpose of finding a practical solution for the structural design of this type of flat plate floors.

This paper summarises one part of the joint BCA-NTU research and focuses on punching shear design of slabs with openings and supported on rectangular columns. A number of researchers had conducted studies on punching shear behaviour of slabs with openings, including: Moe [1], Hognestad, Elstner and Hanson [2] and others. More recent research includes those by El-Salakawy, Polak and Soliman [3]. However, they concentrated on slabs with essentially square columns. So, their results may not be suitable for direct extrapolations to cases involving larger openings and rectangular columns.

The ACI Code [4] and the Eurocode-2 include the design procedure for treating the effect of column rectangularity. However, they were based on limited number of experimental data and their accuracy should be checked further when more data become available. Hawkins, Fallsen and Hinojosa [5] conducted experiment on punching shear strength of slabs supported on rectangular columns. More data are obviously needed. This paper presents the data of twenty full-scale slabs with openings and various column sizes that had been tested to failure. The parameters included in the investigation are the effect of different locations of holes, rectangularity of columns and the different loading ratios in the x and y directions of the slabs.

TEST PROGRAMME

Specimen Notations

Figure 1 shows the typical details of two slab specimens. The overall size of the slab is either 2.2 x 2.2 x 0.15 metres or 2.2 x 2.7 x 0.15 metres. Concrete of Grade 40 was used and the deformed steel bars have yield strength of about 460 N/mm^2. The columns were monolithically cast with the slabs. The column stubs has a height of 200 mm. Three column cross-sections were used: 200x200mm (or 1x1 column), 200x600mm (1x3 columns), and 200x1000mm (1x5 columns). Thus, the notation used to identify the slabs can either be OC11, OC13, or OC15. The size of the opening is 200x400mm and its location in each specimen is given with respect to the centre of the column. "V23" indicates that the hole is oriented vertically (on plan) with its centre located at a coordinate of (200mm, 300mm) from the column centre. A typical notation would be "OC13V23-0.63", which refers to a slab with 1x3 (200x600mm) column, having a vertically oriented opening whose centre is at the coordinate of (200, 300mm) from the column centre. The last numerals "0.63" indicates that the loading along the x-direction (producing moment about the y-direction) is equal to 0.63 times the loading along the y-direction.

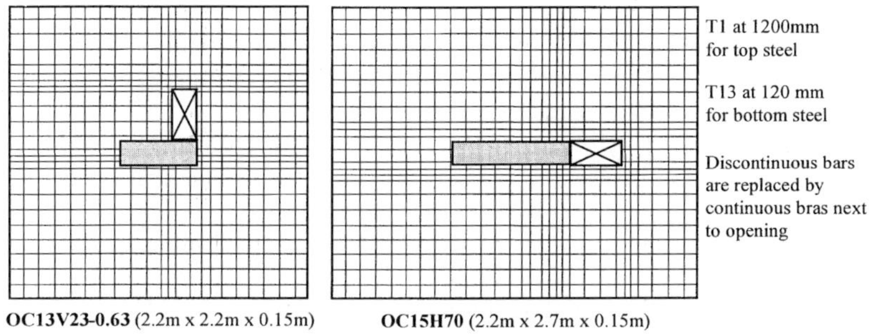

Figure 1 Specimen detail

Test Procedure

Each slab was loaded with a system of eight point loads of equal magnitude on the topside. For all the slabs, the first crack was observed carefully and crack patterns were drawn on the white washed surface of the slabs. Throughout the testing, data from all the instruments such as displacement transducers, steel strains, and load cells were recorded at every load increment.

Crack Propagation and Deflection

As expected, deflection increases with an increase in the applied load. Once the peak load is reached, the deflection increases suddenly, indicating the abrupt nature of punching shear failure. Normally, the final shear-crack inside the slab is inclined at an angle of between 30 to 40 degrees to the horizontal. The presence of the opening reduces the overall stiffness of the slab. Figure 2 shows the crack patterns in three slab specimens after failure.

Punching Shear Strength

All the slabs failed in brittle sudden punching mode. The ultimate punching shear loads were recorded as the peak loads during the tests. The slabs without opening (OC11, OC13, and OC15) had the maximum punching load capacity in its series. A summary of all the ultimate punching strength of all the slabs is given in Figure 3. As can be expected, a large part of the shear stress is concentrated at the two end of the elongated column. This means that the punching shear strength of the slab is determined by the strength of slab around the end portion of the rectangular column. This also means that an opening located at the end zone, such as those in slabs OC11V20 and OC13V40, will have very serious effect on the punching shear strength. Note that different ratio of applied load in the x and y directions have some effect on the punching strength. However, this effect tends to be limited to those cases where the loading on the shorter side of the column is larger. That is, when the shear force on the shorter side of the column is higher by 60% compared to that applied on the longer side, the punching strength reduces by only about 10%. On the contrary, no reduction in strength can be seen if the shear force on the longer side is the one that is higher.

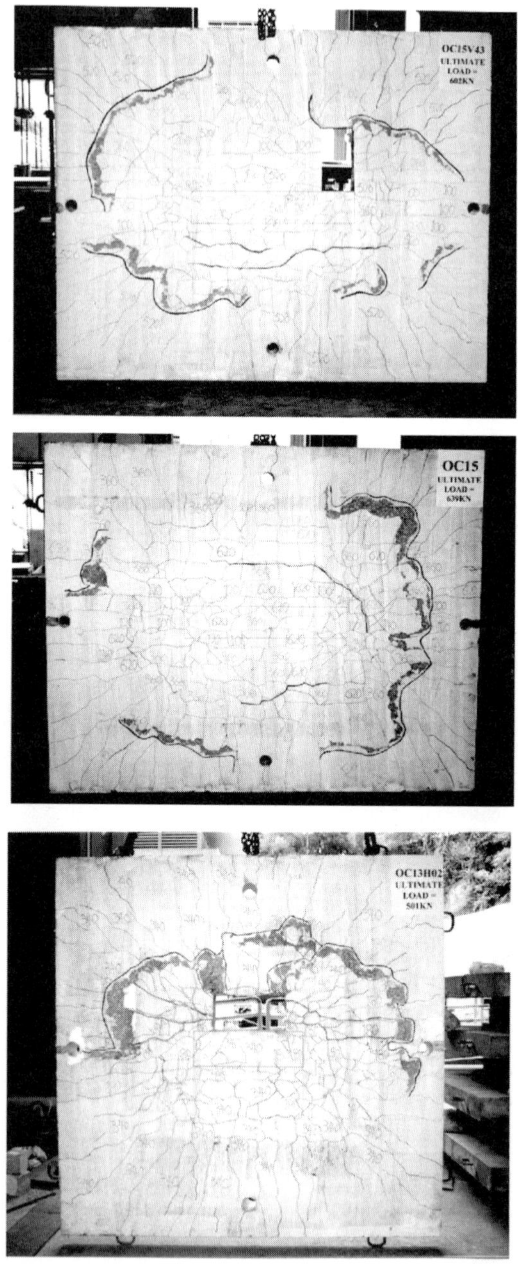

Figure 2 Crack patterns in slabs OC15V43, OC15 and OC13H02

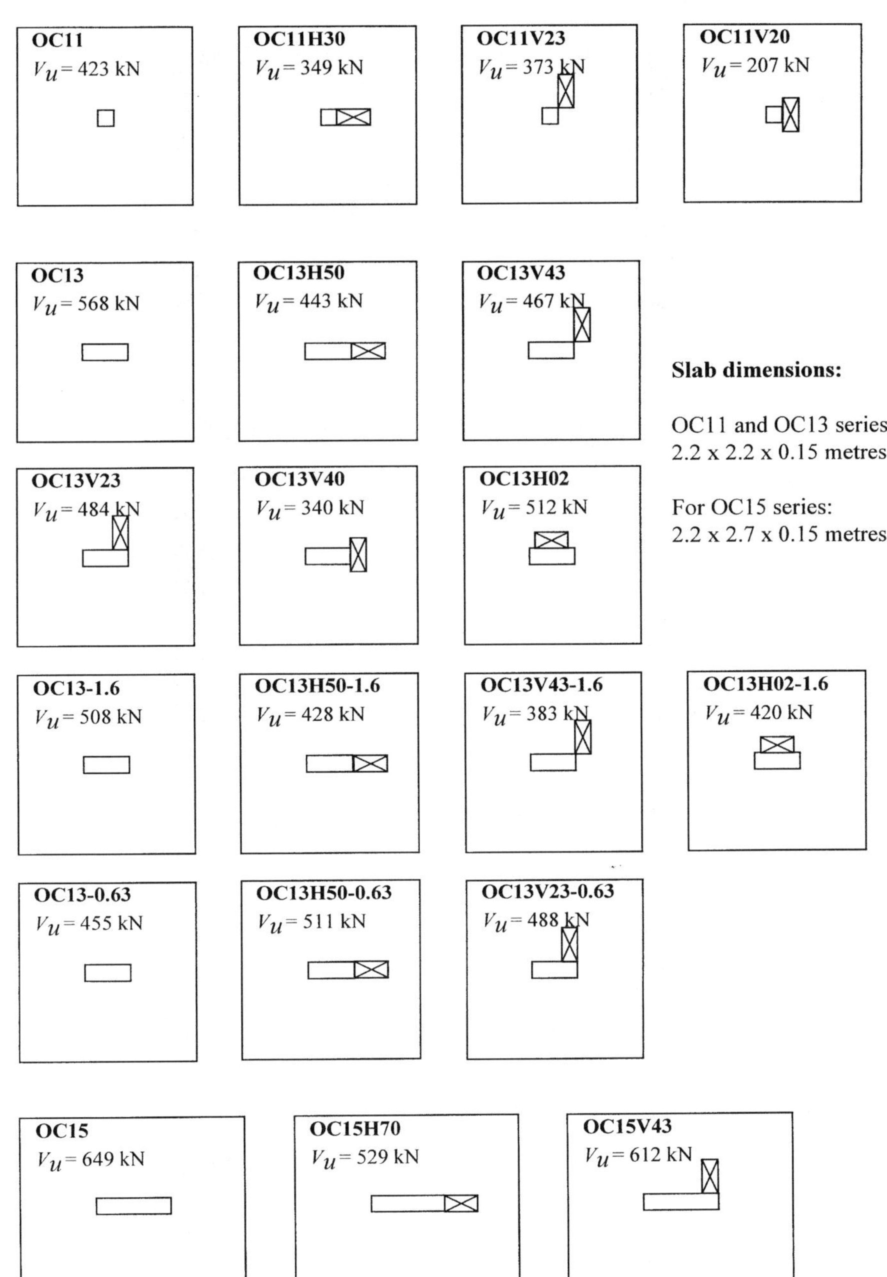

Figure 3 Punching shear strength of tested slabs

PUNCHING SHEAR DESIGN

The punching shear strength of concrete slab is obtained by calculating a nominal shear stress on an effective shear perimeter at some fraction of the slab depth away from the loaded area. Various codes of practice include design methods for punching shear, such as the ACI Code the British BS8110 Code, the Eurocode-2.

Those codes of practice, however, treat column rectangularity, shear perimeter, and opening quite differently from each other. To treat column rectangularity, the ACI Code reduces the nominal shear stresses while the Eurocode-2 reduces the shear perimeter and the BS8110 is silent. The ACI considers the shear perimeter to be located at $0.5d$ away (d is the effective depth of slab) from the column face while the BS8110 and Eurocode-2 take it to be at $1.5d$ away. BS8110 and Eurocode-2 neglect the effect of an opening that is located farther than $6d$ from the column while the ACI will neglect such an opening only if it is located farther than $10d$.

In this joint research, it is found that the BS8110 method is very reliable and with a simple extension described below, the method can be use for cases involving rectangular columns and openings.

ACI Code

According to the American Concrete Institute ACI 318-99, the punching shear capacity of slabs without shear reinforcement can be determined from the lowest of the following expressions (in US Customary units)

a) $$V_c = \left(2 + \frac{4}{\beta}\right)\sqrt{f_c'}b_o d \quad \text{(kips)} \qquad (1)$$

b) $$V_c = \left(\frac{\alpha_s d}{b_o} + 2\right)\sqrt{f_c'}b_o d \quad \text{(kips)} \qquad (2)$$

c) $$V_c = 4\sqrt{f_c'}b_o d \quad \text{(kips)} \qquad (3)$$

where β is the ratio of the long side to the short side of the concentration load (or columns), α_S is 40 for interior column, 30 for edge columns, and 20 for edge columns. b_O is the critical shear perimeter taken at a distance of $0.5d$ away from the column face and has square corners for square columns and rounded corner for circular columns.

Equation (1) shows that the ACI 318 method treats the rectangular column case by reducing the nominal shear stresses. From Equations (1) to (3) above, it is clear that the ACI method ignores the influence of flexural reinforcement and size effect on the punching shear capacity. When an opening in the slab is situated at a distance less than 10 times the slab thickness to the column or loaded area, the ACI method recommends a reduction in the critical shear perimeter in the same manner as that used in the BS8110.

BS8110 Method

The existing BS8110 method seems to be able to predict the punching failure load of slabs with square column fairly accurately compared with other codes. For reinforced concrete slabs, the available nominal shear stress can be calculated from

$$v_c = 0.79\left(\frac{100A_s}{bd}\right)^{1/3}\left(\frac{400}{d}\right)^{1/4}\left(\frac{f_{cu}}{25}\right)^{1/3} / \gamma_m \quad \text{(N/mm}^2\text{)} \qquad (4)$$

where A_s/bd is the percentage of flexural reinforcement, $400/d$ is the size effect factor which should not be taken less than 1.0, $f_{cu} \leq 40$ N/mm^2, $\gamma_m = 1.25$. The ultimate punching shear, v, due to a concentrated ultimate load can then be obtained from

$$v = V_u/ud \text{ (should not exceed } v_c\text{)} \qquad (5)$$

where V_u is the ultimate design shear force, u is the effective shear perimeter, and d is the effective depth of slab. The BS8110 uses a shear perimeter at $1.5d$ away from the column face and has square corners regardless of the shape of the column. For slabs with an opening located within a distance of $6d$ from the column, BS8110 considers that that part of the perimeter enclosed by the radial projections from the centroid of the loaded area (column) to the opening to be ineffective. BS 8110 also recommends an upper limit to the shear resistance expressed in terms of the shear stress at the periphery of the loaded area. The shear stress v at the periphery of the loaded area has to be less than the smaller of $0.8\sqrt{f_{cu}}$ or 5 N/mm^2.

The Code is silent on the issue of rectangular columns and a large opening near a rectangular column. The design procedure described below is recommended for those cases where the BS8110 does not apply.

Eurocode-2

According to the Eurocode-2, the punching shear strength of slabs without shear reinforcement is given by

$$v_{R1} = \tau_R k(1.2 + 40\rho) \quad \text{(N)} \qquad (6)$$

where τ_R is the basic shear strength of members without shear reinforcement, k is a size effect coefficient which also depends on the percentage of reinforcement, and ρ is the flexural reinforcement ratio. The punching shear load of slabs is then given by

$$V_{R1} = v_{R1}ud \qquad (7)$$

where u is the shear perimeter, which has rounded corners and is located $1.5d$ from the column face. For a typical column having a dimension of a by b with a not longer than $2b$, the shear perimeter u is equal to $2(a + b + 1.5d\pi)$. For a more rectangular column, the Eurocode-2 considers the perimeter in the middle of the longer side of the column to be ineffective as shown in Figure 4. For slabs with openings that are located at a distance less than 6 times the effective depth of slab from the column, the shear perimeter is reduced in the similar manner as that in the BS8110. That is, the length of the perimeter enclosed by the radial projection lines from the centre of the column to the opening is considered ineffective.

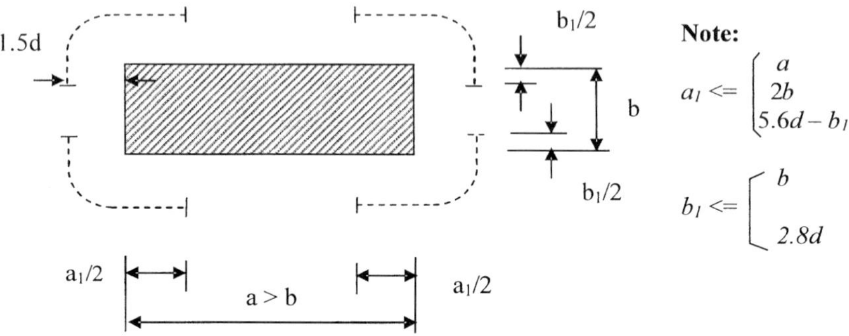

Figure 4 Shear perimeter according to Eurocode-2

RECOMMENDED PROCEDURE

This design procedure is essentially an extension to the BS8110 to make it applicable to punching shear design involving rectangular columns with an opening near the column. To accommodate column rectangularity, a reduction in the shear perimeter is introduced in the similar way as the Eurocode-2 procedure, except that a longer effective shear perimeter will result compared to that recommended by the Eurocode-2. When an opening is presence, the radial projection lines will not necessarily originate from the centre of the column but from the centre of the end portion of the column. When the column cross section is square, the procedure reduces exactly to the BS8110 method.

Note that the recommended procedure is intended for slab/interior column connections where unbalanced moment is negligible. The complete procedure is described below.

Design Procedure for Slabs Without Shear Reinforcement

The punching shear capacity should be checked on a shear perimeter 1.5d from the face of the column. For a rectangular column, the centre portion on the longer side of the shear perimeter may not be effective in transferring shear stresses, and therefore, the shear perimeter u should be computed according to that shown in Figure 5.

If the calculated shear stress, v, does not exceed, v_c, then no shear reinforcement is needed. The nominal design shear stress v is calculated from Equation (5), that is $v = V_u/ud$, where V_u is the design shear force due to ultimate loads, u is the shear perimeter as defined in Figure 5, and d is the effective depth of the slab.

For applications to flat-slab structures, the design shear force should be calculated using the design effective shear force, V_{eff}, in accordance with Clause 3.7.6 of the BS8110 Code.

Thus, V_u should be replaced by V_{eff}

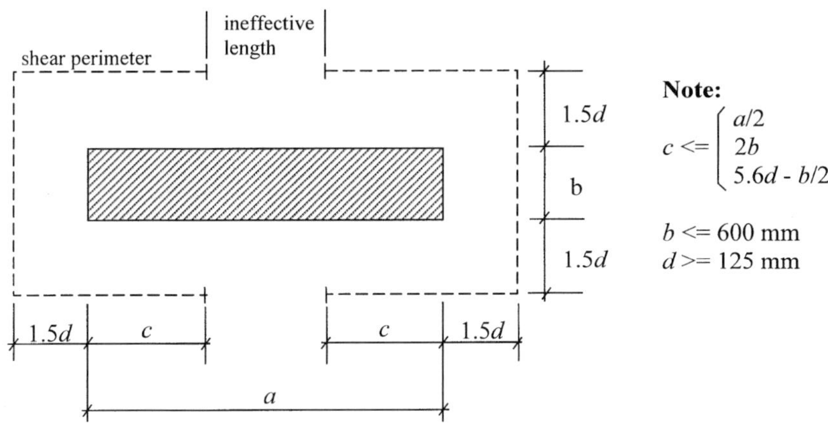

Figure 5 Recommended effective shear perimeter

Modification of Effective Shear Perimeter to Allow for Holes.

When an opening in the slab is located at a distance less than six times the effective depth of the slab from the edge of a column or a concentrated load, then the shear perimeter u should be reduced. That part of the shear perimeter, which is enclosed by the radial projections from the centroid of the end portion of the column to the edge of the openings, should be considered ineffective (see Figure 6). The projection lines from one end of a column should not cross the centre line of the column.

COMPARISON WITH EXPERIMENTAL RESULTS

The accuracy of the recommended design method, the Eurocode-2 method, and the ACI method had been compared with experimental data and the results are shown below. For the purpose of this comparison, all the safety factors for the code procedures are set to 1.0. Table 1 shows the average values of the predictions over experimental results, their standard deviations, and coefficients of variation (COV).

Note that for slabs with rectangular columns and openings, taking the radial projections from the centre of the column did not result in good prediction, so in Table 1, the radial projections for all methods were taken in accordance with Figure 6. From Table 1, the COV of the recommended method, that is, the extended BS 8110 Equation is 0.128 for the 20 slab specimens. These COV's are the lowest among the methods discussed above, indicating that the extended BS8110 procedure is reliable.

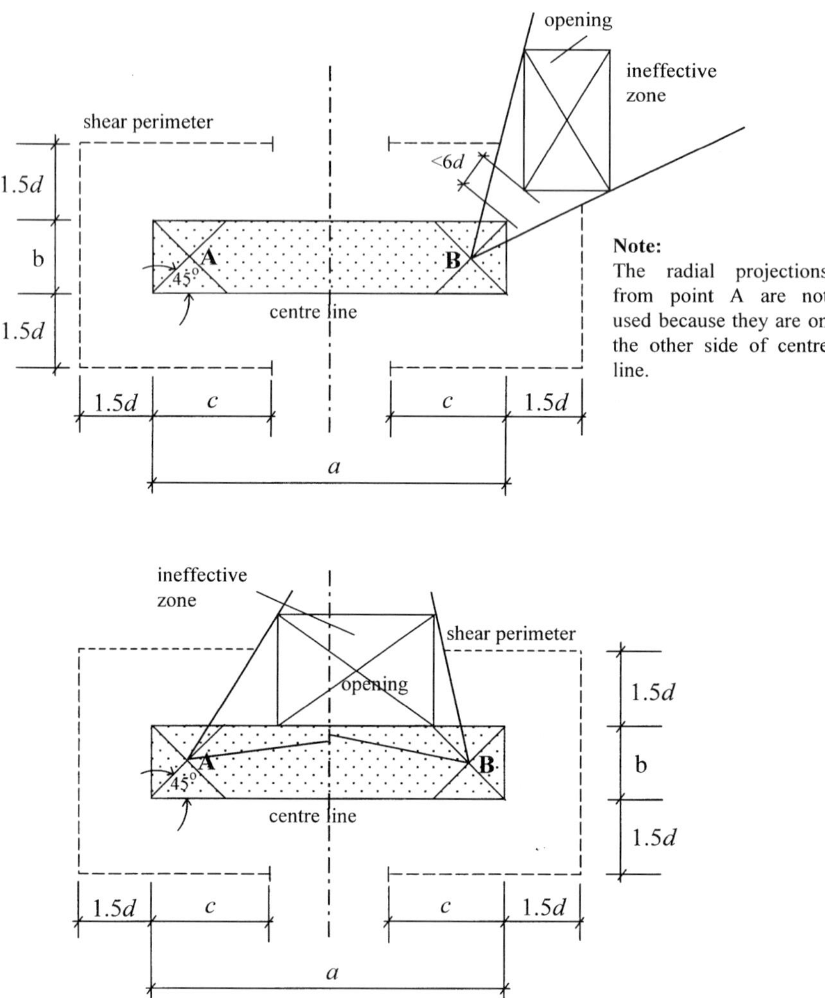

Figure 6 Recommended effective shear perimeter of slabs with openings

CONCLUSIONS

A recommendation on the punching shear design of concrete slabs with opening and supported on rectangular columns has been presented. The 20 new slab specimens tested under BCA-NTU joint research programme have been used to verify the accuracy and consistency of the recommended procedure. The recommended procedure, which is essentially an extension to the BS8110 procedure for cases involving rectangular columns with large openings, has been shown to be reliable and accurate.

Table 1 Comparison with experimental results

METHOD		BCA-NTU 20 SPECIMENS
current BS8110	Ave of Vpre/Vexp	0.898
	Standard of Deviation	0.133
	Coef. of variation	0.148
Recommended	Ave of Vpre/Vexp	0.891
(Extended BS8110)	Standard of Deviation	0.114
	Coef. of variation	0.128
ACI-318	Ave of Vpre/Vexp	0.692
	Standard of Deviation	0.098
	Coef. of variation	0.142
Eurocode-2	Ave of Vpre/Vexp	0.695
	Standard of Deviation	0.115
	Coef. of variation	0.165

Note: Vpre = predicted punching strength (safety factors = 1)
Vexp = punching strength from experiments

ACKNOWLEDGEMENT

This research is part of the joint BCA-NTU research on flat plate structures. Research grants from the Building and Construction Authority (BCA), Singapore and the Nanyang Technological University (NTU) are gratefully acknowledged.

The following individuals provided significant supports for the joint research project:

BCA Staff: Lam Siew Wah (Deputy CEO), Tan Tian Chong (Director), Ong Chan Leng (Director).

Advisory Panel: Tan Ee Ping (Tan Ee Ping & Partners), Tan Guan (T.Y. Lin - South East Asia), Tan Kheng Hwee (BBR Construction), Soh Seng Siong (KTP Consultants).

Thanks also due to Paul Regan (University of Westminster, UK) and Chris Morley (Cambridge University, UK).

REFERENCES

1. MOE, J, Shearing strength of reinforced concrete slabs and footings under concentrate loads, Development Department Bulletin D47, Portland Cement Association, Skokie, Illinois, April 1961, p 130.

2. HOGNESTAD, E, ELSTNER, R C, HANSON, J A, Shearing strength of reinforced structural lightweight aggregate concrete slabs, ACI Journal, Vol 61, No 6, June 1964, p 29-58.

3. EL-SALAKAWY, E F, POLAK, M A, SOLIMAN, M H, Reinforced concrete slab-column edge connections with openings, ACI Structural Journal, Vol 96, No 1, January-February 1999, p 79-87.

4. ACI COMMITTEE 318, Building code requirements for reinforced concrete (ACI 318-99) and commentary (ACI 318R-99), American Concrete Institute, Detroit, Michigan.

5. HAWKINS, N M, FALLSEN, H, HINOJOSA, R C, Influence of column rectangularity on the behaviour of flat plate structures, ACI Special Publication, SP-30-6, American Concrete Institute, 1971, p 127-146.

6. BRITISH STANDARDS INSTITUTION, Design of concrete structures – Part 1, General rules and rules for buildings (Eurocode 2), European Prestandard ENV 1992-1-1:1991, Comité Europeén de Normalisation, Brussels, p 253.

7. BRITISH STANDARDS INSTITUTION, Structural use of concrete, Part 1, Code of practice for design and construction, (BS 8110:Part 1:1995), 1985, London.

DEFLECTIONS OF REINFORCED CONCRETE SLABS: CASE STUDIES

W H Boyce

Halliburton KBR

Australia

ABSTRACT. The move to limit states design philosophy has, in principle, clarified the need to satisfy serviceability requirements. Nevertheless there are still many examples of suspended reinforced concrete slabs that have excessive deflection. The paper considers a flat slab floor with 8.4 m panel dimensions and deflections exceeding 80 mm at mid-panel. The building was the subject of a legal dispute.

The paper discusses:

- The legal proceedings
- Deflection issues
- The findings of the court
- Defects and rectification work.

The paper shows that engineers cannot rely solely on design codes, nor can they argue that they are too busy to keep themselves informed. The paper highlights the deficiencies of some aspects of deflection calculation and suggests improvements.

Keywords: Flat slabs, Serviceability, Deflections, Span-to-depth ratios, Design codes.

W H Boyce is a Technical Adviser Structural with Halliburton KBR in Brisbane and concurrently held a fractional appointment as Reader in Civil Engineering at the University of Queensland for 10 years until 2001. Over his 40 years experience, he has been involved in the design of a wide range of civil engineering and building structures. He has investigated many structural failures for insurance and litigation purposes.

INTRODUCTION

The move to limit states design philosophy has, in principle, made more explicit the need to satisfy serviceability requirements. Nevertheless there are still many examples of suspended reinforced concrete slabs that have excessive deflections. The paper considers one such example.

The building comprises a two-level basement car park, ground floor and three upper floor levels. It has insitu reinforced concrete floors, columns and basement walls; piled foundations; masonry walls to lift and stair shafts; and structural steel roof. The structural design was carried out in the period 1987/88 and the building was constructed in 1988. The building was purchased by the present owner (the plaintiff) in 1989.

The floor system is a flat slab with drop-panels at internal columns only and typical panel dimensions of 8.4 m. Deflections at mid-panel exceeded 80 mm.

The structure was the subject of legal proceedings during which the adequacy of the original design came into question.

The paper discusses:

- defects in the building and rectification works;
- the legal proceedings;
- the design issues involved;
- the need for realistic definition of the serviceability requirements;
- the findings of the court appointed referee.

The paper shows that engineers cannot rely solely on design codes when the literature has identified shortcomings in the code rules. Nor can they argue that they are too busy to read the literature and keep themselves informed.

The paper highlights the deficiencies of calculation procedures for slab deflections and suggests improvements.

BUILDING DEFECTS & RECTIFICATION

The author's employer was engaged during the period 1993 to 1999 by the property managers on behalf of the building owners to investigate and report on various matters relating to the structural condition of the building and to provide advice on rectification.

During these investigations, certain deficiencies were found in the structure and rectification works were carried out as follows:

- External stairs, wheelchair access ramp and landings: There had been excessive movements due to inadequate compaction of the fill under these exterior slabs. The defective stairs, landings and ramps were demolished and reconstructed with appropriate footings to control settlement.

- Piles: Piles under certain pile caps were not adequate to support the design load. One pile in each group of three was overloaded due to uneven distribution of the applied load because of poor geometry of the pile group/column layout. Rectification comprised the construction of engaged piers alongside the existing columns between the pile cap and the floor above, in order to spread load more uniformly between all three piles.

- Floor Slab/Column Junctions: The floor slabs did not have adequate punching shear strength at certain columns. Rectification comprised the installation of steel column capitals (collars) at the head of the relevant columns bearing tight against the underside of the floor slabs.

- Floor slab deflections: Floor slabs had deflected excessively such as to cause serviceability problems. The floor slabs had been determined to have adequate flexural strength, and rectification comprised the laying of a topping screed to restore the floor surface to acceptable levels.

The external works rectification was carried out in 1995 and the remaining rectification in 1998.

CHRONOLOGY OF BUILDING CASE

In this section an outline of the events relating to the legal action in the Supreme Court is provided, from the viewpoint of the structural 'expert'.

February 1994: Preliminary report based on walk through inspection and providing recommendations for further investigation.

February–June 1994: Meetings with solicitor agreeing on floor level survey, comparison of measured and theoretical deflections, consideration of options for rectification with indicative costs. Levels taken on underside of slab, more detailed inspection with photographs, structural adequacy checked.

July 1994: Report concluded most of the damage is the result of floor deflections, the measured deflections are in excess of the maximum permissible deflections, the slab thickness is less than that required to satisfy AS 1480 or AS 3600.

August 1994 Onwards: Preparation of Plaintiff's Statements of Claim for Supreme Court action, responding to requests for further and better particulars, meetings with solicitor and barrister.

December 1995–January 1996: External stairs, landings, ramps, tiles rectified.

March 1996: Further report with same conclusions as July 1994 report, noting that rectification is feasible by using screed and levelling compound, with bonded plates to strengthen and stiffen floor a necessary adjunct to this.

July 1996: Supplementary report, covermeter readings show excess top cover, cores show low strength, column capitals needed to increase punching shear capacity, some piles overloaded because of unfortunate arrangement in original design.

August–October 1996: Meetings of experts.

October 1996: Three day mediation conducted by QC.

January 1997: Consolidated report with some modification to previous views in light of discussions with other experts and matters raised during mediation, levels on top of slab confirmed deflections calculated based on levels to underside, requirement for bonded plate stiffening removed.

January–February 1997: preparation of questions to be determined by Special Referee appointed by the court, preparation of answers as part of evidence before Referee.

March 3–7, 1997: Hearing before Referee.

April 1997: Report of Referee, in almost all respects the findings were in favour of the plaintiff.

May 1997 Onwards: Interpretation of Referee's findings in relation to details of rectification proposals.

December 1997: Amended defence claiming defects should have been observed by purchaser.

June 1998 Onwards: Rectification of internal works; columns, shear capitals, slab topping.

December 1998: Reports (a) how the defects found by the Referee may affect the safety of the building in its use as a commercial office building, (b) in the event of being asked in 1989 to provide to a prospective purchaser a structural report on the building what measures would have been taken and what likely observations would have been made.

In *Tod Group Holdings Pty Limited v Fangrove Pty Limited* (1 December 1998, Queensland Court of Appeal, unreported) the court unanimously found that a firm of engineers was not liable to a subsequent purchaser of commercial premises for a claim relating to a defect in the structure itself.

As a result of this Appeal Court decision the Supreme Court action was abandoned.

BUILDING DEFLECTIONS

Observations

Floor sags calculated from levels taken on the underside of Level 5 in August 1995 are shown in Figure 1.

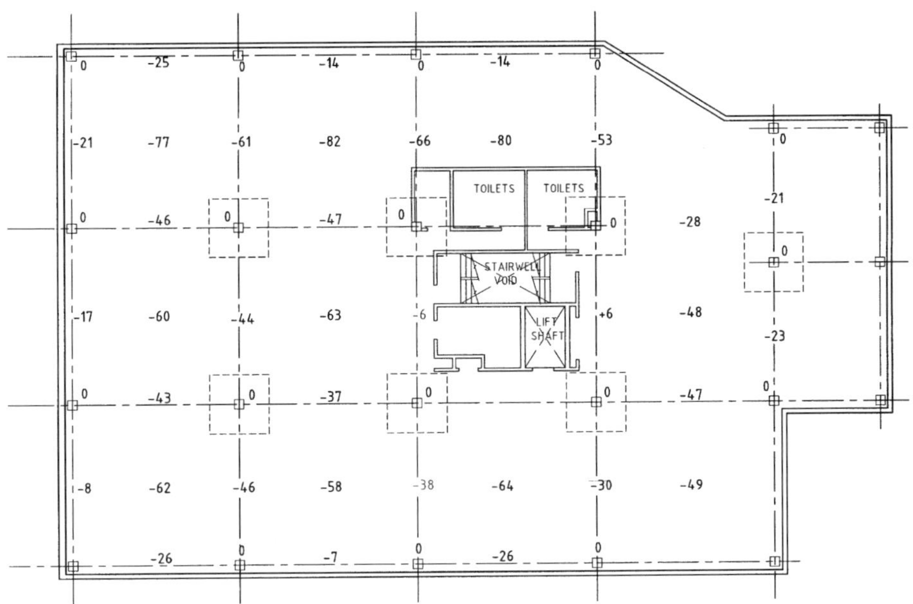

Figure 1 Floor sags (mm), Level 5, August 1995. (Seven years after construction)

These sags are also representative of levels 3, 4 and part of 6.

The findings of the referee, based on the evidence presented to him, were that the floor slabs have deflected to such an extent as to cause:

- Desks, tables, shelves, cupboards, drinking fountains etc to lean on a slope unless they are chocked.
- Ceiling tiles and ceiling grids to be misaligned and tiles to fall out.
- Office doors to jam or not to close completely.
- Toilet walls to rotate or crack, opening joints and cracking tiles.
- Fitted partitions to be visually disturbed.
- Visually apparent floor deflections.

In summary, there is a significant diminution in the serviceability of the office floors.

Documents discovered during the trial indicate that troublesome movements (40–55 mm) were of concern to tenants within 12 months of occupancy of the building.

Deflection Limits

The relevant concrete code at the time of design was AS 1470–1982 [1]. Under the building legislation, the use of AS 1480 was mandatory.

The deflection to be considered is that part of the total deflection which occurs **after** the attachment of non-structural members and the limits are set at:

- Span/500 where the non-structural members are likely to be damaged by deflection.
- Span/250 where the non-structural members are not likely to be damaged by deflection or the slab does not support non-structural members.

It is noted that the Span/500 limit may be inadequate for brittle finishes or partitions and that in two-way slab construction, the deflection to which the above limits apply is the theoretical deflection of the line diagram representing the equivalent frame.

The limits as set are very difficult if not impossible to correlate with actual deflections in the finished structure since (a) it is not usual to survey the floor prior to adding partitions etc so there is no defined datum for future deflections and (b) inherently in a two-way system, the deflection across the width of the equivalent frame varies. In addition there is no limit on total deflection or on mid-panel deflection.

The revised concrete code AS 3600–1988 [2] modified the limits to set a total deflection limit for all members of Span/250 with separate limits for the deflection which occurs after the addition of masonry partitions. For two -way slabs the limits are again applied to the theoretical deflection of the line diagram representing the idealized frame. The most recent edition of AS 3600 [3] has maintained the limits set in the 1988 revision.

It was argued, and found by the Referee, that an acceptable mid-panel total deflection in a building of this nature was Span/250.

Deflection Calculations

No evidence was discovered, or presented by the defendant, to indicate that deflection calculations had been carried out as part of the design process. Calculations were carried out during the court action by all parties and, based on the evidence presented, the Referee determined that the floor slabs as designed did not satisfy the calculated deflection limits of AS 1480. A note in AS 1480 indicated that Standards Australia was not prepared to recommend a formula for effective moment of inertia but suggested that the Branson formula may be used. It was also suggested that the cracked section moment of inertia would be more appropriate than the effective moment of inertia in some instances. A calculation using the effective moment of inertia satisfied the AS 1480 requirements. A similar calculation using the cracked section moment of inertia did not. The use of the effective moment of inertia does not correctly reflect the cracked state and nature of the slab.

AS 3600–1988 repeated the procedure of AS 1480 except that the use of the Branson formula was fully incorporated into the procedure. AS 3600–2001 has modified the procedure by incorporating a shrinkage-induced tensile stress which has the effect of reducing the bending moment to cause cracking of the section. While this most recent provision is an improvement, it does not address the problem that the calculated deflection cannot be related to an actual deflection of the structure.

Rangan [4] developed a simple expression to calculate deflections of two-way slab systems. His method leads to the values shown in Table 1, where observed sags are also shown.

Table 1 Final (long-term) deflection. Comparison of calculated and observed values

LOCATION		DEFLECTION (mm)	
		Calculated (Rangan)	Observed Range
External panel	mid-panel	74	49–82
	grid-line	53	30–66
Internal panel	mid-panel	46	48–63
	grid-line	39	28–47

It may be seen that Rangan's expression leads to realistic values and had this simple calculation been attempted at the design stage, the existence of a potential problem would have been recognised.

Span-to-depth Ratios

An alternative means of satisfying the permissible deflection requirements of AS 1480 was to comply with span/effective depth ratios. These ratios were based on work by Beeby [5] and are almost identical to the requirements of B.S. CP110 [6].

Slab thicknesses calculated on this basis are shown in Table 2. It was known at the time of the design that the span-to-depth criteria of AS 1480 were unreliable and the Referee found that the design engineer in 1987 ought to have been aware of the inadequacies. AS 3600 introduced more stringent requirements and the use of its span-to-depth ratios led to slab thicknesses shown in Table 2.

In 1988, the Cement and Concrete Association of Australia, in collaboration with the Steel Reinforcement Institute of Australia, published 'Design Guide for Long-span Concrete Floors' [7]. This provided span to depth information for different floor systems based on a survey of concrete buildings under construction or recently built in all main cities of Australia. It would thus represent design practice of the early to mid-1980's. Based on information in this publication, floor thicknesses would be as shown in Table 2.

Table 2 Slab Thickness (mm). Based on span/depth ratios

CONDITION	AS 1480	AS 3600	CACA	DESIGN REPORT
Without drop panels				
General floor area:				
• end span	240	285	260–280	290
• interior span	240	240	260–280	290
Toilet area	310	370		
With drop panels				
General floor area				
• end span	200	260	260	220
• Interior span	200	220	230	220
Toilet area	260	340		

Also shown in Table 2 are slab thickness values presented in the designer's preliminary design report.

The slab thickness used was 220 mm and drop panels were not used on the external columns. The slab thickness does not satisfy the span-to-depth ratios of either AS 1480 and AS 3600 and it may be concluded that the slab thickness is not adequate to prevent deflections exceeding the deflection limits in either code. The actual thickness of 220 mm is less than the general practice of the period. It had been recognised as early as 1977 [8] that the use of the AS 1480 span-to-depth criteria produced slabs that were too thin to satisfy deflection limits. A prudent engineer would have used thicker slabs than obtained from AS 1480 span-to-depth criteria.

One witness argued that engineers rely on codes exclusively and are too busy to be aware of other problems set out in the technical literature. The Referee rejected this evidence, stating that as professionals they should not be too busy to be aware of problems in their field of expertise. They build up their expertise from experience and knowledge, not by relying on codes exclusively.

Rectification

The Referee found that the slab deflections required rectification so that the floor could be made serviceable. The plaintiff argued that this should be done by providing a level topping on the slab. The Referee did not accept this argument and found that the extent of rectification required is to top the slabs with additional material so as to bring the deflections to 27 mm. The 27 mm is based on the reasonable deflection of 35 mm found previously and allowing for a further deflection of 8 mm due to the effects of additional weight of topping and the expected long term creep. This puts the owner in the position he reasonably would have been in had the design defect not occurred.

CONCLUSIONS

While the move to limit states design philosophy has highlighted the need to consider serviceability requirements more explicitly, there are still many examples of suspended reinforced concrete slabs with excessive deflections.

This paper has considered one such building which was the subject of a court action and the following conclusions are drawn from this example.

The deflection limits set by the concrete code of the time were not capable of being related to actual deflections in the finished structure since they were based on that part of the deflection occurring after attachment of partitions and also on the theoretical deflection of a line diagram. The current concrete code has set a limit on total deflection while maintaining a separate limit on the deflection occurring after addition of masonry partitions. However, the limits are still based on the theoretical deflection of a line diagram. The limits should be amended for two-way slabs to be the deflection at mid-panel. This deflection is readily calculated using the Rangan [4] procedure. An acceptable mid-panel limit for commercial buildings is Span/250.

Deflection calculations are inherently difficult because of the number of variables involved, including cracking in the slab due to shrinkage restraint. The use of effective moment of inertia based on the Branson formula markedly underestimated the deflection in service. The procedure introduced into AS 3600–2001 incorporating a shrinkage-induced tensile stress should improve accuracy but it is still preferable to have the calculated deflection relating to an actual location in the structure.

The span-to-depth ratios of AS 1480 (based on Beeby) produced slabs that were too thin to satisfy deflection limits. While it might be possible to modify the ratios, it seems that the modified and simpler procedure of AS 3600 will produce acceptable results.

It is not acceptable for professional engineers to rely exclusively on design codes. They need to be aware of current literature particularly if it has identified shortcomings in code procedures.

REFERENCES

1. STANDARDS ASSOCIATION OF AUSTRALIA, The Use of Reinforced Concrete in Structures, AS 1480–1982, Sydney, 100 pp.

2. STANDARDS ASSOCIATION OF AUSTRALIA, Concrete Structures, AS 3600–1988, Sydney, 156 pp.

3. STANDARDS AUSTRALIA INTERNATIONAL, Concrete Structures, AS 3600–2001, Sydney, 175 pp.

4. RANGAN, B.V., Deflections of Reinforced Concrete Flat Slab and Flat Plate Floors, Institution of Engineers, Australia, Civil Engineering Transactions, 1986, Vol. CE28, No. 1 January, pp 28–31.

5. BEEBY, A.W., Modified proposals for controlling deflections by means of ratios of span to effective depth, Cement & Concrete Association Technical Report 42.456, London 1971.

6. BRITISH STANDARDS INSTITUTION, The Structural Use of Concrete, CP110 : Part 1 : 1972, London, 154 pp.

7. CEMENT & CONCRETE ASSOCIATION OF AUSTRALIA/STEEL REINFORCEMENT INSTITUTE OF AUSTRALIA, Design Guide for Long-span Concrete Floors, Sydney, 1988.

8. WALSH, P.F., Deflections of Reinforced Concrete, Institution of Engineers, Australia, Civil Engineering Transactions 1977, Vol. CE19, No. 2, pp 146-152.

WHAT`S NEW IN FLOORS – AN OVERVIEW OF THE KEY CHANGES IN TR34

T Hulett

Concrete Society

United Kingdom

ABSTRACT. This paper reviews the processes and research underpinning the new edition of The Concrete Society's Technical Report TR34, Concrete Industrial Ground Floors – A guide to their Design and Construction. The rationale for the contents and design of the new edition are described. Key changes are identified.

Keywords: User's needs, Design, Construction, Materials, Guidance, Limit state analysis, Plastic analysis, Process modelling.

T Hulett is a principal engineer with the Concrete Society specialising in industrial floors. He was project manager and lead author for the third edition of Technical Report TR34, Concrete Industrial Ground Floors – *A guide to their Design and Construction.*

INTRODUCTION

Concrete industrial floors are playing an increasingly important role in modern economies. They are the ‘working platform’ for most manufacturing, warehousing and distribution activities. They are also to be found in retail premises and specialist applications such as sports halls and other leisure facilities. The annual value of this form of construction in the UK is now approximately £100 million with some 6 million square metres of floor being constructed each year.

In August 2000, The Concrete Society began a review of its Technical Report TR34, Concrete Industrial Ground Floors – A guide to their Design and Construction[1]. The review was a DETR (now Department of Trade and Industry) funded project as part of the PiI - Partners in Innovation scheme. The Association of Concrete Industrial Flooring Contractors (ACIFC) provided half of the funding.

THE NEED FOR NEW GUIDANCE

Existing guidance has developed from the work at the Cement and Concrete Association in the late 1970s and early 1980s. The Concrete Society produced its first edition of TR34 in 1988. This document and the next edition of 1994 referred extensively to other earlier publications and standards and consequently did not provide complete, integrated, guidance. The result is that guidance has developed piecemeal throughout a period of rapid expansion of floor construction particularly in the distribution and retail trades.

This period has also been one of significant development in design philosophy and material technology. Construction methods have also changed to make more use of mechanisation. These developments have been driven by and have in turn contributed to this rapid growth in floor construction. These changes have been economically beneficial and it has been possible to “have more for less”, however, such change has led to greater risk of failure to perform and so it is appropriate to consolidate and update the guidance. The objective is to provide a document which enables a floor user to obtain required performance at acceptable cost with an appropriate level of risk i.e. value for money.

Although The Concrete Society is primarily concerned with the UK, the pace of development has also been apparent in the rest of Europe and particularly in the former Eastern European countries. It has been noticeable that the UK property development and logistics sectors have been heavily involved in Europe. As the project developed the European dimension became increasingly apparent.

GETTING STARTED

Towards the end of 1998, The Association of Industrial Flooring Contractors and The Concrete Society began discussion on the development of a new edition of TR34. Two key objectives were agreed. TR34 should provide complete design and material selection guidance, with only minimal requirement to refer to other documents and it should be non-prescriptive. This is in common with the trend away from prescriptive guidance in the construction industry as a whole. Where it is possible to give performance requirements there is no requirement to be prescriptive although underpinning guidance is often needed.

The target audience for the report was established and divided into two categories:

- **End Users:**
  - The owner of the property
  - The user of the property
  - Supplier of equipment for use on the floor

- **Suppliers:**
  - Consultant/Designer
  - Main contractor
  - Specialist flooring contractor
  - Construction equipment suppliers
  - Materials suppliers
  - Other service providers

The benefits to be gained were seen as follows:

- **For the end users;**
  - Better whole life value
  - More consistent quality
  - Compatibility with international standards
  - Flexibility to specify to performance criteria

- **For the suppliers;**
  - Flexibility to deliver economic, fit-for-purpose, solutions
  - Better design practice to reduce material costs and wastage

On the basis of these discussions, the Society made a submission to the Department of Transport and the Regions for funding from the Partners in Innovation scheme-PiI. This was successful and was matched by an equal contribution from the flooring industry. The project formally commenced in August 2000 and was completed in June 2002.

PROJECT AND REPORT FRAMEWORK

From the early pre-project discussions it became apparent that the research for the project could be broken down into four primary areas: User's needs; Design; Construction; and Materials. Working groups were formed to guide the research in each of these areas. The framework proved to be robust and is reflected in the completed report. This is less obvious in relation to construction but reflects that the report is intended to be non-prescriptive and as a result does not give detailed information floor construction methods. There was considerable overlap in terms of skills and experience between the four main working groups and their activities were coordinated through the project manager and the steering group.

Before starting detailed research for the project and to ensure effective coordination of the project, it was recognised that there was a need to understand how the many interested sectors within the flooring industry interacted. This resulted in a suggested model of the function of the flooring market as shown in figure 1.

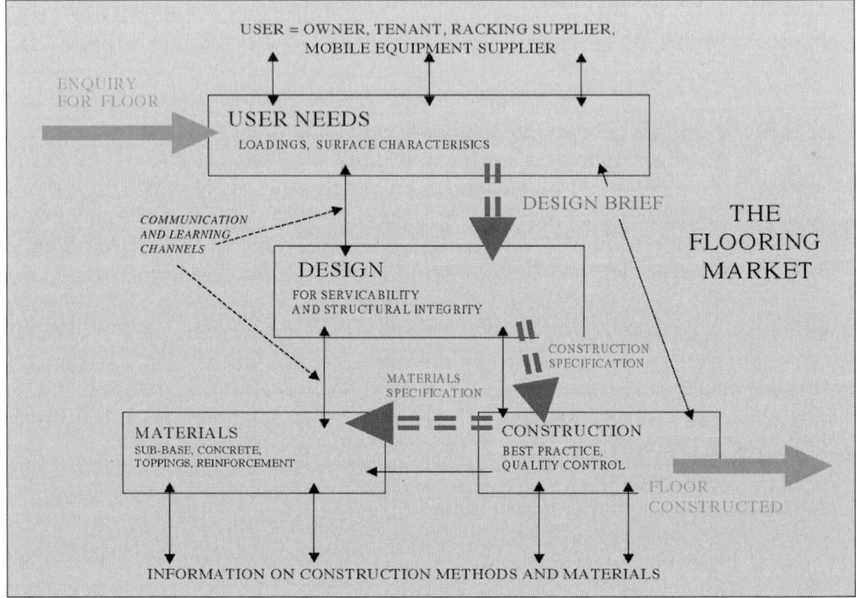

Figure 1 The market for floors

Based on the model, the primary activities for working groups were identified and are described here:

User's Needs

The objective was to understand how a user's planned activities interact with a floor. These activities are typically storage and mechanical handling; manufacturing and retail activities. There are other minority uses. The features of a floor, which are of interest, are load carrying capacity, flatness, joints, durability-particularly abrasion resistance and appearance.

It is perhaps not appreciated that the performance requirements for a floor with pallet stacking or other low level use can be quite different to those of a high bay racked warehouse. For the former, the durability of, or reduction in the number of, joints is a high priority whereas in the latter, flatness is a priority. To some extent these can be opposing needs.

Considerable emphasis has been placed on helping a user to understand the interactive nature of the decisions taken in relation to the above features and also the iterative nature of the design process. The report also attempts to encourage realistic expectations of what can be expected from a floor, for example the significance of cracks.

A key output of this section is a model design brief, which provides a disciplined approach to a thorough analysis and appraisal of the planned use of the floor.

Design

The design group conducted a state of the art review of slab load/stress analyses and slab early age movement data, along with associated material properties. The outcome is a limit state design based on a plastic analysis predominantly following the analysis of Meyerhof. Serviceability checks for deflections are included. The design section is broadly in line with the proposed Eurocode 2[2].

Shrinkage induced strains and associated stresses are an important factor in restrained ground bearing slabs. The design group were unable to identify a satisfactory analysis of the stresses involved across a wide range of load and construction scenarios, without incurring significant and potentially unjustified economic impact. They did however identify situations where greater caution might be adopted and placed great emphasis on the need to minimise shrinkage in all situations.

The group has updated the guidance on assessment of soils and construction of sub bases.

A spreadsheet based model was developed for testing and establishing the economic impact of the design proposals. The cost effect of the new load capacity analysis is broadly neutral by comparison with the plastic analysis in the appendix of the 1994 edition.

Construction

It is not intended that the report should describe or dictate the construction method as is the case in some codes of practice such as ACI 302[3]. To include such information was considered to be prescriptive and likely to hinder innovation. Although there is no specific construction section in the report, this group made a significant contribution in ensuring that the report, as a whole, reflected current economic construction practice. An important result of this groups work is a section on good practice in construction.

Materials

The materials group has reviewed the guidance on concrete constituents and mix design and other materials to be found in floors. Particular emphasis has been placed on minimising concrete shrinkage by reducing water contents. The report takes note of the recent work at University of Dundee on the role of cement in concrete. Guidance on material and construction factors affecting abrasion resistance has been updated using data from work commissioned by the Society at The University of Aston.

PROJECT COORDINATION

To ensure that the report contains all the information required for floor design and construction, a process modelling technique was adopted using the ADePT methodology [4, 5]. The technique involves the identification of the activities and information flows that make up a process, such as the design of the slab.

It is crucial that any process model is validated to eliminate any erroneous logic. The model is not concerned with which party to the process carries out each activity, only that the tasks are performed. This has proved to be a useful stimulus in breaking down perceived contractual barriers between the elements in the design and construction process. The work-breakdown structure of the model was used to challenge the structure of the new TR34 and highlight any missing sections. It has also identified communication barriers particularly in relation to the need for establishing common terminology.

THE KEY CHANGES IN THE NEW EDITION

General

The report layout attempts to reflect a typical route which might be taken through the processes of appraising the needs of the user and designing the floor. It is recognised that this is an iterative process with the potential for multiple exit and re-entry points, and therefore there is no absolutely correct way of setting out a written document, which is essentially a linear medium. However, it is hoped the extensive cross-referencing and comprehensive index will help users to find their way around the document efficiently. It has been suggested that a digital version of TR34 will facilitate this cross-referencing more effectively than a paper based document and this is being considered.

Soils and sub bases

Following recommendations in a 1995 Transport Research Board (USA)[6], sub bases are no longer considered to enhance modulus of sub grade reaction values. Greater emphasis is placed on the performance of a sub base as a working platform for construction in terms of stability and level tolerances.

Materials

TR34 has been updated to reflect BS 8500[7]. Particular emphasis has been placed on minimising early thermally induced shrinkage and drying shrinkage. The new guidance on materials and mix design for achieving abrasion resistance is not linked directly to cement content or compressive strength.

Design

The report reflects Eurocode 2[2] with some modifications. The main design method now uses plastic analysis based on Meyerhof's original equations and includes checks for punching. Serviceability checks for deflections are included. Steel mesh fabric is now included along with steel fibres as suitable means of achieving ductility. Guidance is given on load transfer capacity for dowelled joints.

Standard floor designs

New standard designs and classifications replace earlier BRE classifications [8].

Surface regularity

An alternative method has been provided for measuring floors in defined movement areas along with associated classification for use. The present edition specifies that only the front axle of trucks be measured. The alternative, in the appendix in the new edition, will broadly follow US practice of measuring the relationship between the front axle and rear wheels to take into account any font to rear motions of a truck. This anticipates work, which has started on a CEN standard and possibly an ISO. This may increase costs for some floors, however, commensurate performance benefits can be expected.

Model design brief

The design brief is intended to discipline the designer and client to ensure that all of the users requirements and construction factors are fully evaluated. Each item will be referenced to the appropriate guidance in the report. The process of design is iterative and it would be expected that there would be a number of changes to the design brief as the design progresses. To facilitate this, it is anticipated that the design brief pro-forma will be in digital format. An abstract from the model design brief is shown in Table 2.

Table 1 Abstract from model design brief

LOADINGS	
Uniform loads	Load (kN/mm^2) Aisle width (m)
Point loads	Load (kN) Minimum spacing (mm) Maximum spacing (mm) Loaded area (L x W mm) Wheel type (eg rubber, pneumatic, polypropylene)
Wheel loads	Load (kN) Wheel spacing (mm) Wheel contact area (L x W mm) Number of repetitions Min distance between wheel and point loads (mm)
Mezzanine	Mezzanine leg load (kN) Spacing (mm) Base plate size (mm x mm)
Other loading information	How long after casting will the floor be loaded? What is the intended temperature regime?
FLATNESS	
Free movement	FM 1 FM 2
Defined movement	Superflat Category 1
APPEARANCE AND SURFACE CHARACTERISTICS	
	Surface abrasion resistance Colour variation

REFERENCES

1. CONCRETE SOCIETY TECHNICAL REPORT 34, *Concrete Industrial Ground Floors, A guide to their design and construction*, Second Edition, The Concrete Society, 1994.

2. EUROCODE 2: Design of concrete structures – Part 1: General rules and rules for buildings, prEN 1992-1, 2nd draft 2001.

3. ACI MANUAL OF CONCRETE PRACTICE, PART 2, Guide for Concrete Floor and Slab Construction, ACI 302.1R-96.

4. AUSTIN S A, BALDWIN A N, LI B AND WASKETT P R, 'Integrating design in the project process', *Proceedings of the ICE: Civil Engineering,* Vol. 138, No. 4, November 2000, pp 177-182.

5. See (www.adeptmanagement.com) for more information on ADePT (the Analytical Design Planning technique).

6. TRANSPORTATION RESEARCH BOARD (USA), Report No.372, Support under Portland Cement Concrete Pavements, 1995.

7. BRITISH STANDARDS INSTITUTE 8500-2:2002, Concrete-Complementary British Standard to BS EN 206-1.

8. BUILDING RESEARCH ESTABLISHMENT, Classes of imposed loads for warehouses, Information Paper, 1987.

ADVANCES IN STEEL-CONCRETE-STEEL COMPOSITE DESIGN

N K Subedi

University of Dundee

N R Coyle

CORUS

United Kingdom

ABSTRACT. Steel-concrete-steel double-skin composite elements have applications in seawalls, underwater tunnels, blastwalls, aircraft hangars and shear walls in tall buildings. Previous designs of s-c-s sections suffered from the lack of full composite actions between the steel plate skins and the concrete core. This problem has been completely overcome by the provision of special surfaces to the plates at the interface. Expamet and Wavy wire surfaces provide high resistance to interface shear and enable the elements to achieve full composite behaviour. This paper discusses the development of the interface resisting surfaces and some results of the recent study.

Keywords: Concrete structures, Steel-concrete-steel, Composite elements, Interface shear, Expamet surface, Wavywire surface.

N K Subedi is a Reader in the Department of Civil Engineering at the University of Dundee. His main areas of research are concrete and composite structures, tall concrete buildings, deep beams and panel walls, concrete under hydrostatic pressure and tensile strength characteristics of concrete.

N R Coyle is an Engineer with CORUS. His main areas of interest are concrete and steel structures, composite structures.

INTRODUCTION

A practical steel-concrete-steel (s-c-s) double-skin element is shown in Figure 1(a). Such elements have applications in seawalls, underwater tunnels, blast walls, aircraft hangars, building components, wave-energy generating structures and bridges. The basic element consists of outer steel plates as skins and concrete in the middle forming the core. The elements are also provided with either crossbars or overlapping studs, Figure 1(b), at certain intervals in both longitudinal and transverse directions. With modular design the elements are highly suitable for factory manufacture in units requiring minimum amount of in-situ welding for assembly.

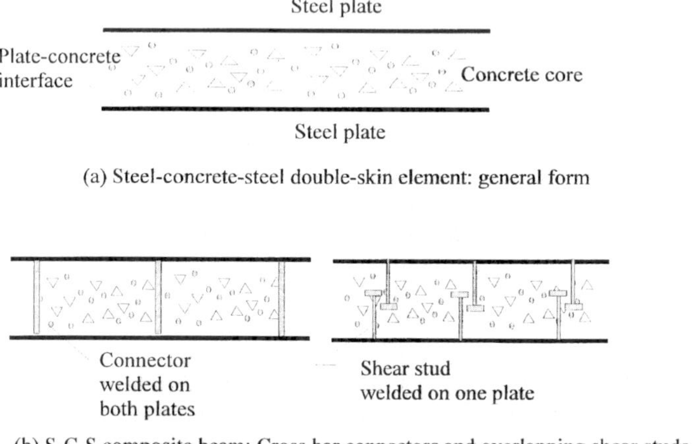

(a) Steel-concrete-steel double-skin element: general form

(b) S-C-S composite beam: Cross bar connectors and overlapping shear studs

Figure 1 Steel-concrete-steel double-skin construction

In s-c-s elements composite behaviour is vital if full benefit is to be derived of the sections. The composite behaviour depends on the bond characteristics at the interface between the concrete and the steel plate. In the case of a plain steel plate the bond relies entirely on the adhesion between the concrete and the plate. In such elements the obvious mode of failure would be due to the buckling of the compression plate at some stage of loading. Such failure happens rather early and quite often the plates do not reach their full strength capacity. This had been the case in earlier detailing [1, 2] of this type of elements, Figure 2, resulting in somewhat inefficient design.

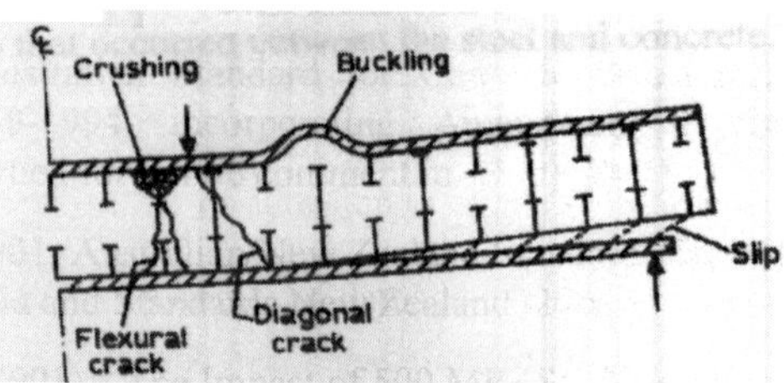

Figure 2 Modes of failure after Oduyemi and Wright [1]

In this paper a recent study leading up to the development of a fully composite section is described. With the development of full composite action the steel plates are utilised to their full capacity with no buckling problem of the compression plate. There is now a good prospect for manufacturing plates with special surface treatment for full composite action.

DEVELOPMENT OF SURFACED PLATE

In order to achieve the full composite behaviour between the concrete core and the steel skins the interface bond between the two materials must be improved. The most practical solution is to detail the plates by creating special surface acting as keys providing mechanical anchorage between the concrete and the plate.

Horizontal sine wire (H)

5 mm square bar (S)

Expamet (E)

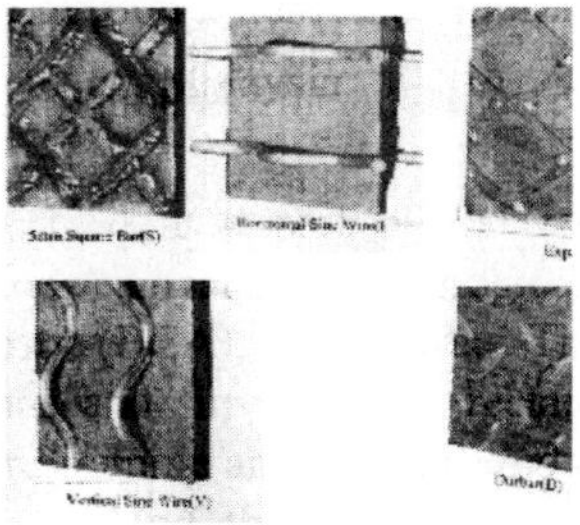

Vertical sine wire (V)

Dunbar (D)

Figure 3 Five different surface preparations for initial study

For the initial study eight different surface preparations were selected. Views of the surfaces 3-7 are shown in Figure 3 as examples.

1. *Plain (P)* : Plain plate cleaned by shot blasting. This will act as control.
2. *Roughened (R)* : Additional molten steel was sprayed on the plain surface to produce a roughened surface.
3. *Durbar (D)* : Standard Durbar floor plate with elliptical shaped embossed surface.
4. *Expamet (E)* : 2.16 mm thick expanded metal mesh welded on to the steel plate.
5. *5 mm square bars (S)* : Small lengths of 5 mm square bars welded on to the plate to produce a similar diamond shape pattern as the Expamet, but significantly thicker.
6. *Vertical sine wire (V)* : Lengths of 6mm diameter wire bent into a sinusoidal shape and welded on to the plate in the longitudinal (to the pull) direction.
7. *Horizontal sine wire (H)* : Similar to (V), but the sinusoidal shape welded on to the plate in the transverse (to the pull) direction.
8. *Air-shot Studs (A)* : Studs approximately 10 mm in height fired on to the steel plate under air pressure.

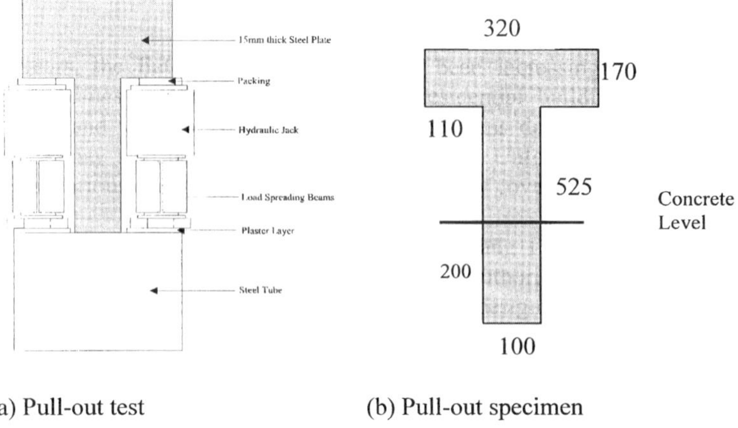

(a) Pull-out test (b) Pull-out specimen

Figure 4 Pull-out test details

For the preliminary study a simple pullout test, Figure 4, was carried out in order to determine the resistance capacity of the various surfaces. The plate thickness was 15 mm with the surface treatment being on both faces. In the case of *Durbar (D)* detail two plates were welded back-to-back to make up the 15 mm thickness. The test specimens consisted of 100 mm wide T-shaped plates inserted into 300 mm diameter steel tubes with 200 mm of the plate cast inside the concrete. Three strengths of concrete, 40, 60 and 80 N/mm^2 were used in the pull out test. The pull out was effectively by two hydraulic jacks mounted on spreader beams reacting on the top surface of the concrete. Two displacement transducers were also incorporated in the test to measure the movement of the plate relative to the concrete.

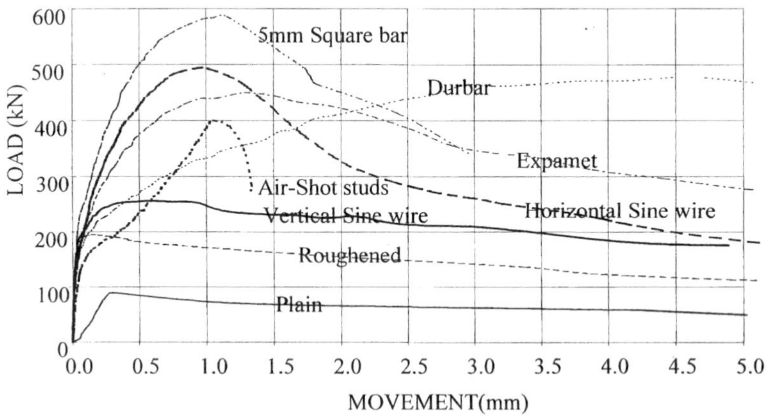

Figure 5 Load-displacement characteristics with C40 concrete

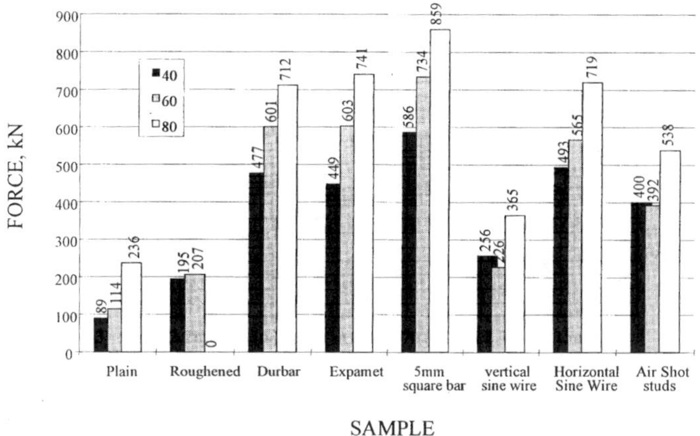

Figure 6 Comparison of pull-out forces (forces in kN)

Behaviour under load

The failure occurred typically by the pull out of the plate from the concrete with a wedge of concrete near the surface remaining attached to the plate. The load versus displacement plot for the plate at the bottom for C40 concrete is shown in Figure 5. The comparison of the pull out forces for all surfaces is shown in Figure 6. As a result of the pull out test a total of three surfaces were selected to be used in the beam-testing programme. These were *Durbar (D), Expamet (E) and Wavy wire*. The Wavy wire surface was a new idea to make the best surface represented by the *5mm square bar (S)* practical. It consisted of sinusoidal bent 6 mm diameter wire laid flat and welded on to the steel plate. This produced a surface with a similar depth and layout to *5 mm square bars*, but was much more practical to manufacture in large scale. The small individual lengths of 5 mm square bar had been replaced by long length of 6 mm diameter wire. The sinusoidal wave shape could be easily produced in a factory shop.

BEAM EXPERIMENTAL PROGRAMME

Following from the initial pull out test results an experimental programme consisting of 32 beams in 5 series was devised. The 32 beams were designed, constructed and tested to investigate the major mechanisms of failure, such as, flexure, shear and bond-slip. The plate surfaces chosen were *plain*, *Durbar, Expamet* and *Wavy wire*. The *Wavy wire* was as a practical substitution to *5 mm square bar* although no pull out test had been conducted using this detailing. Beams with a wide range of shear span/effective depth ratio were studied. The combination of the overall dimensions varied in the range 160 – 320 mm wide, 160 – 268 mm deep and 1500 – 4000 mm long. For the resistance of shear forces the beams contained either dowels welded on to the opposite faces of the plates at regular intervals or over lapping studs welded on one end only. Three strengths of concrete were used, 40, 80 and 150 N/mm^2.

The details of the experimental work will be published separately. Here only typical examples of the test results are discussed.

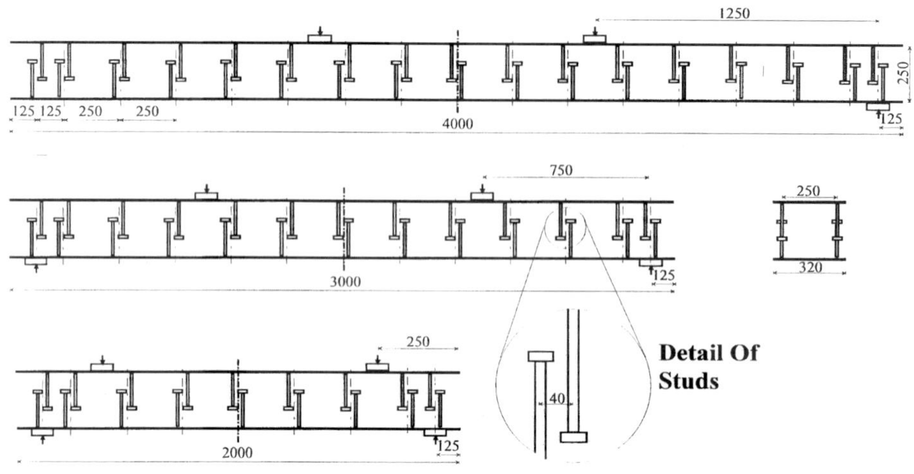

Figure 7 Dimensions of Series 5 beams

Experimental results: Series 5: Typical example

Results for shear span/concrete depth ratios 1 and 5 are presented for four beams in each category, 5P1, 5D1, 5E1, 5W1 and 5P5, 5D5, 5E5, 5W5. The symbols represent P – *plain* (control), D – *Durbar*, E – *Expamet* and W – *Wavy wire*. The concrete used for this series was C80. In this series overlapping studs were provided for the resistance of shear forces. The dimensions of the beam and the plate surfaces are shown in Figure 7 and a view of test arrangement is shown in Figure 8.

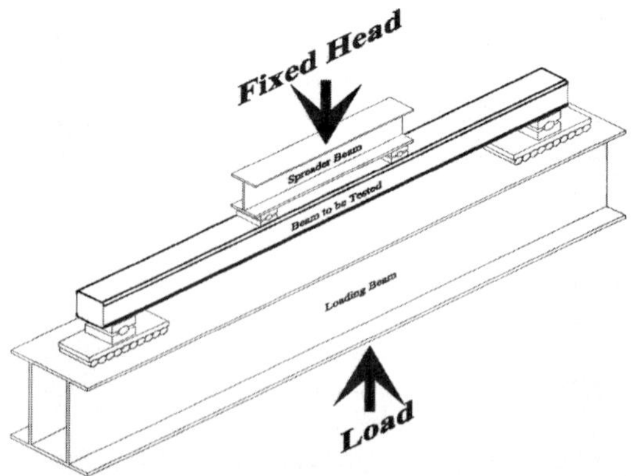

Figure 8 Test arrangement for beams

Shear span/concrete depth ratio 1

For these beams, as expected, the failure mode was shear, Figure 9 and Table 1, except for 5P1, in which the lack of interface resistance caused slip at much lower load of 1403 kN. In the case of the surfaced plates the failure loads increase between 122 and 133 percent compared to that of the *plain* surfaced beam. This highlights the importance of surfaced plates in enhancing the resistance capacity in shear. The inadequacy of the interface bond is much more pronounced when examining the service loads. With the plain plate the service load is only 17% of the ultimate load, but with *Expamet* and *Wavy wire* surfaces the load values are 60% representing a much improved serviceability requirement. The load-deflection plot in Figure 11 shows a more ductile behaviour for the *Expamet* and *Durbar* surfaces and less so for the *Wavy wire* case. The beam with *plain* surface failed suddenly due to slip.

Table 1: Series 5 results for shear span/concrete depth 1 and 5

BEAM	SERVICE LOAD, kN	SERVICE LOAD AS PERCENTAGE OF ULTIMATE LOAD, %	ULTIMATE LOAD, kN	ULTIMATE LOAD AS PERCENTAGE OF CONTROL	MODE OF FAILURE MoF
5P1	240	17	1403	100	Interface shear (slip)
5D1	589	32	1864	133	Vertical shear
5E1	1079	60	1795	128	Vertical shear
5W1	1030	60	1707	122	Vertical shear
5P5	69	28	245	100	Interface shear (slip)
5D5	157	48	343	140	Flexure
5E5	215	53	408	167	Flexure
5W5	225	63	355	145	Flexure

Figure 9 Typical mode of failure in shear : Beam 5E1

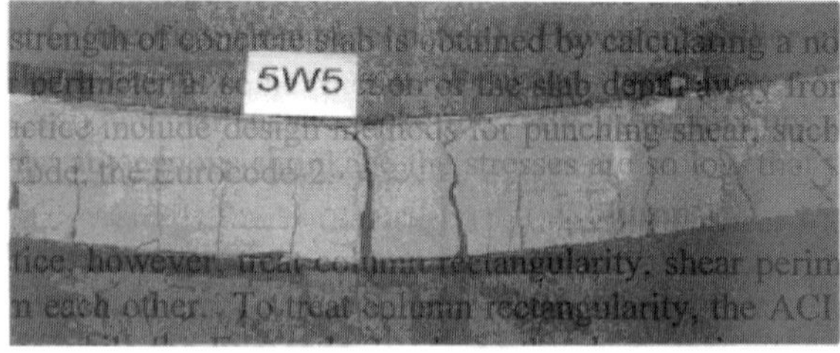

Figure 10 Typical mode of failure in flexure : Beam 5W5

Shear span/concrete depth ratio 5

This configuration illustrates the case of a normal flexural behaviour. Here in all but *plain* surface case, the failure was due to flexure as shown in Figure 10. There is clear evidence of uniformly spaced flexural cracks within the zone of maximum bending moment, full composite action between the concrete and the plates and the formation of a large curvature before the failure. The surfaced plates provide full resistance to the interface shear keeping the plates and the concrete core to act compositely well up to the failure. The enhancement to ultimate load in flexure varies from 140 to 167 percent, Table 1. The serviceability loads for all surfaced beams, ranging from 48% to 63% of the ultimate load, are also satisfactory. The load-deflection plots, Figure 12, show good ductile behaviour for the *Expamet* and *Wavy wire* surfaces and less so for the *Durbar* surface. The *plain* surface case is vulnerable to sudden slip at failure with little warning.

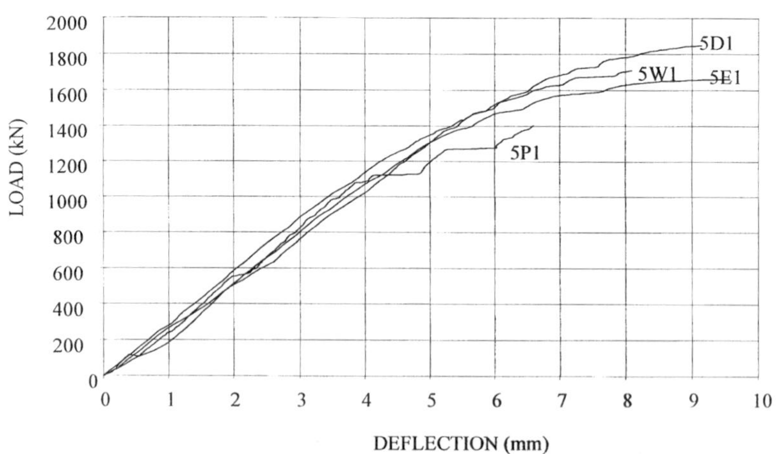

Figure 11 Load-deflection behaviour for shear span/depth 1

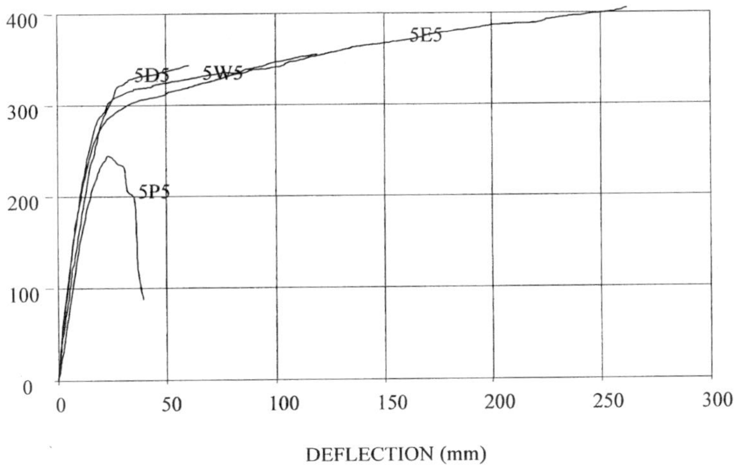

Figure 12 Load-deflection behaviour for shear span/depth 5

From the observed overall behaviour it is clear that *Expamet* surface provides an excellent resistance against the interface shear followed closely by the *Wavy wire* detail. Although *Durbar* surface has shown a good overall resistance it is prone to larger slips due to the particular type of surface indentation. The serviceability load is much lower with the *Durbar* surface.

CONCLUSIONS

1. Surfaced plates have been developed for use in steel-concrete-steel double-skin beam elements.
2. Of the eight surfaces examined here, *Expamet* and *Wavy wire* surfaces provide the best keys for resisting the interface shear and provide full composite behaviour.
3. Although *Durbar* surface also showed an overall good performance, but due to the configuration of the surface embossment, it has a tendency for larger slips and thus bigger cracks in concrete. The performance of *Durbar* surface with respect to serviceability load is much lower.
4. For the efficient use of steel-concrete-steel elements full composite action is essential. Full composite action enhances both the shear and flexural capacities of the section.

ACKNOWLEDGMENTS

The work described in this paper was carried out with a grant from the Department of the Environment, Transport and the Regions under its PiI programme. The industrial collaborator was The British Steel (CORUS), who also provided part of the funds. The authors gratefully express their gratitude for the contributions.

REFERENCES

1. ODUYEMI T O S, WRIGHT H D, An Experimental Investigation into the Behaviour of Double-Skin Sandwich Beams, *Journal of Constructional Steel Research*, Vol 14, 1989, 197-220.

2. NARAYANAN R, ROBERTS T M, NAJI F J, Design Guide for Steel-Concrete-Steel Sandwich Construction, Vol1: general Principles and Rules for basic Elements, *CSI Publication P 131*, The Steel Construction Institute, 1994.

SLAB-COLUMN JUNCTION DETAIL – A NEW CONCEPT

N K Subedi

P S Baglin

University of Dundee

United Kingdom

ABSTRACT. NUUL system is a new concept of detailing for slab-column junctions in concrete structures. It consists of a steel plate; solid steel cross bars and staples forming a prefabricated composite construction. The development of the system is aimed at eliminating the problem of punching from vulnerable sites such as flat slab-column junctions. Preliminary study shows that the system produces a robust design with very high resistance to punching force and it is suitable for prefabrication with minimal amount of steel fixing at the site. The paper describes the detailing system and discusses the preliminary results. A comparison with ACI recommendations for shearhead design is also discussed.

Keywords: Slab-column junction, Flat slab, Shearhead, Concrete structures, Composite construction, Punching resistance.

N K Subedi is a Reader in the Department of Civil Engineering at the University of Dundee. His main areas of research are concrete and composite structures, tall concrete buildings, deep beams and panel walls, concrete under hydrostatic pressure and tensile strength characteristics of concrete.

P S Baglin is a Research Fellow in the Department of Civil Engineering at the University of Dundee. His main areas of research are concrete and composite structures and lightweight structures.

INTRODUCTION

The aesthetic value and the cost benefit by having clear space between the floor and the ceiling and without the provision of column head in a flat slab construction are always appreciated by the clients and the architects alike. The attraction however creates a major problem to the structural engineer with regard to ensuring adequate and guaranteed safety against possible punching at the slab-column junctions. In the wake of punching failures of flat slabs, including a multi-storey carpark in Wolverhampton, England in 1997, the design provisions and detailing methods at the slab-column junctions and the safety issues were once again rekindled. This led to further appraisal of latest studies on punching resistance of various details and the conclusions were published as a Best Practice Guide (Ref 97.376 – 2000) by the British Cement Association (BCA). The Guide suggests that the use of proprietary prefabricated punching shear reinforcement systems, such as stud rails and shear ladders appear to be almost always worthwhile as justified by their long history of successful application.

Using traditional links is considered time consuming and expensive. The Best Practice Guide, however, does not address the main structural issue and that is 'do the recommended systems provide extra safety to avoid the vulnerability of slab-column junctions with respect to punching?' It is considered that the first and foremost criterion for deciding a particular system should be the safety and complete elimination of punching from such vulnerable sites as the slab-column junctions. The mode of failure based on flexural capacity, but with remote possibility of punching should be the design criterion for flat slabs. With this as the main objective this research looks at the development of a new detailing system called NUUL, which comprises a steel plate, cross bars and staples, Figure1. Test so far suggests that this is a robust system, which provides a very high degree of punching resistance and thus eliminates the possibility of punching in the normal design of flat slabs.

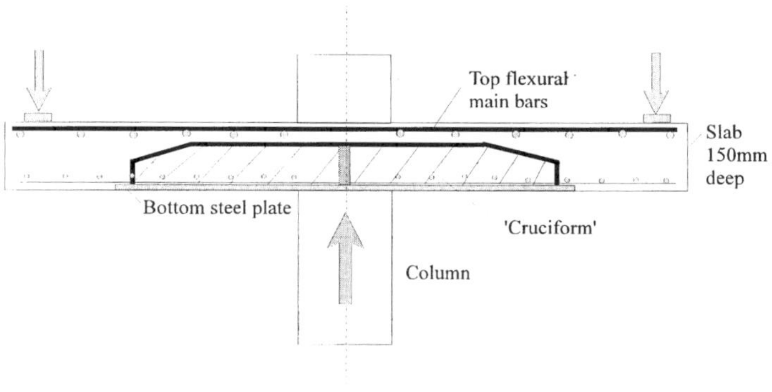

Figure 1 Punching resistance: NUUL system (staples not shown for clarity)

This paper presents the development of the NUUL system of joint detail for slab-column junctions. Initial results are discussed and compared with the ACI provisions for shearhead design.

PRELIMINARY INVESTIGATION – SLAB STRIP TEST

A series of four beams, 100mm wide x 150mm deep, representing a strip of slab was tested for preliminary investigation. The details of the strip are shown in Figure 2. The main shear-resisting element over the column is a steel member 400 x 90 x 16mm thick welded on to a base plate flushed with the soffit of the beam. From the preliminary test it was concluded that (a) the shear failure in each case occurred outside the steel section characterised by the inclined crack or cracks and the local buckling of the base plate (b) the composite section itself is very strong in shear and therefore failure is characterised by other secondary causes, such as the anchorage failure of the base plate or the main bars, Figure 3. The results from the preliminary test are given in Table 1.

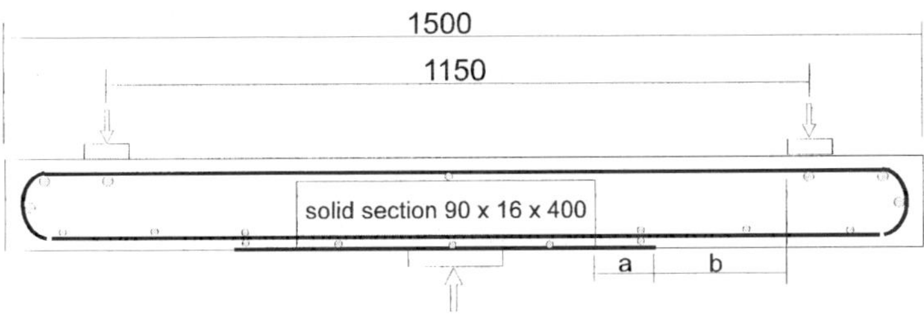

Figure 2 Typical beam detail representing slab strips for preliminary investigation

Table 1 Beam – test results

BEAM	DIMENSIONS			TEST RESULTS		
	a, mm	b, mm	Plate length, mm	Failure load kN	Shear stress N/mm^2	Failure mechanism
A1	100	250	600	89	3.6	Shear failure at end of plate
A2	200	150	800	118	4.8	Plate anchorage failure – followed by buckling of plate
A3	300	50	1000	119	4.9	Shear induced dowel failure of tension reinforcement
A4	400	-50	1200	86	3.5	Shear induced dowel failure of tension reinforcement

Beam cross-section 100 mm wide x 150 mm deep

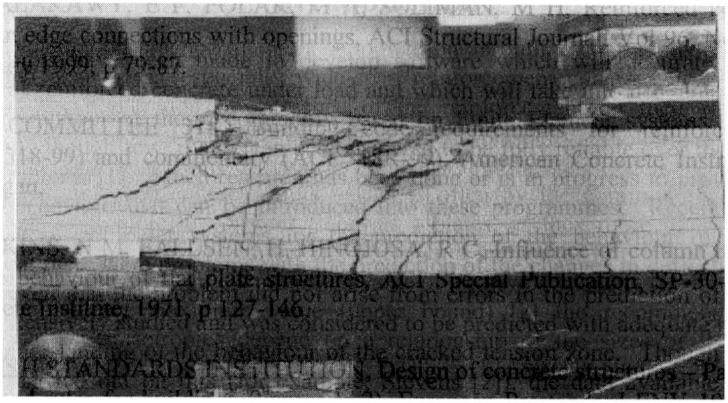

Figure 3 Typical beam after failure

PUNCHING RESISTANCE TEST

NUUL System – Detail

The NUUL system consists of solid steel cross bars, steel plate and U-shaped staples with a hole cut out at the centre of the plate for the column bars to pass through. The details are shown in Figure 4. The plate will form the soffit of the slab around the column. The level of the plate can be lifted to accommodate a fire protection layer flushed with the slab soffit. The steel cross bars forming a 'cruciform' will pass over the column. The plate provides the anchorage for the 'cruciform' and confinement to the soffit of the slab resisting local crushing around the column. In the figure, part of the opening for the passage of the column bars through the slab can be seen. The inverted U-shaped staples tie the top reinforcement of the slab preventing separation.

Figure 4 Plate, 'cruciform' and staples detail

Test Specimens and Procedure

The test specimens consisted of 1500 mm square slabs 150 mm deep with a square column in the centre. The lower part of the column varied in length either150 mm or 320 mm. A small part of the column also protruded through the top of the slab. The thickness of the control slab was 138 mm. The main reinforcement in the slab consisted of T16 at 100 mm centres. The bottom reinforcement was T12 at 100 c/c. A typical configuration of the plate and the cross bar 'cruciform' detail is shown in Figure 1. Other details and concrete properties are given in Table 2. The test specimens represented a typical internal panel of a flat slab construction.

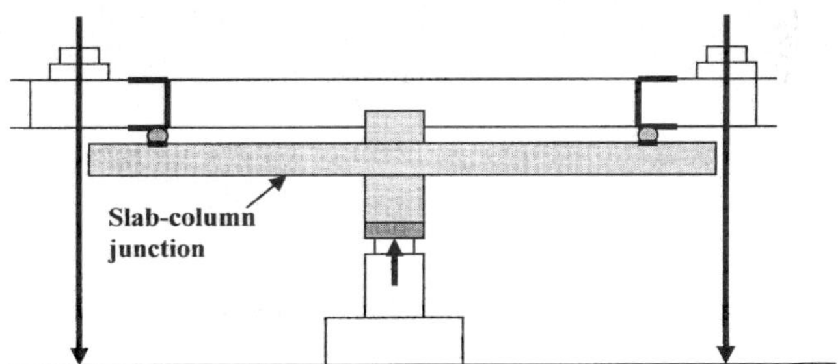

Figure 5 Test arrangement

Table 2 Detail of test specimens and concrete properties

SLAB	PROPERTY					
	Concrete	Dimensions, mm		Flexural Reinforcement		Shear Reinforcement
	f_{cu}, N/mm^2	Slab	Column	Top	Bottom	
S1	72	1300 square 138 deep	320	T16 100c/c	T12 100 c/c	Conventional links
S2	62	1300 square 150 deep	320	T16 100c/c	T12 100c/c	800 sq plate 800 mm cruciform
S3	74	1500 square 150 deep	150	T16 100c/c	T12 100c/c	Trimmed plate, 400 mm cruciform
S4	73	1500 square 150 deep	150	T16 100c/c	T12 100c/c	Trimmed plate, Stapled tie 400 mm cruciform

The test was conducted by applying the load to the column from the underside of the slab, Figure 5 and effecting punching to the slab. The slab was supported on roller supports at a square perimeter of side 1150 mm by means of rigid beams tied down to the strong floor of the laboratory. Typically the response of the slab under the punching load started with radial cracks from the perimeter of the column at the top tension surface. The cracks continued to progress radially whilst cracks parallel to the column perimeter also started to form. At any stage the radial cracks were more prominent than the perimeter ones. The upward deformation of the slab was monitored using an array of transducers. The progress of cracks both in terms of width and length continued until the spalling of concrete was observed around the perimeter on the soffit of the slab. The spalling and further deformation of the slab continued until full punching with the formation of large crack around the plate limited the load carrying capacity of the slab. The failure was completed. Views of failure are shown in Figure 6. The results of three of the slabs plus control are presented in Table 3.

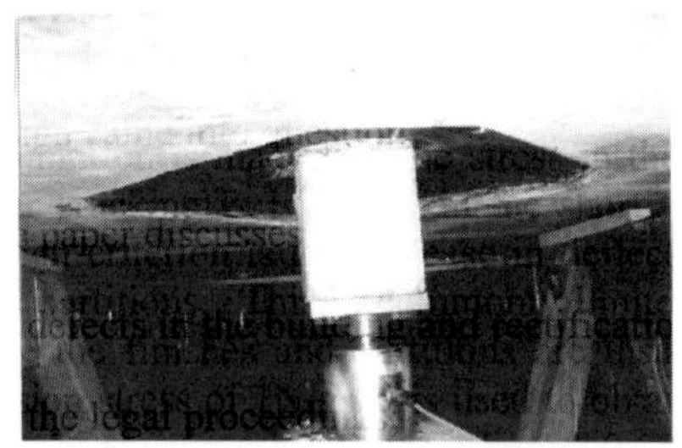

(a) S4: Failure around plate perimeter

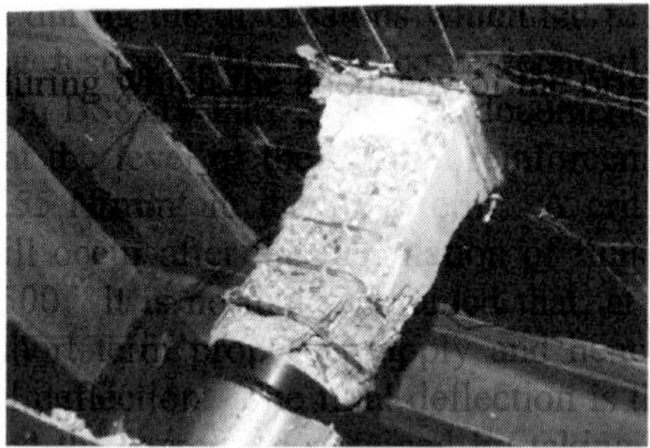

(b) S2: Failure of column by bursting

Figure 6 Views of slab-column junctions after failure

Table 3 Test results

SLAB	ULTIMATE CAPACITY	SHEAR STRESS	FAILURE MODE
	kN	N/mm^2	
S1	395	Column: 4.54	Shear failure around column perimeter
S2	800	Column: 9.0	Punching column burst; slab still intact
S3	767	Column: 8.5	Separation of top reinf mat followed by failure of cruciform/plate weld
S4	879	Column: 9.8 Plate:2.7	Shear failure around plate perimeter

CODE PROVISIONS AND COMPARISON

The use of solid bars or structural sections for the resistance of punching force is not covered in BS 8110. But, the ACI Building Code Requirements for Reinforced Concrete (ACI 318M-89) does have the provision of structural sections, I and Channel, as shearheads in slabs. A typical small interior shearhead is shown in Figure 7.

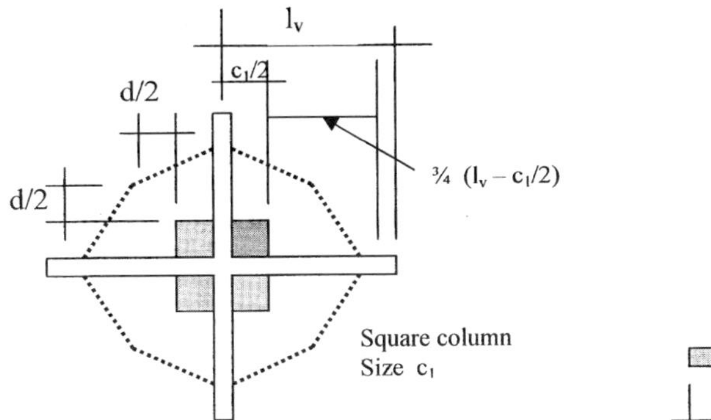

Figure 7 Small interior shearhead

Figure 8 Effective section for shearhead

The relationship between the plastic moment M_p required for each arm of the shearhead and the ultimate shear V_u is given by

$$\phi M_p = V_u \left[h_v + \alpha_v (l_v - c/2)\right] / 2\eta \ldots\ldots \qquad (1)$$

in which,

h_v	is the depth of the section
α_v	is the ratio between the stiffness of each shearhead arm and that of the surrounding composite cracked slab section of width (c + d)
l_v	is the minimum length of each shearhead
ϕ	is the strength reduction factor for flexure (= 0.9)
η	is the number of arms of shearhead

The critical slab section for shear is assumed at ¾ $(l_v - c/2)$ from the face of the column. In the case of a known shearhead detail, M_p, h_v, l_v and c will be known. α_v can be calculated and the equation (1) can be used to evaluate the punching shear capacity, V_u, of the joint.

For the test specimens reported here the Equation (1) was used to calculate the punching shear capacity of the NUUL system. In the calculations the cross bars forming the 'cruciform' and the bottom plate was assumed to act compositely and an effective section as shown in

Figure 8 was assumed for each arm of the shearhead. The parameters and the results are compared in Table 4.

Table 4 Punching shear resistance according to ACI and comparison with test results

SLAB	α_v	M_p	V_u		RATIO	RATIO
			ACI	Test	ACT/Test	Test/Control
		kNm	kN	kN		
S1*	-	-	303	395	0.77	1.00
S2	0.26	16.57	644	800	0.81	2.02
S3	0.24	16.57	842	767	1.10	1.94
S4	0.24	16.57	842	879	0.96	2.22

* based on $Vu = (\sqrt{f_c'})/3\ b_o\ d$, $f_c' = 57.6\ N/mm^2$, $b_o = 1056$ mm, d = 114 mm

Comparison of Results

The preliminary calculations show that the ACI approach using the assumed inverted T-section gives acceptable assessment of the punching resistance capacity for the proposed detail. Except in the case of S3, in which the over-estimation is 10%, other two cases, S2 and S4, show reasonable and safe predictions. The ratio, test/control, highlights the general enhancement of punching resistance for the proposed NUUL system, increasing the resistance by as much as 2 times that of the conventional detailing using shear stirrup reinforcement.

CONCLUSIONS

1. A new system of shearhead comprising of a steel plate, solid steel cross bars and staples for slab-column junctions has been proposed.

2. The preliminary test shows that such a system is robust and has very high resistance to punching force.

3. The ACI approach for the design of shearheads gives a good assessment of the punching resistance capacity of the proposed system.

REFERENCES

1. RATIONALISATION OF FLAT SLAB REINFORCEMENT, Best Practice Guide, BCA, 2000.

2. AMERICAN CONCRETE INSTITUTE, Building Code Requirements for Reinforced Concrete (ACI 318M-89) and Commentary – ACI 318RM-89.

3. WINTER, G., NILSON, A.H., Design of concrete structures, McGraw-Hill International Book Company, 1981.

RESEARCH UPON LONG SPAN CONCRETE SLABS, POST-TENSIONED BY EXTERNAL CABLE NETWORKS

C Mircea

Z Kiss

Technical University of Cluj-Napoca

Romania

ABSTRACT. The paper deals with testing and theoretical research made upon orthotropic concrete slabs, prestressed by external cable networks. The tendons are disposed on a parabolic surface, placed at the inside of the building. The structural solution can be highly efficient for slabs with spans up to 60.0 m. Tests were made upon a model made at scale 1:10, by simulating loading and unloading cycles between the long-term load, the total service load and the cracking load respectively, followed by loading until structural failure. Comparison of test data with theoretical results obtained on a Finite Element Model, based on an innovative modelling technique using virtual bitmaps in non-linear analysis of reinforced concrete structures, in terms of stresses and displacements is presented. Finally, general guidelines of simplified structural analyses, according to the Limit States Design, are presented according to the conclusions of the testing and theoretical investigations.

Keywords: Reinforced concrete, Slab systems, Large spans, Post-tensioning, Finite element.

C Mircea is currently an associate professor in the Faculty of Civil Engineering at the Technical University of Cluj-Napoca. His research interests comprise innovative reinforced concrete and prestressed concrete structures, non-linear modelling of concrete structures and rehabilitation of structures.

Z Kiss, actually is an associate professor in the Faculty of Civil Engineering at the Technical University of Cluj-Napoca and General Manager of Plan 31 Ltd, a consulting company. His research activity is orientated on reinforced concrete and prestressed concrete structures.

INTRODUCTION

Initial stress states are appropriate for slender spatial structures covering large spans. Inducing them by outside mechanical systems in respect with the concrete structure, the main advantages are obvious: increased stiffness of the whole structural ensemble, lower steel consumption and improved pouring conditions, and not the least the simple replacing of the post-tensioning systems. Hence, the beginning of the last decade were marked by intensive research on external prestressing, essentially aiming to expand its use at bridge structures. When referring to outside mechanical systems, two solutions can be adopted face to the main concrete body. The first is by cables stayed on pillars, connected to the structure in certain points, forming supports controlled by post-tensioning systems. Tendons with plane layouts or cable networks scattered on advantageous paths, anchored in rigid boundary members, with post-tensioning systems ensuring the interaction with the main structure. Testing and theoretical investigations were made in Romania on linear, plane and spatial structures, in order to develop useful tools for practical design. Next, the primary stages of the research, made on slab systems, are emphasised.

TESTING OF A CONCRETE ORTHOTROPIC SLAB MODEL POST-TENSIONED BY A PARABOLIC CABLE NETWOFK

As Figure 1 exhibits, the model, built at 1:10 scale, is a square ribbed slab of 3.00 m spans, made of 36 precast open box panels and poured in situ concrete. Stiff girders form the boundaries, the entire ensemble being pinned supported at the corners. The post-tensioned tendons, anchored in the boundary girders have the tracks laid on a parabolic surface, placed below the concrete slab (see Figure 2). The equation of the cables surface related to the centre of the mid surface of the slab is:

$$z = 300 - \frac{300}{1{,}500^2}(x^2 + y^2) + \frac{300}{1{,}500^4}x^2 \cdot y^2 \text{ (mm)} \tag{1}$$

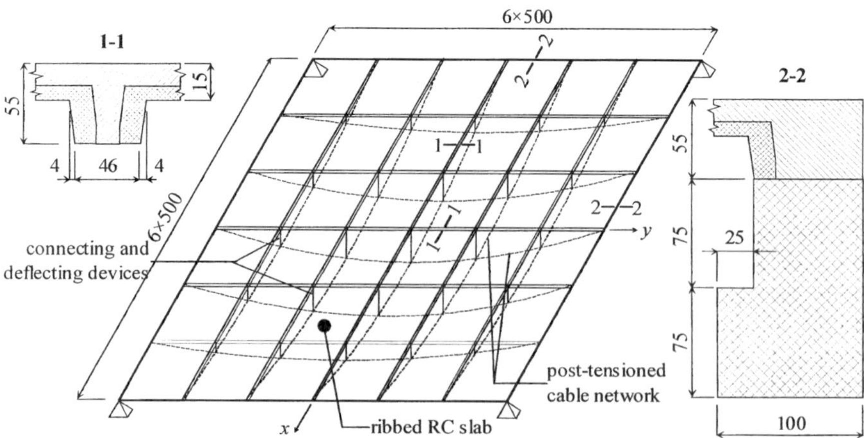

Figure 1 Post-tensioned slab model

Figure 2 Parabolic cable network

The model was built of micro-concrete with a specific weight of 22.0 kN/m^3. The witness specimens shown a cylinder strength of concrete of 24.3 N/mm^2 for the precast concrete with a corespondent strain of 1.99 ‰, and an ultimate strain of 2.98 ‰, while for the cast in situ concrete the cylinder strength was of 22.0 N/mm^2, with the correspondent strain of 2.03 ‰ and an ultimate strain of 3.04 ‰. The passive reinforcement with 3 mm in diameter was made of mild steel, while the post-tensioned steel wires were of 4 mm in diameter and made of high strength steel. The characteristics of the reinforcement are presented in Figure 3.

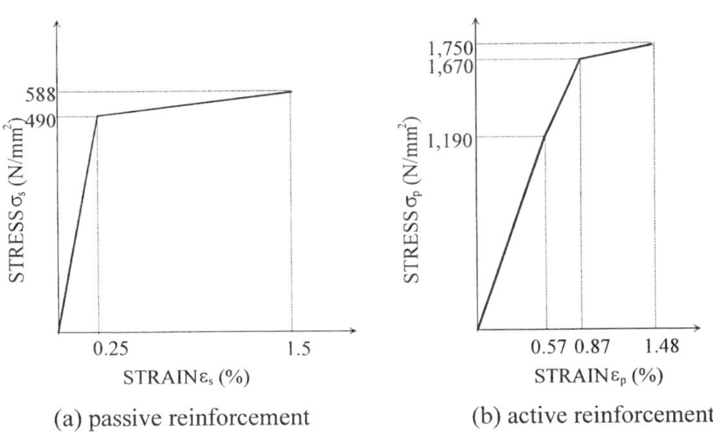

(a) passive reinforcement (b) active reinforcement

Figure 3 Stress-strain relations for steel

Steel devices (see Figure 2) made the connection between the slab and the post-tensioned network, and ensured the deviation of the cables. The testing loads carried on the model had three natures:

Figure 4 Loading device

i. The dead load of the full-scale slab, including the weight of the testing device;
ii. The live load with variable intensity, increased until the collapse;
iii. Forces produced by the post-tensioned cables.

Two hydraulic jacks (see Figure 4) induced gravitational loads, each acting on a lever system of bars to uniform the applied loads, concentrated in the nodes of the grid. Figure 5 shows the testing stages. Initially, the cables were tensioned to σ_0=250 N/mm^2 until the vertical load reached the dead load value, corresponding to the full-scale slab. Then, the cables post-tensioning stress was increased to 750 N/mm^2 and the model was loaded and unloaded in several steps, as shown in Figure 5. Stresses were induced by an adjustable anchoring system with recoil of 2 mm, shown in Figure 6, the tensioning order being from the middle to the borders.

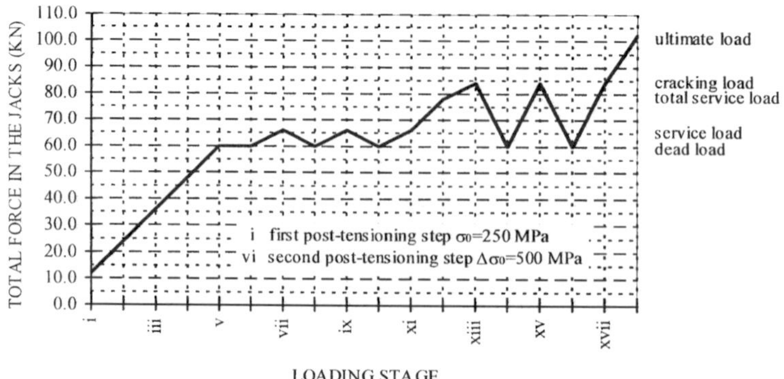

Figure 5 Loading program

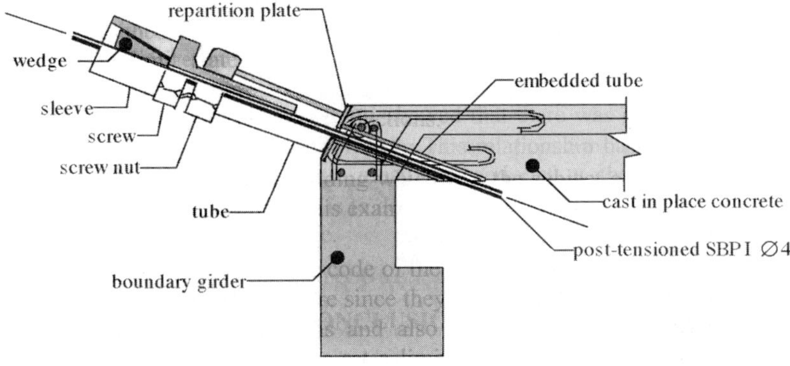

Figure 6 Anchorage system

FE MODELING AND COMPARISON WITH TEST DATA

Several FE models were made and analyses were performed. The most accurate proved to be a complex physically non-linear FE model, that was conceived in order to certify once more a few numerical procedures and to perform studies on full-scale FE models. The general principles of it are given hereinafter.

The concrete slab, modelled as a grid, was divided in interconnected beam elements. The sample cross-sections were analysed like virtual bitmaps, an innovative procedure that allows the implementation of different constitutive models to the virtual pixels, assimilated to the material points of a cross-section, as shown in Figure 7.

$$\frac{\sigma}{f_{ck}} = 2\frac{\varepsilon}{\varepsilon_{c1}}\left(1 - \frac{\varepsilon}{2\varepsilon_{c1}}\right) \quad \text{for } 0 < \varepsilon \leq \varepsilon_{c1} \tag{2.a}$$

$$\frac{\sigma}{f_{ck}} = 1 - 0.15\frac{\varepsilon - \varepsilon_{c1}}{\varepsilon_{cu} - \varepsilon_{c1}} \quad \text{for } \varepsilon_{c1} < \varepsilon \leq \varepsilon_{cu} \tag{2.b}$$

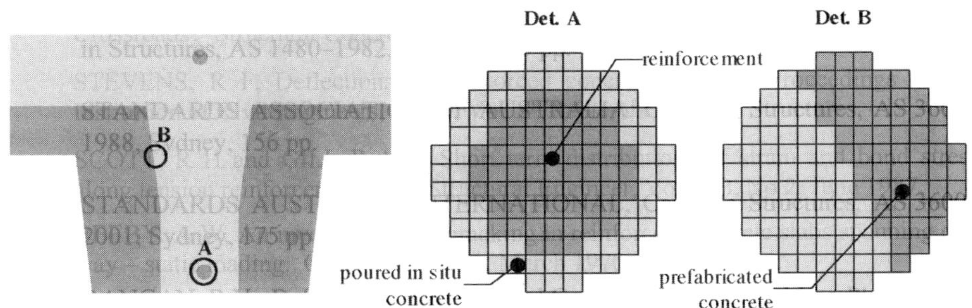

Figure 7 Virtual bitmap of a typical cross-section and types of material points

An algorithm based on the Bernoulli's hypothesis achieve the equilibrium of the section. A bilinear stress-strain relation was taken into account for the passive reinforcement (Figure 3(a), while for compressed concrete, the Hognestad [1] constitutive model (Figure 8) was considered with both ascending and descending (ie, softening) branches:

Tensioned concrete was modelled by a linear stress-strain relation, the characteristic strength being calculated by the formula specified in the Romanian standard [2]:

$$f_{tk} = 0.22 f_{ck}^{2/3} \left(0.3 + 0.7 \frac{22.0}{24.0} \right) \tag{3}$$

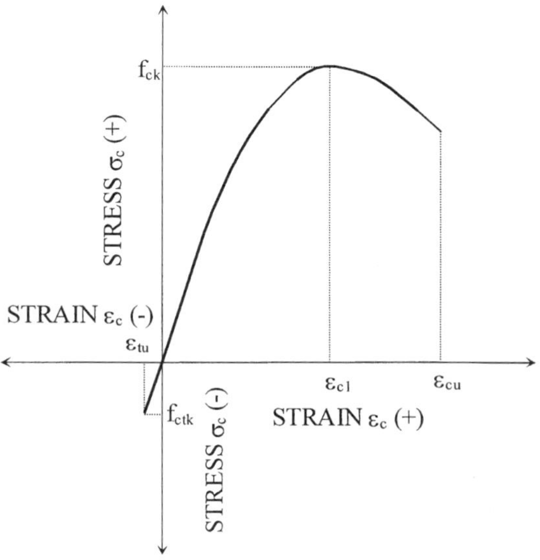

Figure 8 Uniaxial stress-strain diagram for concrete

Cables were modelled as truss elements, the equilibrium along them being computed with an iterative algorithm developed by Mircea [3], taking into account the friction and the slips in the deflecting devices. The procedure is based on the nodal compatibility equations with stress redistribution in respect with the variation of the potential energy of deformation, as depicted in picture below.

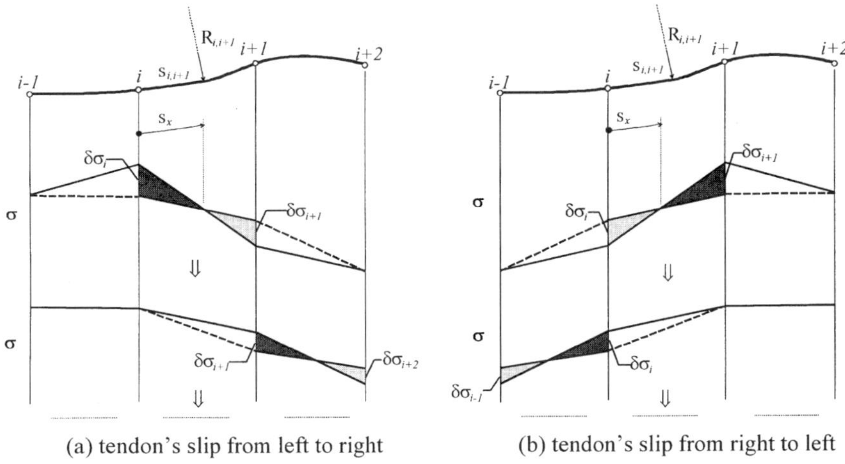

Figure 9 Unbalanced stress redistribution to adjacent segments

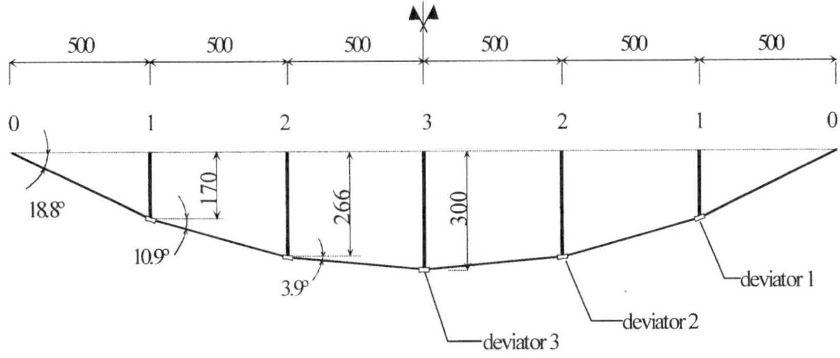

Figure 10 Layout of the median cables

The biographical FE analysis was organised in the traditional Newton-Raphson incremental scheme, using original algorithms developed by Mircea [4] for concrete modelling. Next, a part of the results accompanied by commentaries will be exposed. For the cable network, results will be summarised for the median cables, with the geometry and symbols given in figure above. Figure 11 presents the variation of the stress in the central segments of the median cables. During the first post-tensioning step, the theoretical values of the stress losses in the median cables were 4.6 % in deflector 1, 4.2 % in deflector 2 and 0 % in deflector 3. At the second tensioning stage, the values were 7.8 % in deflector 1, 6.9 % in deflector 2 and 0.0 % in deflector 3. Starting with the first tensioning operation, many slips, accompanied by stress gradients, were registered. The directions estimated within the FE analysis, and the friction factors, starting with an initial value of 0.35 and subsequent increases of 30% at every change in the slips direction, are shown in Table 1. Measured stresses confirm the behaviour of the tendon found with the numerical procedure.

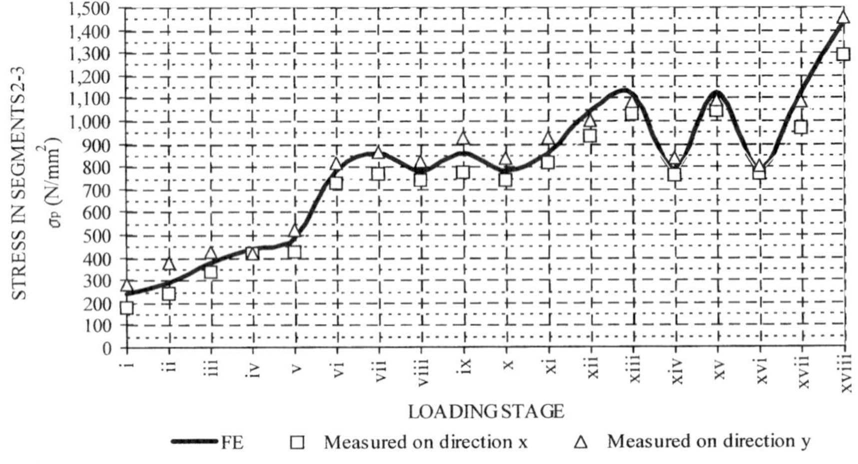

Figure 11 Stress variation in the central segments of the median cables

Table 1 Theoretical direction of slips (support ←, ◆ blocked, → central deviator)

LOADING AND TENSIONING STAGE	DEVIATOR 1 θ_1=7.9°			DEVIATOR 2 θ_2=7.0°			DEVIATOR 3 θ_3=7.8°		
	μ_{LEFT}	DIRECTION	μ_{RIGHT}	μ_{LEFT}	DIRECTION	μ_{RIGHT}	μ_{LEFT}	DIRECTION	μ_{RIGHT}
i during first tensioning	**0.35**	←	0.46	**0.35**	←	0.46	0.35	◆	0.35
i after tensioning	0.59	→	**0.46**	0.35	◆	0.46	0.35	◆	0.35
ii	0.59	→	**0.46**	0.59	→	**0.46**	0.35	◆	0.35
iii	0.59	→	**0.46**	0.59	→	**0.46**	0.35	◆	0.35
iv	0.59	→	**0.46**	0.59	→	**0.46**	0.35	◆	0.35
v	0.59	→	**0.46**	0.59	→	**0.46**	0.35	◆	0.35
vi during second tensioning	**0.59**	←	0.77	**0.59**	←	0.77	0.35	◆	0.35
vi after second tensioning	1.00	→	**0.77**	1.00	→	**0.77**	0.35	◆	0.35
vii	1.00	◆	0.77	1.00	◆	0.77	0.35	◆	0.35
viii	1.00	◆	0.77	1.00	◆	0.77	0.35	◆	0.35
ix	1.00	◆	0.77	1.00	◆	0.77	0.35	◆	0.35
x	1.00	◆	0.77	1.00	◆	0.77	0.35	◆	0.35
xi	1.00	◆	0.77	1.00	◆	0.77	0.35	◆	0.35
xii	1.00	→	**0.77**	1.00	→	**0.77**	0.35	◆	0.35
xiii	1.00	→	**0.77**	1.00	→	**0.77**	0.35	◆	0.35
xiv	**1.00**	←	1.30	**1.00**	←	1.30	0.35	◆	0.35
xv	1.69	→	**1.30**	1.00	◆	1.30	0.35	◆	0.35
xvi	1.69	◆	1.30	1.69	◆	1.30	0.35	◆	0.35
xvii	1.69	◆	1.30	1.69	◆	1.30	0.35	◆	0.35
xviii	1.69	→	**1.30**	1.69	→	**1.30**	0.35	◆	0.35

Obviously, the slips produced many stress gains and losses in the segments of the cables, in accordance with the deformation state of the concrete slab. All these events caused the reducing of the work performed by the cable network. n indicator of the performance of the cable network, is the variation of the potential energy in the segments of the network. Practically, the energy losses occurred in the deflecting devices were insignificant for the general equilibrium of the structure (see Table 2). However, these losses have a local importance that can influence the safety of the global structure. In the case of the tested model, energy dissipation in the deflecting devices caused local deformation of the wires and small incrustation in the deflectors. These changes were materialised in the increased friction factors, but practically had no effect on the bearing capacity of the wires, due to the small number of applied loading-unloading cycles.

In general, energy losses that occur during the slips of the tendon in the deflecting devices give rise to:

i. Temperature gradients in the contact area;
ii. Growth of the penetration of the wires of the strand into the duct;
iii. Increased local irregularities at the contact surface.

Table 2 Energy losses associated to the slips (Erg)

LOADING STAGES	DEVIATOR 1: θ_1=7.9°	DEVIATOR 2: θ_2=7.0°	DEVIATOR 3: θ_3=7.8°
i-v	4,930	3,070	0
vi-xi	13,210	9,820	0
xii-xviii	49,930	41,360	0

Temperature gradients modify the physical properties of the two materials in contact (naturally, in the case of steel-HDPE contact, that most affected is the duct). Softening of the duct increases the penetration of the strands in the duct, increasing the friction properties of the contact. Thus, the slips will be obstructed during the service of the structure, and this problem has to be considered in practical design when simplified analyses are performed in the conditions of the Ultimate Limit States. Nevertheless, the entire penetration of the duct will lead to an intimate contact between steel tendons and deviators, which can cause the failure of the strand wires, reducing the bearing capacity of the cable. This last situation must be avoided by selecting the appropriate thickness of the duct, problem that is already fixed in the common practice. Figures 12 and 13 present the midspan deflections of the slab and of the boundary girders.

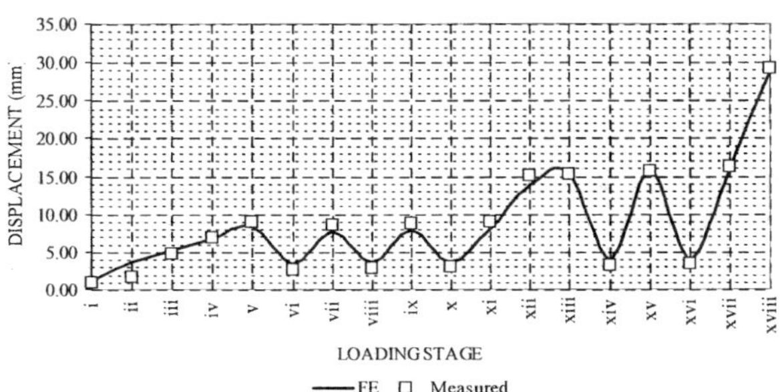

Figure 12 Midspan slab deflection

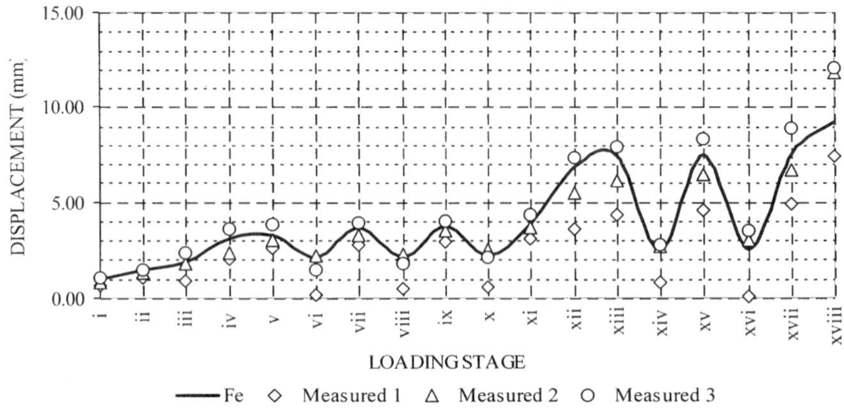

Figure 13 Midspan boundary girders deflection

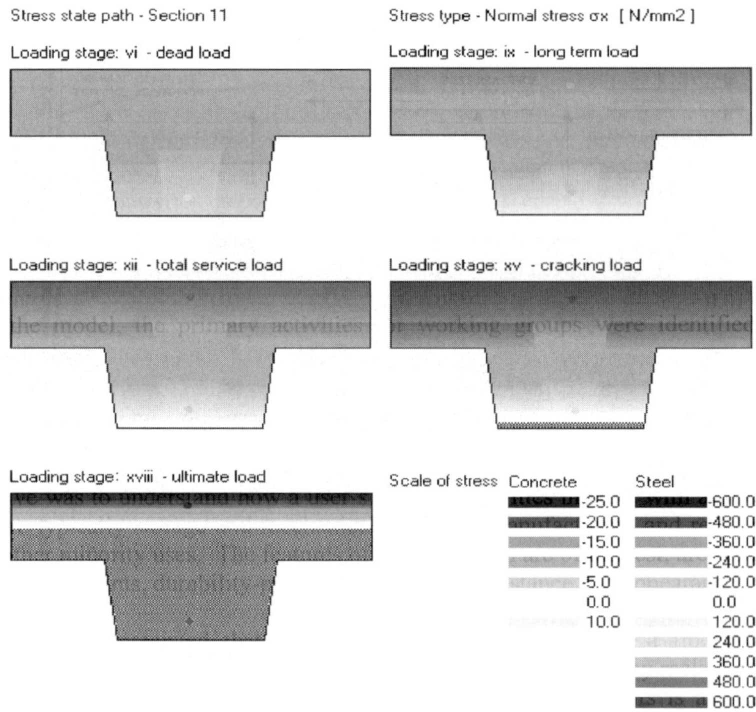

Figure 14 History of the stress state in cross-sections adjacent to the middle point

FE analysis predicted a decompression load of 13.45 kN/m^2 and a foreseen cracking load equal with 13.64 kN/m^2, while practically, macrocracking was registered under the equivalent load of 14.20 kN/m^2. The nodes of the grid being strongly confined by the triaxial compression state, the theoretical analysis shows that the macrocracks begin to develop near the central ribs on both sides, progressively, starting from the middle point. Figure 14 emphasises the stress states corresponding to the main loading stages in the middle sections, considered as sample sections in the FE analysis. Figure 15 reveals the same cracking pattern registered within the test. Due to the symmetry of the loads, the initiation and developing of the normal cracks are not practically influenced by the shear stresses (due to shear forces and torsion moments), plastic hinges being formed in these sections due to the normal compressive stresses. The redistribution of the stress caused by them does not lead to the initiation of other significant macrocracks. Structural collapse was the consequence of the crushing of concrete in the centre of the slab, while the passive reinforcement was still yielding. The anticipated failure load was 18.39 kN/m^2, while the collapse of the structure occurred at an equivalent load of 17.20 kN/m^2.

Other simplified analyses were made in accordance with Limit States Design. A linear FE analysis and a simplified analysis in the plastic range, in respect with the cinematic method of theory of plasticity, proved to be significant for the common practice.

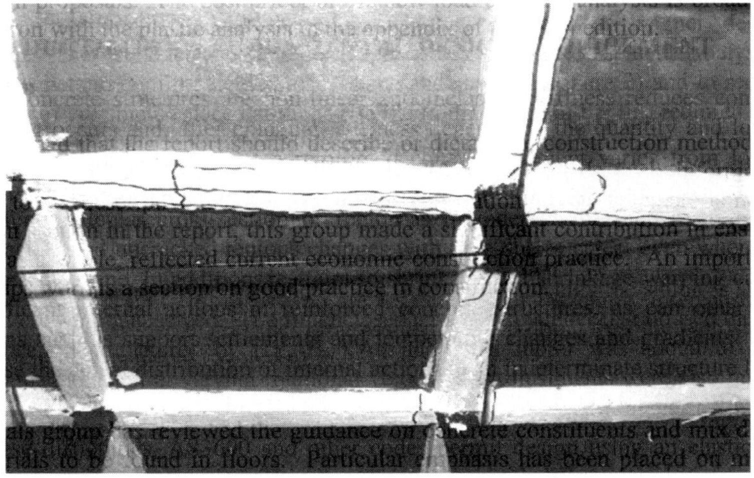

Figure 15 Cracks opening near failure

CONSIDERATIONS UPON THE LIMIT STATES DESIGN

Evidently, Limit States Design should take into account all significant stages. Tensioning steps should be programmed in order to avoid concrete cracking by compensating the desired/necessary correspondent load. Structural analyses performed in the conditions of the Serviceability Limit States, could be performed following the assumptions:

i. Loads are concentrated in the nodes of the grid;
ii. Torsion stresses are insignificant for rectangular grids;
iii. The tension losses due to friction and slip in the deflecting devices and anchorage zones, should be related only to the tensioning stages;
iv. Stress increases due to structural displacements may be calculated considering only the vertical components of the grids displacements;
v. Prestressing forces can be considered like an external system of forces acting on the nodes of the concrete grid, applying the load balancing method developed by Lyn;
vi. The influence of time dependent behaviour should be introduced by considering the rheological properties of the prestressing steel and adjusting the values of concrete Young module with the creep coefficients, according with the simplified procedures accepted by the codes of practice;
vii. The stiffness of the grid segments should be calculated on the ideal cross-section, using equivalence factors for passive steel related to the concrete Young module.

In the opinion of the authors, a simplified plastic analysis is much more appropriate for structural analyses performed in the conditions of the Ultimate Limit States. It offers a better understanding of the failure mode and the proper control of the safety. In accordance with the cinematic method of plasticity theory, the directions of the plastic analyses are given below:

i. The failure mechanism is set up by the development of the plastic joints on the concrete grid segments, near the central nodes, as shown in Figure 16;
ii. The grid segments between the plastic joints are inflexible;
iii. The energy losses due to slip and friction are negligible;
iv. The work done by the external forces will be found in the internal work, accomplished by the rotation of the plastified cross-sections and the post-tensioned tendon network;
v. Special attention must be accorded to the rotation capacity of the plastic hinges, which must be correlated with the limit stress in the central cables of the post-tensioned network.

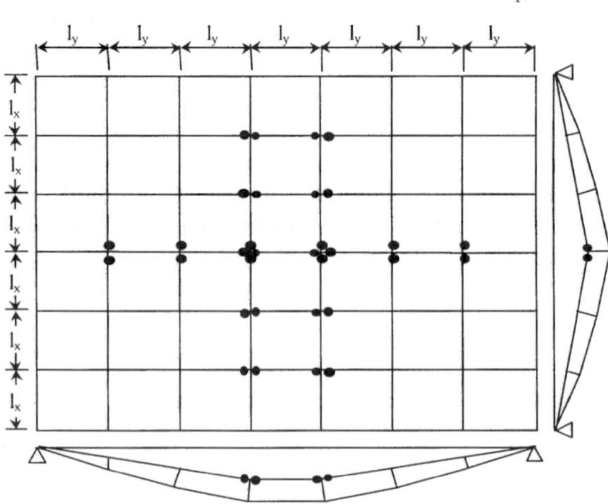

Figure 16 Failure mechanism

FINAL REMARKS

The composite slab system that was briefly presented here improves the strength potential of concrete and steel, in a harmonious ensemble. The solution can satisfy many necessities of the modern design. High stiffness, easy maintenance and facile rehabilitation - by interchangeable post-tensioning systems, complete original aesthetic, adequate and flexible functionality.

As a result of the research, non-linear modelling procedures, simplified methods and information on different supports were provided to structural engineers.

REFERENCES

1. HOGNESTADT E., A Study of Combined Bending and Axial Load in Reinforced Concrete Members, Engineering Experiment Station, University of Illinois, 1951, Bulletin no 399, vol 49, November, 22 pages.

2. STAS 10107/0-90, Calculul si alcatuirea elementelor structurale din beton, beton armat si beton precomprimat, 1990, MLPAT.

3. MIRCEA C., BUCUR I., Finite Element Modelling of Unbounded Post-Tensioned Tendons, Proceedings of FIB Symposium 1999, vol 2, p 617-622.

4. MIRCEA C., Thin Reinforced Shells – Finite Element Approach, UT Press 2000, 245 pages.

EFFECT OF TENDON LAYOUT ON STRESSES AND DEFLECTIONS IN POST-TENSIONED SLABS

A J Schokker

S C Lee

A Scanlon

Pennsylvania State University

United States of America

ABSTRACT. This paper presents results from an analytical study of various tendon layouts in two-way post-tensioned flat plate slabs. The slab is modeled using plate-bending elements in the software package SAP 2000. Four tendon layouts are considering varying from fully banded distributions to fully uniform distributions. Prestressing loads are applied as equivalent external loads based on cable profiles. Results from gravity loads are superimposed on results from the equivalent tendon loads to determine net results at service load levels. Deflection results and distribution of moments are compared for each tendon layout.

Keywords: Two-way slab, Post-tensioned, Tendon layout, Flat plate.

A J Schokker is the Henderson Chair and Assistant Professor of Civil Engineering at the Pennsylvania State University. Her research focuses on design and materials related improvements for post-tensioned structures and durability of prestressed concrete.

S C Lee is a Research Assistant and Master of Science candidate at the Pennsylvania State University.

A Scanlon is Professor of Civil Engineering at the Pennsylvania State University. His research interests focus on safety and serviceability of concrete structures.

INTRODUCTION

Post-tensioned flat plates are an economical solution for two-way slab systems because the slab thickness can be reduced relative to the thickness required for reinforced concrete slabs. Design is usually accomplished using the well-known load-balancing concept as described in the Post-Tensioning Manual [1] published by the Post-Tensioning Institute. Various distributions of tendons have been employed successfully in practice and tested in the laboratory [2,3]; however there does not appear to have been any systemic study of the effects of various tendon distributions on the distribution of moments or deflections in slabs.

The simplest distribution consists of a uniform distribution in one direction carrying the entire balanced load in that direction, and a banded arrangement along column lines in the other direction. This arrangement mimics the one-way slab supported on beams system to transfer the desired balanced load to the supporting columns. Usual methods of slab analysis are then used to determine the moments due to the unbalanced load. Stresses due to the unbalanced load are added to the uniform compressive stress corresponding to the balanced load condition.

This paper presents results of an analytical study to determine the effects of various tendon layouts on distributions of moments as well as deflections of two-way systems.

METHOD OF ANALYSIS

The general-purpose computer program SAP 2000 was used to perform a plate-bending analysis of a typical interior panel of a two-way slab system consisting of a slab of uniform thickness equal to 165 mm under both dead load and the equivalent vertical load caused by the force in the tendons. These load cases were run separately and the results superimposed to obtain net moments and deflections. The dimensions and boundary conditions of the slab panel are shown in Figure 1. The tendon force was calculated to produce a uniform compression in the slab equal to approximately 1200 kPa in each direction. For tendons spanning in the short direction, the corresponding equivalent load is 0.86 times self-weight, while for tendons spanning in the long direction the equivalent load is 1.25 times the self-weight. The average equivalent load is 1.06 times self-weight.

TENDON LAYOUTS

Four tendon layouts were considered as follows:

1. 100% of tendons banded along the column line in each direction
2. 75% of tendons placed within the column strip with the remainder in the middle strip, in two directions
3. 100% of tendons banded along the column line in one direction, and uniform distribution of tendons across panel in the other direction.
4. Uniform distribution of tendons in both directions

The tendon layouts are shown in Figure 2.

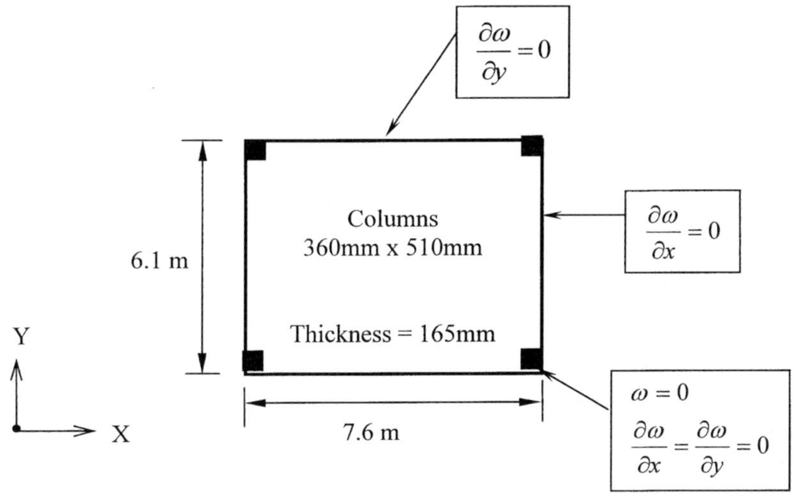

Figure 1 Slab panel dimensions and boundary conditions

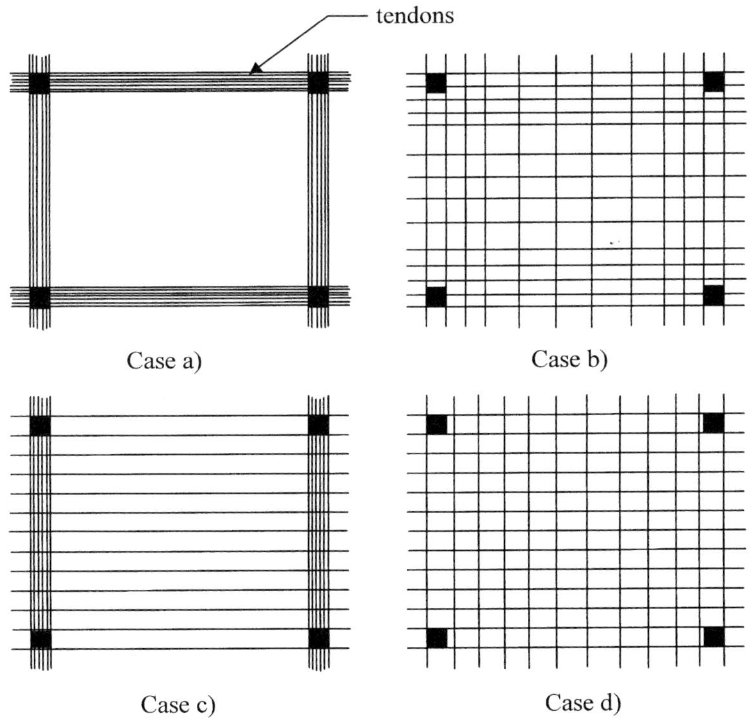

Figure 2 Tendon Layouts

Case A) is not generally considered to be a practical layout because most of the slab is essentially unreinforced, however this case was included to demonstrate the effect of an extreme tendon layout producing essentially line loads between columns in each direction.

Case B) represents a distribution corresponding to the typical distributions of nonprestressed reinforcement in conventional reinforced concrete slabs.

Case C) is a commonly used distribution as described earlier.

Case D) is the opposite extreme from Case A with all tendons uniformly distributed in each direction.

RESULTS

Figure 3 compares the deflections obtained for each case. Numerical values are given in Table 1. Deflections are plotted at mid-span along the column line in the long direction (Point A), at mid-span along the column line in the short direction (Point B), and at mid-panel (Point C). Results are presented for prestressing only, self-weight only, and the combined effect. The most effective layout in terms of producing upward deflection in the slab is Case A while the least effective is Case D, however the mid-panel panel deflections for all four cases are quite similar, ranging from 4 to 4.8 mm. A larger variation can be seen in the deflections at mid-span on the column lines. When combined with the self-weight deflection Case A results in a slight upward net deflection while Case D results in a slight downward deflection.

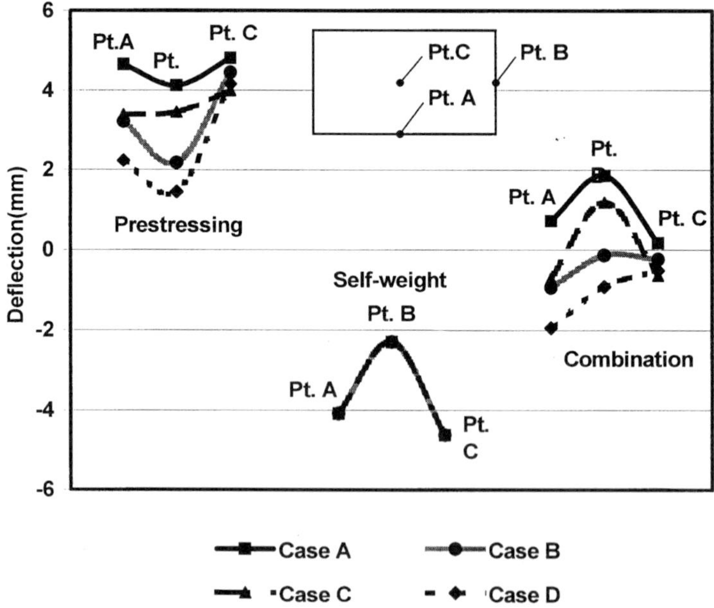

Figure 3 Comparison of the deflections obtained for each case

Table 1 Numerical values of deflections obtained for each case

	PRESTRESSING DEFLECTIONS (mm)		
	Point A	Point B	Point C
Case A)	4.65	4.12	4.81
Case B)	3.22	2.19	4.45
Case C)	3.39	3.46	4.00
Case D)	2.24	1.45	4.16
	SELF-WEIGHT DEFLECTIONS (mm)		
	Point A	Point B	Point C
Case A)	-4.08	-2.30	-4.61
Case B)	-4.08	-2.30	-4.61
Case C)	-4.08	-2.30	-4.61
Case D)	-4.08	-2.30	-4.61
	COMBINATION DEFLECTIONS (mm)		
	Point A	Point B	Point C
Case A)	0.72	1.86	0.17
Case B)	-0.93	-0.13	-0.24
Case C)	-0.70	1.18	-0.63
Case D)	-1.94	-0.92	-0.51

Figures 4-11 demonstrate the effect of tendon layout on bending moment distribution. Moments are plotted across the panel at the column face for each of the tendon layouts in both the long and short span directions. For Case A, prestressing induces high positive moment intensities in the vicinity of the columns. The moment intensity decreases rapidly with distance from the column and becomes negative in the region midway between columns. For this case, the combination of prestressing and self-weight produces net positive moments in the vicinity of the column in both the long and short directions. For Cases C and D, the combination of prestressing and self-weight produces net negative moments near the columns in both the long and short directions. Case B is very effective in balancing the prestressing moments with the self-weight moments in the long direction as shown in Figure 5. In the short direction, Case B behaves similarly to Case A with positive moments near the columns under the combination of prestressing and self-weight.

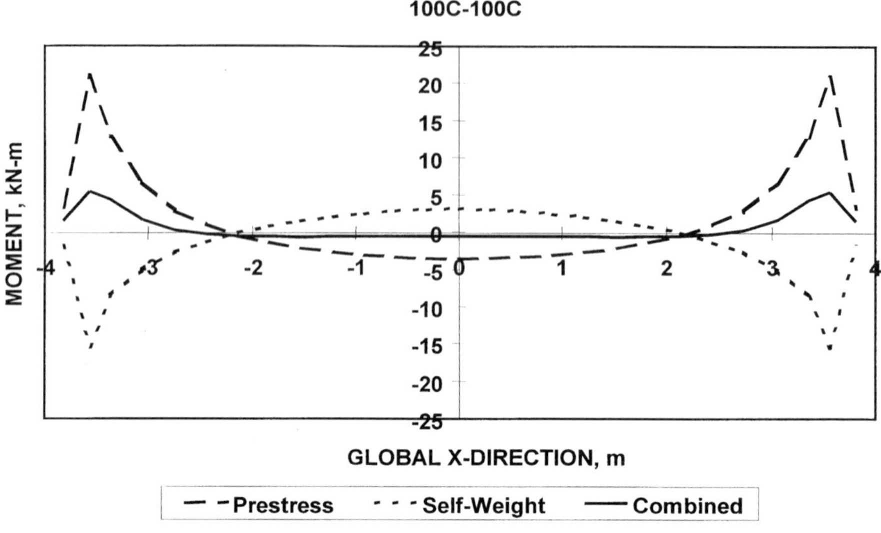

Figure 4 Distribution of moments, long direction, case A (banded two directions)

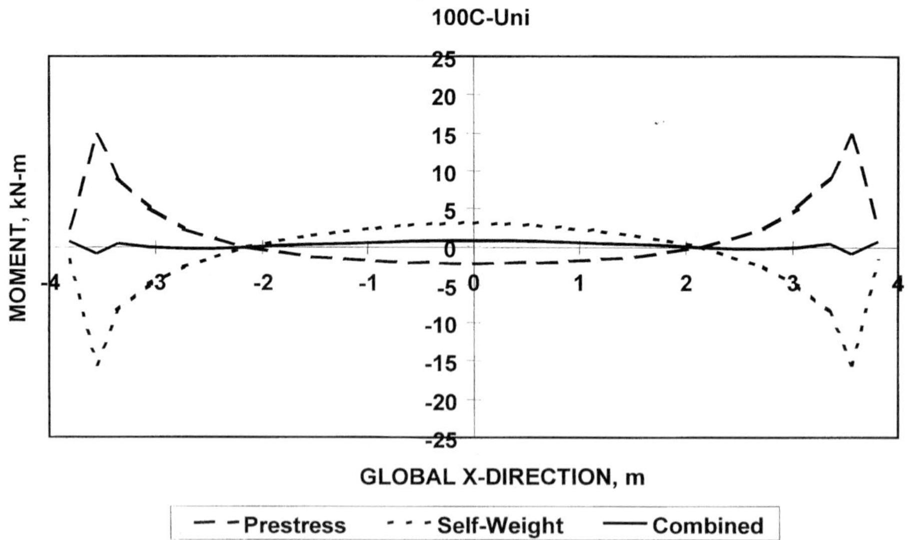

Figure 5 Distribution of moments, long direction, case B (banded one way, uniform one way)

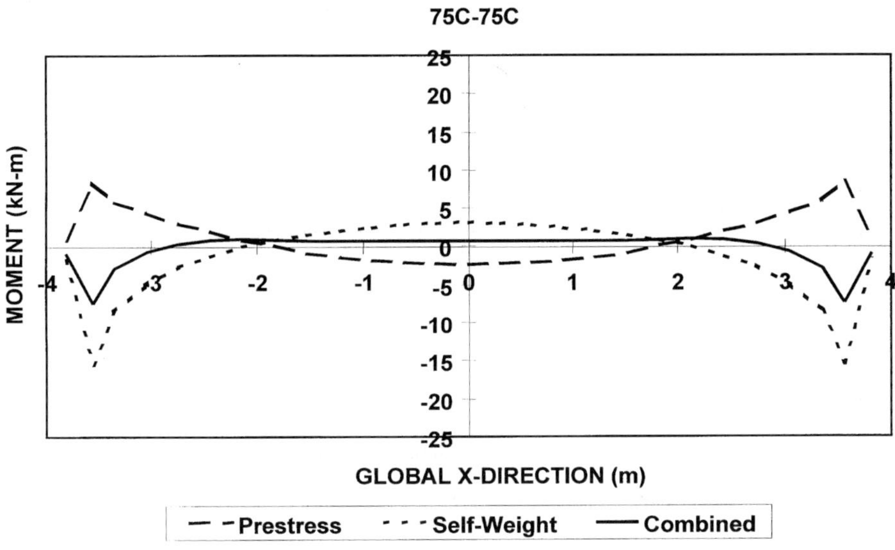

Figure 6 Distribution of moments, long direction, case C (75% in columns two directions)

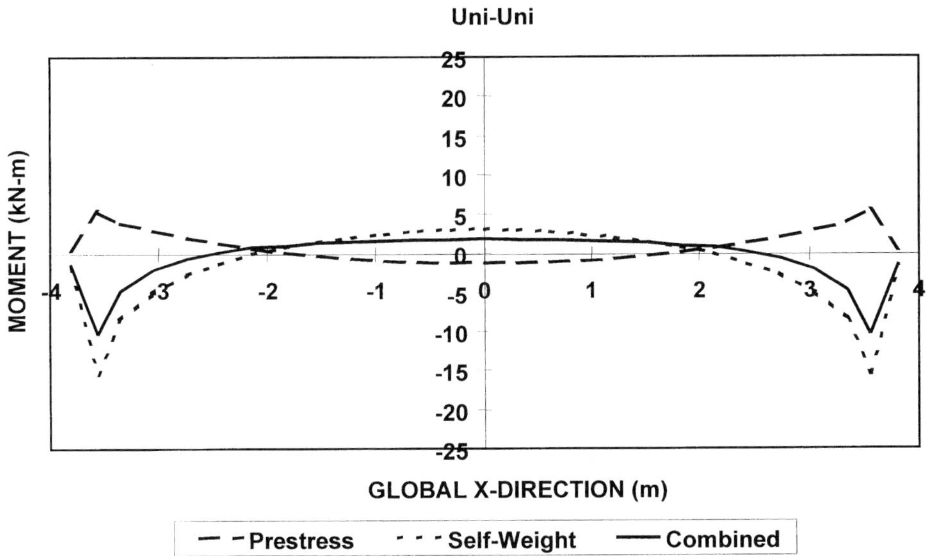

Figure 7 Distribution of moments, long direction, case D (uniform two directions)

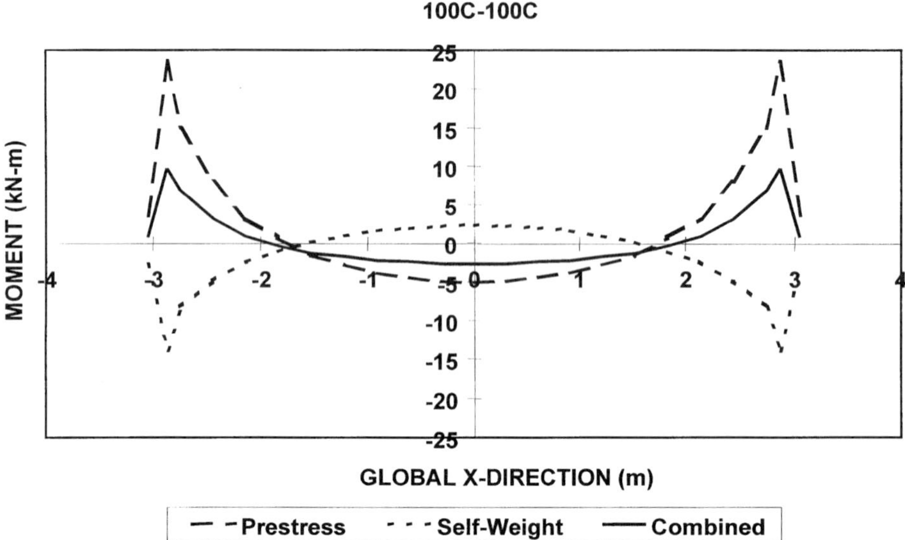

Figure 8 Distribution of moments, short direction, case A (banded two directions)

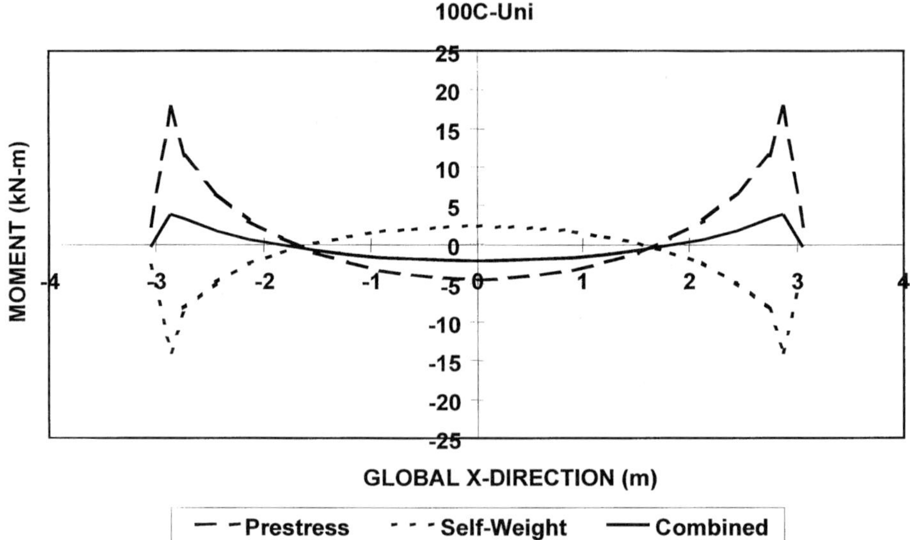

Figure 9 Distribution of moments, short direction, case B (banded one way, uniform one way)

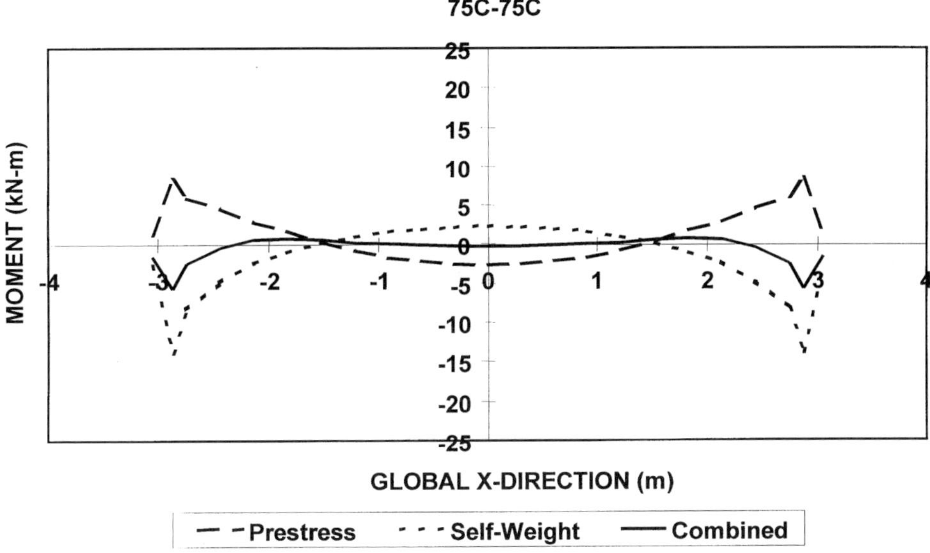

Figure 10 Distribution of moments, short direction, case C (75% in columns two directions)

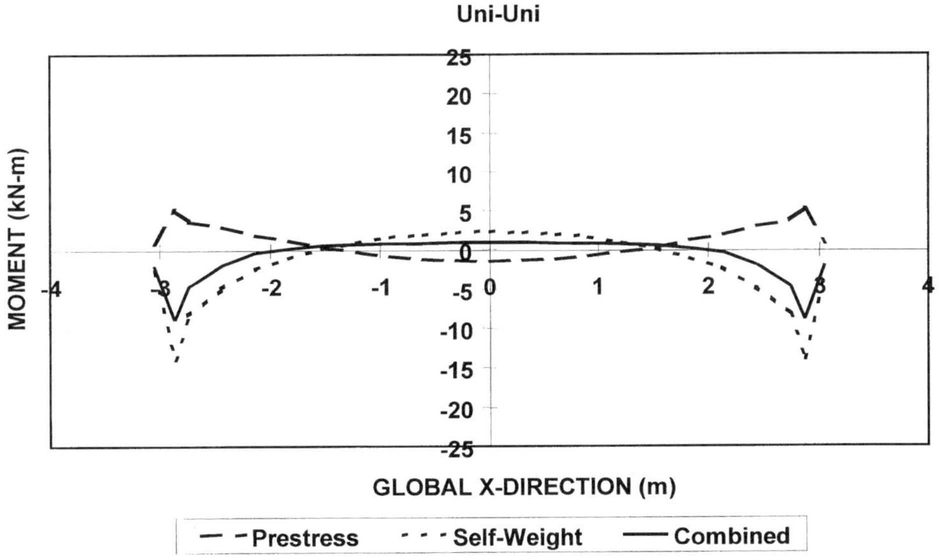

Figure 11 Distribution of moments, short direction, case D (uniform two directions)

CONCLUSIONS

This exploratory study has demonstrated that mid-panel deflections of post-tensioned slabs are relatively insensitive to tendon layout assuming the slab remains uncracked. A preliminary examination of moment intensity distributions indicates significant variations, in some cases resulting in different signs for moments at locations on critical sections. Further study is needed to determine whether these variations could have an impact on potential cracking in the slab as well as expected behavior at ultimate.

REFERENCES

1. POST-TENSIONING INSTITUTE, Guide Specification for Grouting of Post-Tensioned Structures, First Edition, PTI Committee on Grouting Specifications, Post-Tensioning Institute, Phoenix, 2001.

2. HEMAKOM, R, Strength and Behavior of Post-Tensioned Flat Plates with Unbonded Tendons, PhD Dissertation, the University of Texas at Austin, 1975.

3. BURNS, N H, HEMAKOM, R, Tests of Post-Tensioned Flat Plate with Banded Tendon, ASCE Structural Journal, Vol 3, 1985, pp 1899-1915.

CONCRETE FLOORS : THE QUEST FOR MINIMAL SHRINKAGE

W F Price

British Cement Association

United Kingdom

ABSTRACT. Concrete ground floors are a key element of many industrial buildings. Drying shrinkage of concrete can lead to problems with cracking and curling. This paper examines the factors influencing both intrinsic drying shrinkage and shrinkage induced cracking and highlights how concrete mix proportions can be modified to minimise shrinkage. The commonly assumed maxim that high shrinkage results, primarily from an increased cement content is shown to be unjustified.

Keywords: Floors, Shrinkage, Mix design, Fibres, Cracking.

Dr Bill Price is Senior Research Manager – Concrete Performance, for the British Cement Association, Crowthorne, UK. He was previously with Sandberg and Taywood Engineering. His interests include the durability of concrete structures and the application of high strength and high performance concretes. He is currently Vice-President of the Institute of Concrete Technology.

INTRODUCTION

Concrete ground floors are an important element of many industrial buildings and usually perform in a satisfactory manner. The performance of the floor, however, is often a key factor in the efficient use of the building and unwanted problems with the floor are less than welcome.

One property of hardened concrete that can lead to problems with an industrial floor, if not properly controlled, is shrinkage.

The purpose of this paper is to identify the different mechanisms of shrinkage that can occur in a concrete floor, examine the effects on floor performance and consider ways in which shrinkage can be minimised.

Shrinkage *per se* is merely a contraction in volume or in a linear dimension, but the underlying cause of this contraction in concrete may result from a range of factors including:

- Movement of moisture
- Thermal changes
- Carbonation

It is useful to discriminate between uniform and non-uniform shrinkage and also between restrained and unrestrained shrinkage. The most significant mechanisms of shrinkage in hardened concrete floors are (most significant first):

- Drying shrinkage
- Thermal contraction
- Carbonation shrinkage
- Autogenous shrinkage

This paper will, however, consider drying shrinkage primarily and examine means in which both shrinkage itself and the risk of shrinkage-induced cracking can be minimised. It is also restricted in scope to common cements and combinations and does not consider expansive cements, which can also be used to reduce or eliminate concrete shrinkage. It should be noted that a new British Standard covering expansive cements (BS 8402) is now at the public comment stage.

DRYING SHRINKAGE

Basic Mechanism

At its simplest, hardened concrete can be considered as a two-phase material consisting of aggregate particles dispersed in a matrix of cement paste.

Of these two phases, it is generally the cement paste that undergoes drying shrinkage, whilst, with the exception of certain rock types, the aggregate particles act to restrain that shrinkage.

When water is lost from a non-rigid porous material (such as cement paste) by exposure to a dry environment, shrinkage will take place. The magnitude of the shrinkage is not, however, equal to the volume of water lost.

Free water lost from the larger pores does not generate significant shrinkage, but as the capillary pores begin to empty, menisci form in the surface of the remaining water. Surface tension forces in the water exert stresses on the walls of the pore, leading to overall shrinkage of the cement paste. Further drying can then lead to the removal of adsorbed water from the cement hydrate gel resulting in additional shrinkage.

Influencing Factors

The magnitude of drying shrinkage is influenced not only by the severity of the drying conditions (temperature, humidity and air movement) and the length of time the concrete is exposed to drying, but also by the composition of the concrete itself. As might be expected from the two-phase model, the volume fraction of (non shrinking) aggregate exerts the major influence over the magnitude of the shrinkage of the concrete as a whole. It should be noted that certain sources of aggregates (typically some dolerite, greywackes and mudstones[1]) can exhibit considerable drying shrinkage in their own right. If minimal shrinkage is an important goal, then these materials should not be selected for use.

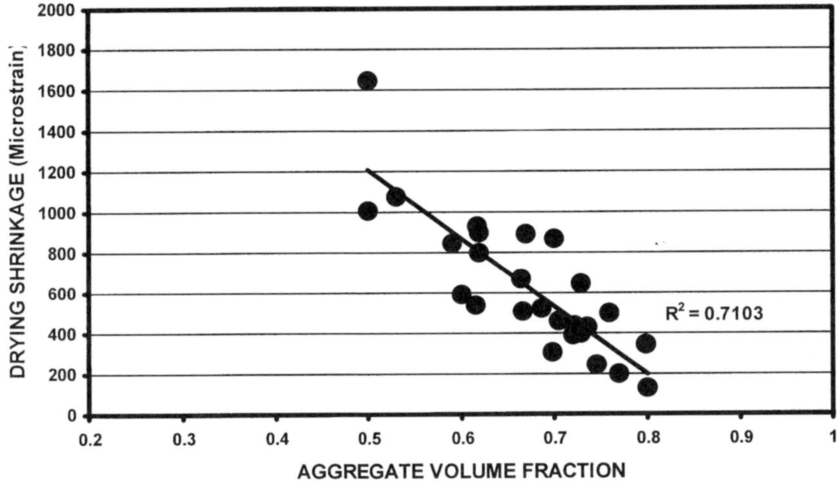

Figure 1 Relationship between concrete drying shrinkage and aggregate volume fractions (based on data from reference 2)

Within the paste fraction, the original water content governs the amount of evaporable water that can be lost and thus contributes to shrinkage. Consequently, at a fixed paste volume, a paste with a low water/cement ratio is beneficial in reducing shrinkage. It is often thought that the cement content of concrete governs its potential for shrinkage directly. Whilst this may be true for concrete at a fixed water/cement ratio, it is not a fundamental relationship for all concretes.

An examination of published shrinkage data[2] shows that there is a strong correlation between measured drying shrinkage and the aggregate volume fraction (Figure 1). This data includes concretes with water/cement ratios of 0.30 to 0.70 and cement contents in the range 200 to 800 kg/m^3 containing a range of different aggregates.

When the same data are examined relative to cement content (Figure 2) there is little, if any, overall correlation. Thus it can be demonstrated that the cement content of a concrete mix alone, is not a reliable indicator of potential drying shrinkage.

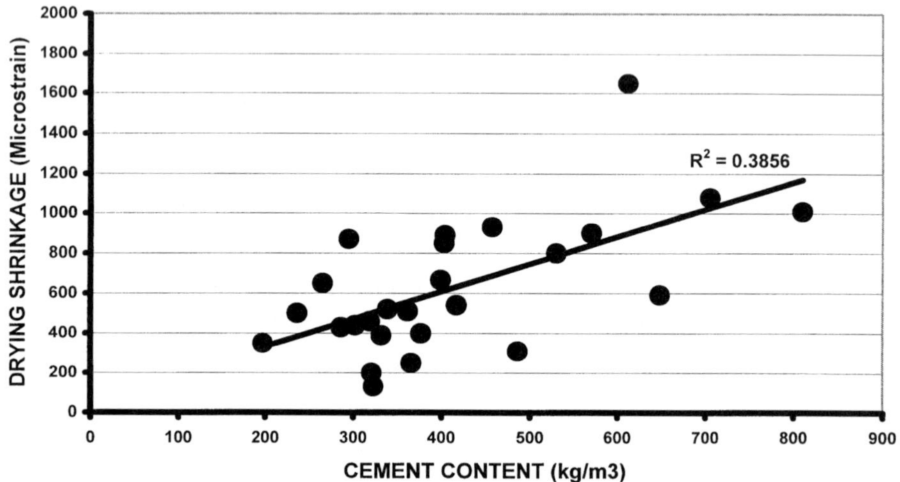

Figure 2 Relationship between concrete drying shrinkage and cement content (based on data from reference 2)

SHRINKAGE EFFECTS

How does shrinkage affect concrete floors?

If shrinkage is both uniform and unrestrained, the effect of shrinkage is merely to reduce the size of a floor panel leading to opening up of the perimeter joints. However, in the real world shrinkage is neither fully uniform nor free from restraint. In such cases a number of different effects can result. (See Figure 3).

Mid Panel Cracking

Even if drying of a concrete floor panel is uniform, the subsequent shrinkage may be restrained by friction between the base of the floor slab and its sub base. If the level of restraint is sufficiently high, tensile forces induced in the concrete can lead to cracking. This usually occurs close to the centre of the panel, perpendicular to the direction of greatest restraint. Such random mid panel cracks are more prevalent in floors where there is no slip membrane below the slab and where the panel sizes are too large.

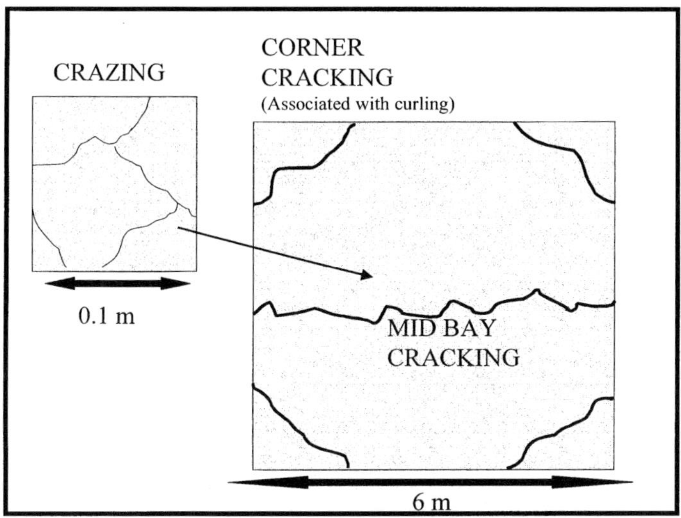

Figure 3 Typical crack patterns associated with drying shrinkage of concrete floors

Curling

In most concrete floors, significant drying only occurs from the exposed upper surface, particularly, when the floor is placed on a slip membrane or gas control membrane that inhibits loss of moisture from the lower concrete surface. The non-uniform drying environment leads to a greater degree of shrinkage in the upper part of the panel. This in turn can lead to upward curling of the panel, particularly at the corners. Loss of contact between the panel and the sub base may cause the panel to rock (which may sometimes be mis-diagnosed as failure of the sub base[3]). Subsequently gravitational forces and trafficking may lead to cracking across the corners of the panel. As with mid-panel cracking, the risk of curling is greater for larger panels or panels with a high length to depth ratio.

Crazing

Crazing is a network of fine cracks on the surface of the concrete floor slab, which whilst unsightly is of no structural significance. Its cause is rapid surface drying, but it is most apparent when there is a discontinuity in the composition or moisture content between the surface zone and the underlying bulk concrete[4]. Crazing is often associated with power-trowelled concrete floors where the surface few millimetres are much more cement rich than the bulk of the slab.

Crazing cracks generally only penetrate a short distance into the concrete (typically only as deep as the power-trowel affected zone).

It is also thought that carbonation of the surface contributes to the formation of crazing.

MINIMISING SHRINKAGE

The factors with the greatest influence over the magnitude of concrete drying shrinkage are the composition of the concrete itself and the environment to which it is exposed.

Concrete Composition

Considering first the concrete composition, it is clear from examination of the mechanisms of shrinkage that it is possible to reduce the potential for drying shrinkage by careful attention to the mix proportions.

The greatest single influence over shrinkage is the paste volume, reductions in which are reflected in lower concrete shrinkage. Consequently, maximising the aggregate size (consistent with the spacing of any reinforcement and the depth of the slab) is beneficial.

The combined coarse and fine aggregates should be proportioned to ensure a smooth grading curve that will minimise the water demand of the aggregate and hence reduce the paste volume needed to provide adequate workability. Aggregates should also be selected that do not display excessive levels of shrinkage[1]. There is some evidence that cements or combinations containing fly ash or blastfurnace slag may cause an increase in shrinkage (compared to Portland cement) at constant water/cement ratio[5]. However, when concrete is proportioned to produce similar strength and workability, the influence of cement type on concrete shrinkage is minor compared to that of aggregate volume fraction.

In order to keep the paste content to an acceptably low level, its water/cement ratio should be kept as low as is practical. This both reduces the total amount of water in the concrete available for loss through drying and also reduces the rate at which it can permeate through the paste and evaporate.

Water reducing and superplasticising admixtures will enable this objective to be achieved without compromising the workability required for effective placing and compaction. Indeed, if minimal shrinkage is a major requirement, then their use is essential to minimise both the paste volume and its intrinsic water content.

However, caution should be exercised in reducing water/cement ratio too far. As the water/cement ratio falls, the risk of autogenous shrinkage increases. This is the reduction in volume of cementitious materials when the material hydrates after initial setting. It is a purely internal phenomenon and does not involve shrinkage due to bulk loss of water or temperature variation. Water is removed from the capillary pore system into the cement hydrates in a form of 'self desiccation'. Under normal circumstances (ie water/cement ratio above 0.45), autogenous shrinkage is negligible in comparison to drying shrinkage, but it may make up as much as 50% of the total shrinkage at a water/cement ratio of 0.30[6].

Recently, a new class of concrete admixture has been introduced to the UK market which claims to reduce drying shrinkage. These shrinkage reduction admixtures (SRAs) function by reducing the surface tension in the water contained in the partially filled concrete pores and hence reducing the forces exerted on the walls of the capillary pores. This in turn reduces the shrinkage stresses within the concrete. When used in combination with an effective superplasticiser, reductions in shrinkage movement of up to 50% have been recorded[7]. This type of admixture will also reduce autogenous shrinkage in low water/cement ratio concrete.

Drying Environment

Control of the environment into which concrete floors are placed is rarely practical. However, where possible, floor construction should be delayed until the roof and walls of the structure have been erected. Windbreaks can also be useful in preventing excessive wind blowing through openings and over freshly placed concrete.

Even in cold weather, windy conditions can cause as much evaporation from exposed concrete conditions as hot weather. Prevention of rapid surface drying of the newly placed slab must be a priority for the constructor.

MINIMISING CRACKING

Shrinkage of concrete in itself does not inevitably lead to cracking and curling in concrete floors. It is generally the restraint to shrinkage that induces the stresses within the concrete that may subsequently result in cracking. Consequently, although reducing intrinsic concrete shrinkage, as described above, will obviously reduce the risk of cracking or curling, other design and construction factors must also be considered, including:

- Panel size and slab thickness
- Curing
- Addition of fibres

Panel size

The tendency for both mid-panel cracking and curling increases as the panel size and the length/depth ratio of the panel increases. A panel size of 6m x 6m is common and is generally considered to prevent problems[8]. However, the author has often observed both cracking and curling in panels of this size. Other authors[3] have suggested a maximum joint spacing of 4.6m as a practical, trouble-free limit.

The provision of a slip membrane below the slab will tend to reduce frictional restraint to uniform shrinkage of the concrete and hence mid-panel cracking. Paradoxically however, this same membrane may increase the tendency to panel curling by inhibiting loss of water from the lower surface of the slab[3]. The resulting non-uniform shrinkage leading to upward curling of the slab as the top surface shrinks more than the base.

Curing

Short-term curing (up to 28 days) does not reduce the intrinsic drying shrinkage of concrete, but prolonged curing may cause a small reduction[9].

The benefits of early age curing are primarily in its effect on increasing the strength and strain capacity of the concrete. This in turn influences the amount of stress or strain induced by restrained shrinkage that the concrete can withstand without cracking. Good curing can also help to reduce the incidence of crazing.

Fibre Addition

Both steel and synthetic fibres are sometimes added to flooring concrete and many claims are made for enhanced performance of fibre concretes.

With regard to drying shrinkage, however, neither type of fibre (when added at typical dosage rates) actually reduces the magnitude of the intrinsic drying shrinkage[10]. The incorporation of fibres can, however, be an effective means of controlling cracking. Due to their low elastic modulus synthetic fibres are not particularly effective in controlling shrinkage cracking in hardened concrete, but they are beneficial in reducing early plastic shrinkage cracking.

Steel fibres, however, when present in sufficient quantities will reduce both the crack width and crack spacing of shrinkage induced cracking, as well as holding the faces of a crack together. This results in a network of fine cracks (which are less visible to the casual observer) as apposed to a prominent single large crack.

CONCLUDING REMARKS

- Problems in concrete floors arising from the effects of drying shrinkage are uncommon but can lead to poor overall performance of the floor. Minimising cracking or curling requires careful attention to the selection of constituent materials and concrete mix design combined with consideration of the effect of design and construction practice. In practical terms the choice of cement type has little significant influence over long-term shrinkage. The oft quoted maxim that high concrete shrinkage is a direct result of a high cement content is, however, not justified.

Concrete mixes intended for flooring, where low shrinkage is a major requirement, should follow the guidelines below:

- Ensuring that aggregates have low intrinsic drying shrinkage.
- Maximising the volume fraction of aggregate in the mix (and hence minimising the paste volume).
- Optimising the aggregate grading curve to minimise water demand.
- Minimising the *water* content of the cement paste fraction.
- Consider the use of shrinkage reducing admixtures

REFERENCES

1. BUILDING RESEARCH ESTABLISHMENT: Shrinkage of natural aggregates in Concrete. *Digest 357*, BRE, Garston, 1991.

2. DHIR R K, TITTLE P A J and McCARTHY M J: Role of cement content in specifications for the durability of concrete – A review. *Concrete*, November/December 2000, pp68-76.

3. WALKER W W and HOLLAND J A: Thou shalt not curl nor crack (hopefully). *Concrete International*, January 1999, pp 47-53.

4. CONCRETE SOCIETY: Non structural cracks in concrete. *Technical Report 22 (3rd Edition)*, Crowthorne, 1992.

5. BROOKS J J and NEVILLE A: Creep and shrinkage of concrete as affected by admixtures and cement replacement materials. *ACI SP-135*, 1992, 19-36.

6. AITCIN P C: Demystifying autogenous shrinkage. *Concrete International*, November 1999, pp 54-56.

7. PERRY B: Admixtures for floors – a help or hindrance. *Concrete*, September 2001, 16-18.

8. CONCRETE SOCIETY: Concrete industrial ground floors – A guide to their design and construction. *Technical Report 34 (2nd Edition)*, Crowthorne, 1994.

9. HOBB D W and PARROTT L J: Prediction of drying shrinkage. *Concrete*, February 1979, 32-37.

10. ASSOCIATION OF CONCRETE INDUSTRIAL FLOORING CONTRACTORS: Steel fibre reinforced concrete industrial ground floor slabs. Pub: Concrete Society, Crowthorne, 1999.

CONTROL OF CRACKING IN UNREINFORCED MASONRY WALLS ON SUSPENDED SLABS

P Mullins

Mullins and Associates

P Dux

University of Queensland

Australia

ABSTRACT. This paper investigates the cracking of unreinforced masonry walls on suspended slabs. It begins with a review of provisions of some codes of practice relating to crack control in walls. The paper then discusses the structural mechanics leading to cracking. The problem is shown to arise from interaction between slab and wall, in which creep and strength properties are of influence. Results of numerical studies are used to quantify the structural effects. These studies relate to a resort building in which several hundred walls developed unacceptable cracking. The studies confirm that codified deflection control provisions for supporting slabs tend to mislead the designer. Practical measures for crack control in walls are discussed in light of the findings.

Keywords: Unreinforced masonry walls, Suspended slabs, Wall-slab interaction, Cracking, Crack control

Dr Peter Mullins conducts a consulting practice and lectures part-time at the University of Queensland and at the Queensland University of Technology in Brisbane, Australia. He is engaged mainly in building design and assessment of concrete structures for durability and repair. His principal research area is strength and serviceability of masonry structures.

Dr Peter Dux is Head of the Department of Civil Engineering at the University of Queensland, Australia. His areas of research and consulting include concrete technology, reinforced and prestressed concrete structures, durability and repair of concrete structures, and stability of steel structures. He is immediate Past-President of the Queensland branch of the Concrete Institute of Australia and is a Fellow of the Institution of Engineers, Australia.

INTRODUCTION

Unreinforced concrete blockwork walls are often used as non-loadbearing firewalls in multi-storey construction, the supporting structure usually being of reinforced or prestressed concrete. The walls are often sheeted in plasterboard, a flexible wall finish. Occasionally, blockwork walls are finished with a coating of render which may be of conventional thickness (10 mm) or may be thin (3 mm). This paper addresses the serviceability of rendered walls in relation to cracking.

The paper reviews the provisions of several codes of practice relating to crack control in masonry walls, supported on suspended concrete slabs. These provisions, where offered, specify incremental deflection limits and do not differentiate between crack control and crack prevention, leaving open the possibility of misinterpretation by designers. The behaviour of walls in relation to attraction of load and cracking is then discussed. Numerical examples illustrate that prior to cracking, storey-height walls without suitable articulation behave as very rigid elements in composite action with the slab and that codified incremental deflection control is largely ineffective in preventing cracking.

INTERPRETATION OF CODES OF PRACTICE IN RELATION TO DEFLECTION AND CRACK CONTROL FOR MASONRY WALLS

Code Requirements On Slab Deflection

Code requirements on slab deflection in the European [1], British [2] and Australian [3] concrete structures codes for slabs supporting masonry walls vary. The European code [1] refers to a maximum deflection of span/300 but notes "deformation sensitivity of non-structural elements ...should be accounted for". The British code [2] states that deflection limits "should normally ensure that the part of the deflection occurring after construction of finishes and partitions will be limited to span/500 or 20 mm. The Australian code [3] requires that the deflection of members supporting masonry walls does not exceed span/500 where provision is made to minimise the effect of movement, otherwise span/1000. These limits refer to the component of the deflection occurring after the construction of the wall. The codes do not give guidance on measures that would minimise the effect of movement. None of the codes indicate that compliance with the incremental deflection limits will prevent cracking, although a designer might assume that this is implied and that adoption of the most severe criterion of span/1000 will prevent serviceability problems.

Neither the British [4] nor Australian [5] masonry structures codes give guidance on limits for deflection of the supporting structure to prevent or control cracking in walls. Masonry walls detailed to tolerate greater foundation movement than normal are referred to in the Australian Residential Slabs and Footings code, AS 2870 [6], as articulated walls. The code requires articulated internal walls to comply with the Cement & Concrete Association of Australia's Technical Note 61 [7], which gives guidance on the location and spacing of vertical control joints. For rendered or painted walls built on slabs or footings on expansive soils, TN 61 [7] recommends a deflection limit of span/800 with vertical control joints at a spacing of 5 m.

AS 2870 [6] specifies a deflection control of span/2000 for full masonry construction, where walls are not articulated. The deflection criteria in the Australian Residential Slabs and Footings Code [6] are more stringent than in the all of the concrete structures codes reviewed.

Acceptance or Rejection of Cracks in Masonry Walls

The codes refer to deflection limits in terms of "controlling or limiting ....cracking" [5], "minimising the effects of movement" [3], being within "tolerable limits" [6], "avoid harmful effects" [1] and "not adversely affect ...appearance" [2]. None of these clauses refer to eliminating or preventing cracks but appear to aim at limiting the extent and magnitude of cracking.

While the British masonry code [4] does not define an acceptance criterion for cracks, the British concrete code [2] states, "in some cases a degree of minor repair work...may be acceptable". The Australian masonry code [5] defines a crack width limit of 1 mm for masonry which is not subject to aesthetic limitations. No guidance is given to acceptable crack widths when aesthetic limitations apply. Interestingly, both codes assume an uncracked condition in determining resistance of a wall to out-of-plane transient loads such as wind loads.

The Australian Residential Slabs and Footings Code [6] is the only code of practice reviewed by the writers, which classifies cracks in walls. The code classifies cracks less than 0.1 mm in width as "hairline" and less than 1 mm in width as "fine cracks which do not need repair". It would be reasonable to assume that this acceptance criterion is related to the deflection limits in the same code. If there is indeed a logical thread between codes, the acceptance criterion implicit in the less stringent deflection limits of the concrete structures codes would deem cracks somewhat greater than 1 mm in width to be acceptable.

The acceptability of a crack depends on the function of the building and the prominence of the crack. How easily a crack is noticed depends on crack location, the level of lighting, the wall colour and the wall finish. A crack at the junction of the wall and the slab is not likely to be noticed, unlike a crack at eye level. Cracks in joints of face brick walls are not obvious, particularly if the joints are raked. However, cracks in walls that are rendered and finished to a smooth surface are much more noticeable. Cracks are more noticeable in rendered walls painted in a light colour than in dark coloured walls. Cracks are more noticeable in areas, which are well lit than in poorly lit areas.

Walls finished in a smooth render therefore require a more stringent criterion on crack acceptance, as cracks are more noticeable than in unrendered walls. A criterion requiring no cracks is unrealistically severe. If the acceptance criterion was that cracks of 0.1 mm width or less are acceptable, the crack control would be an order of magnitude more severe than that implied in the deflection limits of AS 2870 [6] which, in turn, are more stringent than the deflection limits of the concrete codes reviewed [1,2,3]. Unfortunately, the matter of acceptability of cracks and definition of a defect is often omitted at the contractual stage, only to surface once wall finishes in the completed building begin to crack.

WALL LOADING, DEFORMATION AND CRACKING SEQUENCE

To illustrate the loading, deformation and cracking sequence of blockwork walls that are nominally non-loadbearing, consider a typical reinforced concrete flat plate structure in which a slab supports a series of parallel, evenly spaced, unreinforced blockwork walls of storey height. The walls are non-load bearing. The line of every second wall coincides with a row of columns, an arrangement taken from a case study of a resort building in which several hundred walls with thin render (3 mm) developed unacceptable cracking.

The following stages identify a typical sequence of wall loading, deformation and cracking:

1. A suspended slab is constructed. It undergoes immediate deflection when props are removed and increasing deflection over a period of years, principally due to creep. In a singly-reinforced slab, the final slab deflection is typically 2 to 3 times the initial deflection.

2. The dominant sustained loading is self-weight of the slab. A lesser component is introduced by construction of the blockwork walls. Most of the time-dependent component of slab deflection takes place after the walls are constructed.

3. A wall is built over a period of, say, one day. The weight of the wall is carried initially by the slab. The main stress in the wall at the completion of construction is vertical compression due to the weight of blockwork above any particular level. Blockwork should be stacked on the slab prior to construction of the wall so that the initial increment in slab deflection occurs prior to erection of the wall. After erection of the wall, the slab continues to undergo time-dependent deformation under the influence of the total sustained loading.

4. The mortar sets and is able to resist various types of stress up to certain failure stress limits. The limiting stresses of the blockwork wall in tension and shear are typically quite small and vary according to the consistency of the workmanship.

5. The slab continues to deflect. Composite action develops between the wall and a strip of slab centred on the wall, to modify the incremental deflections along the line of the wall.

6. Storey-height blockwork walls are very stiff. After only a very small deflection of the wall-slab composite unit, the slab no longer supports the wall along its full length. Rather, it begins to load the wall-slab composite unit.

7. The potential for loading to be attracted to a slab-wall composite member depends on its location in relation to supporting columns. For example, unless cracking intervenes, a wall-slab composite member extending between columns will attract much the same load as would a row of props along the same line, installed immediately after the wall is constructed. A parallel wall-slab composite unit situated away from columns tends to prevent other than uniform deflection increments along the line of the wall.

8. At the interface between the wall and slab, stresses change from compression along the full length of the interface (ie. the stresses at the completion of wall construction) to tension in some zones, compression in others and varying shear stress. Zones of tensile stress also develop within the body of the wall, introducing the possibility of cracking.

9. Composite behaviour continues until an applied stress either in the body of the wall or at the interface between the wall and slab reaches a limit. The wall may crack in one or more regions. The wall and the slab may separate over much of the interface, leaving the wall as an essentially self-supporting unit, carrying its self-weight by a combination of arching and tension. Some walls may therefore feature cracking in the body of the wall and some may not.

10. Cracks along mortar beds in the wall do not necessarily form where tensile stresses are highest. Rather, cracking occurs where the local strength is first reached, and this local strength can be highly variable.

11. Because deep walls are so stiff, the deflection of the wall is very small at the time that cracking occurs, much less than typical codified incremental deflection limits for the slab.

NUMERICAL STUDIES

Two walls have been analysed. In some instances the numerical model represents a wall acting compositely with a strip of slab (Figures 1, 2). Other models represent free-spanning walls (Figures 3, 4).

Wall Details

The first is a wall of length 6 metres with a door opening. The second is a solid wall also of length 6 metres. The height of the walls is 2.5 metres and thickness is 140 mm. Other data appear in Table 1. The walls are similar to actual walls on suspended slabs at the case study building referred to earlier, and for which the crack patterns were recorded by the authors. Comments in the text concerning observed cracks relate directly to the recorded data.

Table 1 Data used in wall analyses and interpretation of results

PROPERTY	SLAB	BLOCKWORK WALL
Elastic modulus (N/mm^2)	27700	8000
Long term modulus (N/mm^2)	11000	5280
Density (kg/m^3)	2450	1640
Flexural tensile strength (N/mm^2)	-	0.200
Direct tensile strength (N/mm^2)	-	0.070
Shear strength (N/mm^2)	-	0.125
Interface friction coefficient	0.6	0.6

The blockwork wall parameters in Table 1 are estimated values for face-shell bedded construction converted to an equivalent solid blockwork construction. The strength values are characteristic values. Limiting frictional resistance along a cracked wall-slab interface can be estimated using a coefficient of friction of 0.6 [4]. The coefficient of limiting static friction should be at least of the same order. Mullins and O'Connor [8] found the coefficient of sliding friction along the base of storey-height masonry walls ranged from 0.7 – 1.9, depending on the interface details.

Walls Acting Compositely With Slab

Analyses involving walls with composite slabs (Figures 1, 2) investigate changes in wall stresses which accompany increasing deflection. Sub-figures (a) and (b) refer to a wall with an opening and a solid wall respectively. For purposes of the example, the wall-slab composite units are simply supported. The width of the slab strip is 1.34 m [1,3]. At the completion of wall construction, vertical compressive stresses are around 0.04 N/mm^2 at the base, reducing linearly with height. As the composite unit deflects under the action of creep and additional attracted loading, these initial stresses alter. Depending on location, composite wall-slab units in buildings have the potential to gradually share a significant component of self-weight of the unit and also to attract substantial additional loading. Figure 2 shows vertical tensile stresses in the walls under an arbitrarily chosen loading equal to the self-weight of the composite unit

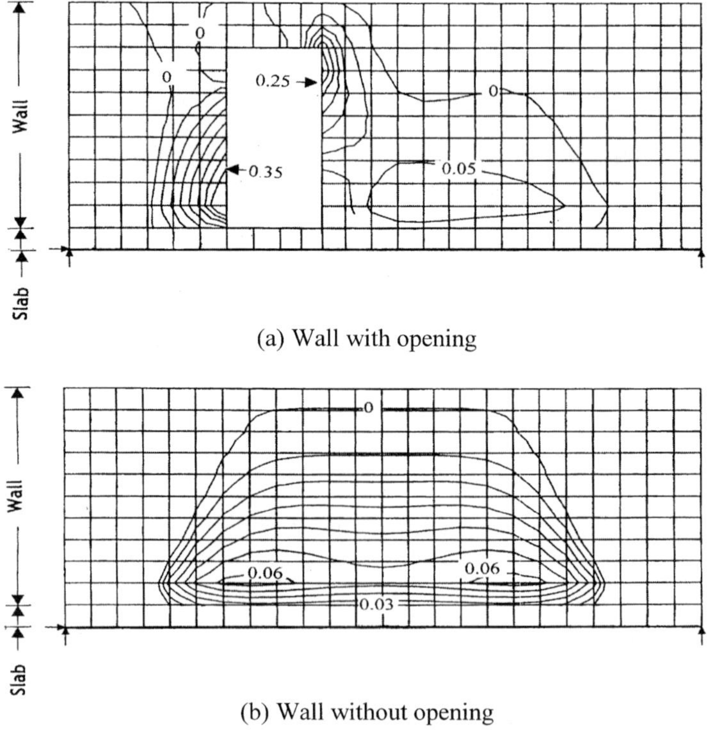

Figure 1 Vertical tensile stresses (N/mm^2) in uncracked composite wall-slab units

In Figure 1(a), concentrations of stress occur in the longer side of the wall at the top of the door opening and towards the bottom of the door opening in the shorter side of the wall. Horizontal cracking was recorded in walls at both of these locations. In the longer portion of the wall away from the door and in the lower half of the wall, the stress contours indicate regions of gradually varying tension. Some case study walls were recorded as having horizontal cracking in this general area, occasionally extending back to the door.

In Figure 1(b), vertical tensile stresses in the wall peak somewhat above the slab, the maximum value being around the direct tensile strength of the wall. As recorded for walls in the resort building, a crack in the body of the wall would be horizontal, it would be expected to coincide with a mortar bed and would not extend to the ends of the wall. Again, while a crack can be characterised by its general location, its precise location is not so amenable to prediction due to the variability of mortar joint strengths. For both walls in Figure 1, horizontal tensile stresses (not shown) are largely confined to the slab, acting as a flange.

Along the wall-slab interface, Figures 2 (a) and (b) show that shear stresses quickly build to critical levels. As loading to the composite units increases, the combination of tension across the interface and shear along it may lead to separation of the wall from the slab. Other analyses indicate that relative shrinkage between wall and slab also causes increased shear stress along the interface and increased tension over much of the interface, again promoting separation of wall and slab to leave the wall to support its own weight over much of its length.

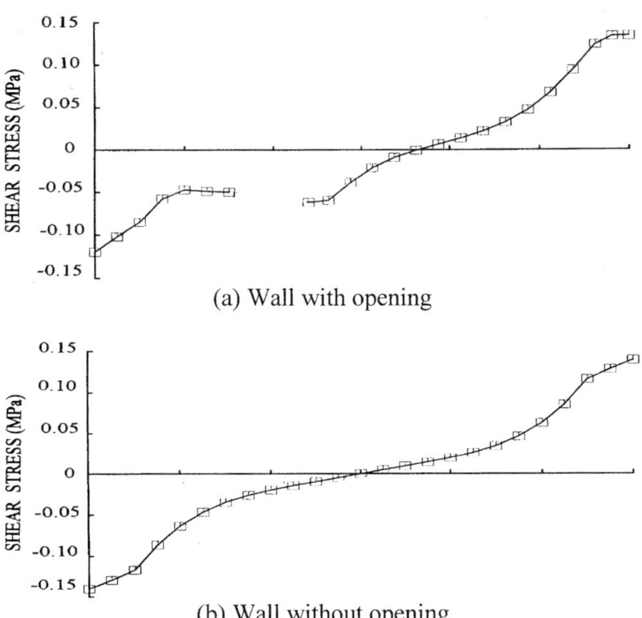

(a) Wall with opening

(b) Wall without opening

Figure 2 Horizontal shear stresses (N/mm^2) along wall-slab interfaces

Figures 1 and 2 describe conditions sufficient for crack initiation. These conditions arise when the deflection of the composite units is still very small, well under 0.5 mm, or less than span/12000. The strength values in Table 1 are characteristic values, so most walls of the type analysed would be able to sustain higher loading and greater deflection in composite action before cracking. However, it is highly unlikely that incremental deflections of order span/1000 or span/2000 could be sustained.

Free-Spanning Walls

Figure 3 shows the long-term behaviour of the walls in the free-spanning state. The necessary horizontal reactions are provided by friction in the end regions of the wall-slab interface. In Figure 3(a), stresses are seen to be highest at the right near the top of the door opening. Cracking will depend on the local tensile strength and the influence of any vertical reinforcement adjacent to the opening. Horizontal tensile stresses (not shown) are also highly localised particularly at the top of the wall above the left-hand jamb. It is therefore quite possible for the wall to span freely without visible cracking, provided that composite action with the slab was terminated by the formation of a crack along the wall-slab interface. The stress plot in Figure 3(b) indicates that the solid wall can freely span without developing horizontal cracks. Other analyses (not shown) indicate that vertical cracking is unlikely.

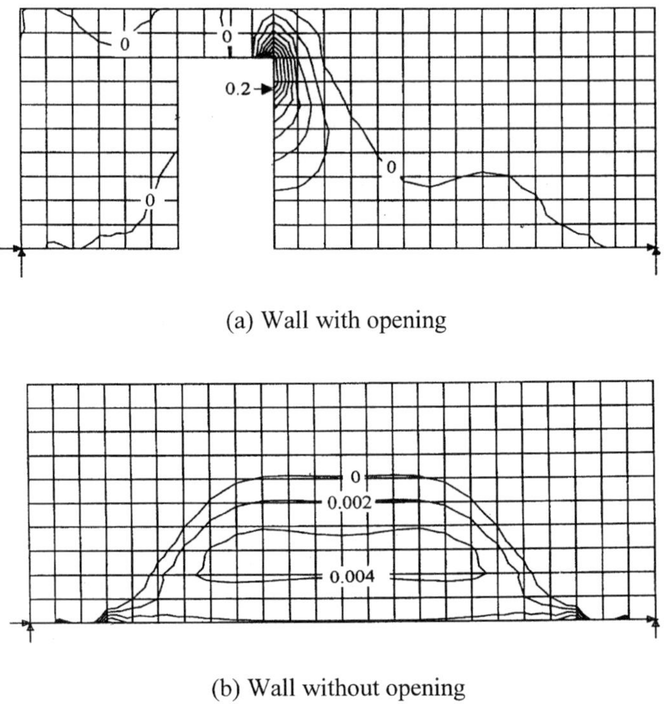

(a) Wall with opening

(b) Wall without opening

Figure 3 Vertical tensile stresses (N/mm^2) in free-spanning walls

Finally, in relation to the free-spanning walls, analysis shows that the maximum long-term deflections are very small. Even if the long-term moduli in Table 1 were in error by a factor of 10, the peak deflections would still be less than span/2000.

Walls Forced To Conform To Codified Slab Deformations

Figure 4 presents vertical tensile stresses in the walls in the long-term with imposed deflections along the base corresponding to a slab deflection of span/1000. In Figure 4(a) the deflection is for a fixed-ended slab. In Figure 4(b) the imposed deflections are for a simply supported slab. The figures show stresses grossly in excess of strength. They dispel the notion that an uncracked wall of normal stiffness and strength can simply follow the slab as it undergoes deformation up to codified deflection limits.

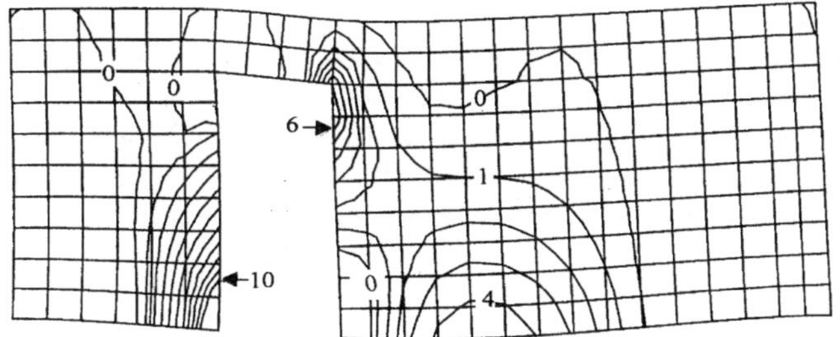

(a) Wall with opening

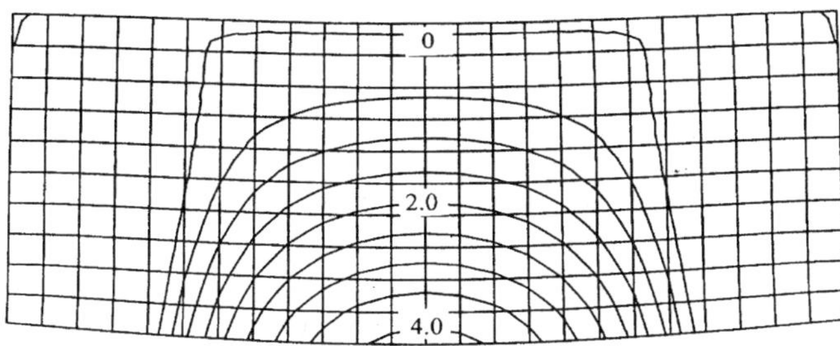

(b)Wall without opening

Figure 4 Vertical tensile stresses (N/mm^2) in walls with imposed deflections of span/1000

CONCLUSIONS AND RECOMMENDATIONS

Principal conclusions are presented below. Included also are recommendations for good practice based on the analyses discussed in the text and on observations from real structures.

1. Code deflection limits for members supporting masonry walls should not be interpreted as providing guidance for the prevention of wall cracking.

2. Masonry walls are able to accommodate only little deflection in the supporting structure before cracking occurs. This may be within the body of the wall or may be at the interface between the slab and the wall. The wall may therefore appear cracked or uncracked.

3. Unless stringent deflection control is exercised, far more stringent than code limits, or unless the masonry wall is articulated, cracking is likely in well-constructed unreinforced masonry walls on suspended slabs.

4. Crack acceptance criteria should therefore be defined in the contractual agreement for a project.
5. Walls that have a potential to develop noticeable cracking, such as walls finished in smooth, brittle render, require a high degree of crack control.

6. Articulation of masonry walls on suspended slabs would include debonding of walls from the slab over much of their length, to ensure separation at the base. The walls can then span freely, typically without cracking in the body of the wall. Debonding should be done in such a way as not to compromise the horizontal frictional capacity at the ends of the wall-slab interface, as this is needed for arching of the free-spanning wall.

7. In regions of sagging slab curvature, wire reinforcement in mortar beds will not prevent cracking nor will it be effective in controlling crack widths, as the majority of cracks develop along bedding planes. In regions of hogging curvature, such reinforcement might assist nominally in controlling crack widths as the cracks will tend to be vertical. However it will be ineffective in preventing the formation of cracks. A far more effective practice would be to articulate the wall by installing vertical control joints in regions of hogging curvature. Articulation of walls on slabs with a large span relative to wall height, by provision of a vertical control joint at mid-span, will reduce tensile stresses in the free-spanning wall.

8. Cracking tends to propagate from upper corners of openings in walls. Articulation by provision of a vertical control joint at the edge of narrow openings closest to the centre of the slab span or on both sides of a wide opening will reduce formation of such cracks.

REFERENCES

1. CEB-FIP Model Code 1990, Design Code, Comite Euro-International Du Beton, Thomas Telford Publishing, London

2. BS 8110 – 1985, Structural use of concrete, British Standards Institute, London.

3. AS 3600-2001, Concrete Structures, Standards Australia.

4. BS 5628 – 1992, Code of practice for use of masonry, British Standards Institute, London.

5. AS 3700-1998, Masonry structures, Standards Australia.

6. AS 2870-1996, Residential Slabs and Footings – Construction, Standards Australia.

7. CEMENT AND CONCRETE ASSOCIATION OF AUSTRALIA, Articulated Walls, Technical Note 61, October 1991.

8. MULLINS, P.J., O'CONNOR, C., Brick Shear Walls: An Experimental Investigation of the Brickwork to Concrete Interface, Proc. 9th Int. Brick/Block Masonry Conf., Berlin, Germany, 1991.

MOISTURE MIGRATION IN CONCRETE SLABS DURING DRYING

N Holmes

R P West

Trinity College Dublin

Ireland

ABSTRACT. This paper presents the profiles of relative humidity (RH) readings taken at various depths in concrete slabs during a time period of over 230 days using probes attached to a humidity-reading device. It also examines the effect ambient conditions have on the drying process. For the experimental work, two sets of slabs were tested, one in a controlled environment with an elevated temperature and the other in a laboratory at room temperatures. It was found, not surprisingly that the slabs in the controlled room appeared to dry out at a much faster rate than those held at room temperature. However, a residual of moisture remained within the slab and the average residual RH over the depth of the latter was as much as 85%, as compared with 80% for the former. This suggests that the industry's standard, specified in BS 8203: 1996, concerning the application of floor coverings when the surface reaches a RH of 75% needs to be treated with caution especially when drying is artificially accelerated. Due to this moisture residue remaining in the slab, any impermeable covering applied to the surface may result in a number of defects occurring, leading to expensive repair work later on. A scheme for modelling the process numerically is presented, employing the finite element method that will eventually account for changing ambient RH, non-linear diffusion coefficients and sealing of the surface at some point in time.

Keywords: Diffusion, Finite element method, Floor slabs, Moisture movement relative humidity.

N Holmes graduated with an Honours degree from Queen's University Belfast in 1998 and a Masters degree from Trinity College Dublin in 1999. He is currently a PhD student at Trinity College Dublin where he is researching finite element modelling of moisture migration in concrete.

R P West is a Senior Lecturer and former Director of the Structural Laboratory at Trinity College Dublin. His research interests in concrete lie in durability, rheology and new materials. He is a former Vice-Chairman of the Council of the Irish Concrete Society. He is currently the Irish liaison Engineer for Eurocode 2.

INTRODUCTION

Establishing the point in time at which a floor covering can be safely applied to a concrete slab is a major concern in the floor construction industry. If a floor covering is applied too early, numerous defects can occur, such as blistering of paint or vinyl, buckling of floorboards or possible rising of tiles [1], [2]. Yet there is a strong economic incentive for a contractor to lay the floor covering as early as possible. The current standard for applying a covering to concrete floors [3] states that it may be applied when the surface reaches a relative humidity (RH) of 75%, usually established using a standard hygrometer test [3].

This so-called safe threshold, however, does not account directly for the moisture condition below the surface and, as shown from the humidity profile in Figure 1, there can still be a considerable residue of moisture in the slab. If a covering is applied at 75% RH, the defects mentioned above may occur due to a vapour pressure forming underneath the covering arising from the trapped residual moisture. If a significant vapour pressure develops under the covering, arising from ongoing gradual non-linear diffusion of the residual moisture to the surface [4], this can lead to expensive repair work later on. This paper reports on the quantification of this effect from experimental work carried out at Trinity College Dublin (TCD), and urges caution when deciding whether the industry standard of 75% surface RH is an acceptable value at which to safely apply such coverings.

The ability to predict the point at which the application of a covering could be safely applied would be a major achievement, reducing the chance of the defects mentioned above from occurring. At present, work is under way at TCD to model the movement of moisture over time using numerical methods and, specifically, to predict the long-term residual vapour pressure that develops under the vapour barrier. The finite element method (FEM) is being employed to represent the mathematical equations that are programmed into a FORTRAN program to model the process. It is anticipated that this model will predict the changes to the humidity profile with depth over time and assist in determining when a covering may be safely applied.

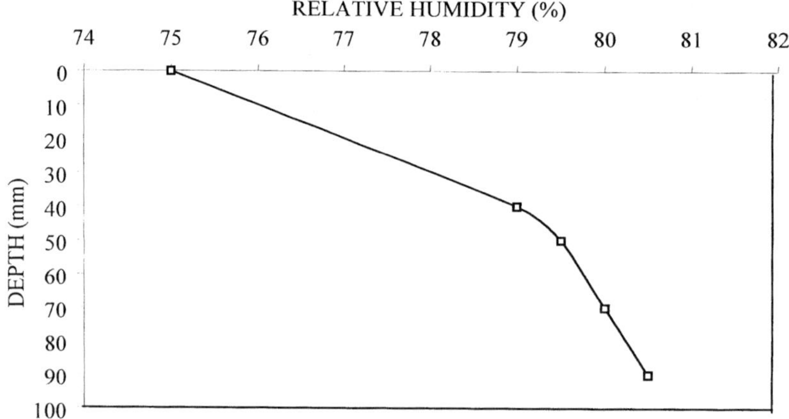

Figure 1 RH profile showing residual moisture in 100mm thick slab when surface is at 75% RH

EXPERIMENTAL SET-UP

A number of concrete slabs were made up each 500 x 500 x1 00 mm with a w/c ratio of 0.5. The slabs were cured for 7 days under wet hessian before the measurements began. Following this, all the slabs were coated with a sealant on five sides, leaving the top surface exposed. This ensured that drying took place in one direction only, namely upwards and this represented actual drying conditions of a slab on grade on site. At this point, four separate holes were drilled into the concrete to depths of 50, 60, 80 and 100 mm. The drilling of the holes did generate some heat but it is expected that this did not affect the moisture condition in the slab too adversely. The holes were 20 mm diameter into which plastic tubes, which had slots 10 mm above the base were inserted, as shown in Figure 2. Therefore, the average heights of the RH readings were, 40, 50,70 and 90 mm. A layer of silicone was applied to the sleeve of the tube to ensure a strong bond developed between the tube and the exposed concrete. In addition, a rubber membrane isolated the slots from the upper part of the hole and developed a ‘chamber’ of uniform RH between the rubber seal and the bottom of the tube.

In order to read the RH at depth, humidity probes were inserted into the tubes and attached to a CE-RH meter that give an instantaneous reading on a display panel [5]. The surface moisture condition is simultaneously revealed by pressing the device onto the concrete [1] where an electrical impulse is imparted to the surface using four transmitting electrodes and the capacitance offered by the concrete is reflected in the reduced signal measured at four receiving electrodes a short distance away. The CE-RH has been developed by a Dublin based company, Tramex Ltd, which design and manufacture moisture meters for concrete and timber materials. The variation in RH was measured over a time period of 230 days. In addition, a number of surface Vapour Emission Tests (VET) and Surface Hygrometer (SH) tests were carried in accordance with the recommended VET [6] and the British Standard [3] procedures to monitor the decrease in the moisture emitted from the surface and to assess the accuracy of the CE-RH meter.

The slabs were placed inside two separate rooms, one at room temperature in a laboratory and the other in a warmer room with a dehumidifier present. The average temperatures and humidities were 14^0 C and 55% and 26^0 C and 35% in the laboratory and control room respectively. he purpose of this was to monitor the effect the ‘forced’ drying condition had on the drying process throughout the thickness of the slab compared with that in the laboratory.

The complete experimental set-up is shown in Figure 3. Surface RH readings were also taken using the CE-RH with an insulated tent mounted onto the surface of the slab [3]. The probe was inserted into a plastic pipe on the tent and the RH was recorded after approximately 2 minutes to allow for any fluctuations in RH to settle. This set-up is shown in Figure 4.

DISCUSSION OF TEST RESULTS

The slabs ‘dry out’ from the surface with time, depending on the ambient conditions, and the concrete’s diffusion characteristics. When a slab is ‘dry’ depends very much on the environment, where an equilibrium is reached between the RH of the air and the residual moisture in the slab. In a previous paper [4], the authors discussed how the trend from numerous VET and SH tests showed the reduction in moisture content over time until they reached their threshold value of 3 lbs/1000ft^2/24hrs and 75% surface RH respectively.

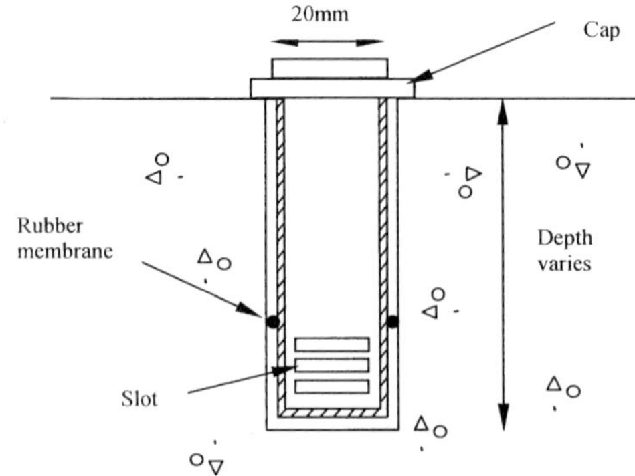

Figure 2 Drilled hole with slotted tube in place for CE-RH readings

Figure 3 The complete experimental set-up

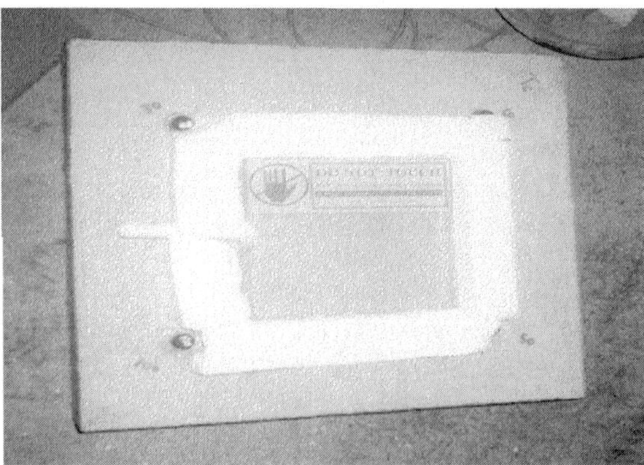

Figure 4 Surface RH apparatus, showing the access tube for the RH probe leading into a sealed plastic tent

However, of course, surface RH reduction tests are usually discontinued at this point in time as the ambient RH has reached the 75% threshold. But, these surface tests say nothing about the moisture condition in the slab deeper down and so continued monitoring of a slab reveals further reduction of the surface RH, assuming suitable ambient conditions.

Figure 5 shows RH profiles at various depths in the concrete as measured using the CE-RH. It is clearly shown from the experimental results that there is still a considerable residue of moisture remaining deep within the slab. For example, at approximately 70 days the surface reaches the BS standard of 75% surface RH to apply coverings. However, it is not until 140 days that the slab had dried sufficiently such that 75% is the maximum RH throughout the thickness of the slab. Figure 6 shows two RH profiles of identical concrete slabs drying in the two different environments; one in a controlled room at an elevated temperature with a dehumidifier present and the other in the laboratory with normal ambient conditions. From Figure 6 it is obvious (and expected) that the slab in the controlled room is drying at a much faster rate than that in the laboratory, and will reach the 75% surface RH in a much shorter time, approximately 40 days. This accelerated rate of drying results in considerable residual moisture remaining in the slab as the surface dries at a faster rate than lower down.

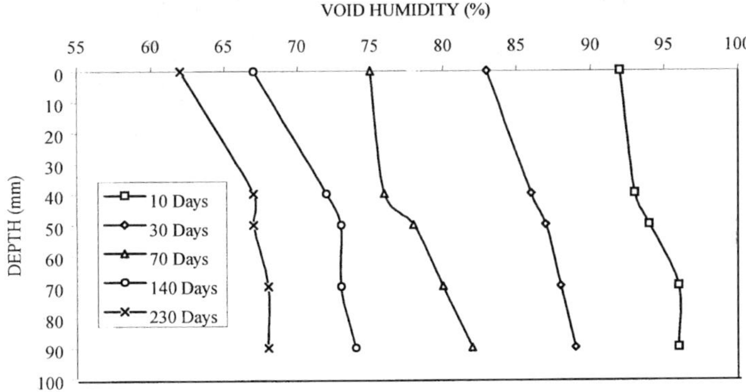

Figure 5 Void RH at various depths over time in the laboratory

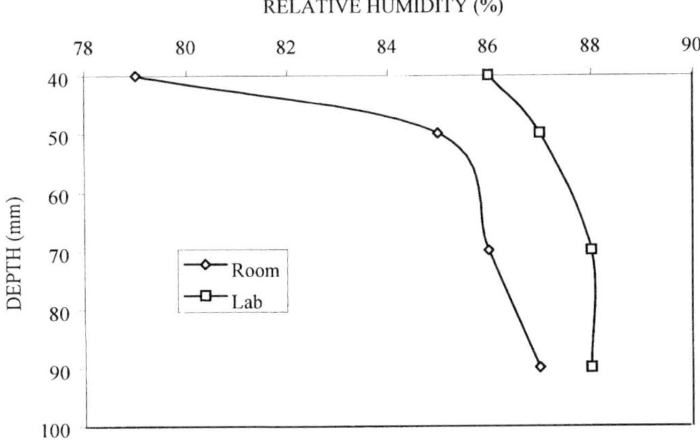

Figure 6 RH profiles for identical slabs at 28 days of drying. One is drying in a heated control room and the other is at ambient laboratory temperature

Following the application of a covering, a vapour pressure will develop underneath the vapour barrier as moisture continues to migrate by diffusion to the surface, under the pressure of a concentration gradient with depth (as seen in Figure 6). An equilibrium RH will be reached some considerable time later between the residual moisture in the slab. This concept is demonstrated in Figure 7.

The concrete, drying in the laboratory, will have residual moisture leading to a long-term average RH below the impermeable covering of, for example, 80%. The moisture condition in the slab drying at an elevated temperature, however, causes a greater average long-term vapour pressure to develop (for example 85% in Figure 7) because the drying conditions have accelerated the moisture loss near the surface, leading to premature achievement of 75% RH at which time the surface may be sealed. This may lead to extensive damage to the covering if this vapour pressure is greater than the adhesion force of the covering to the surface.

It is, therefore, vital that concrete slabs are allowed ample time to dry out sufficiently and naturally before the application of coverings is allowed. At present, the current tests for establishing the 'dryness' of slabs and the construction standards do not take full account of this residual moisture. Indeed, in order for a true reflection of the moisture condition, the RH must be assessed through the full slab depth.

However, it should be noted that previous experience indicates that if slabs are allowed to dry naturally, then residual moisture on the slabs is not usually enough to develop sufficiently high vapour pressure to cause defects. Traditional tests such as the VET and SH take 1-3 days to complete and even then tell one little about the moisture condition within the slab. The CE-RH is a much faster test and does allow an assessment of the moisture condition through the depth of the slab.

It has become apparent that being able to predict the residual moisture in a concrete slab would be a major advantage for the flooring industry. It would mean that coverings could be applied as early as possible with little risk and the defects mentioned above could be reduced or eliminated with substantial savings made in relation to the repair work needed.

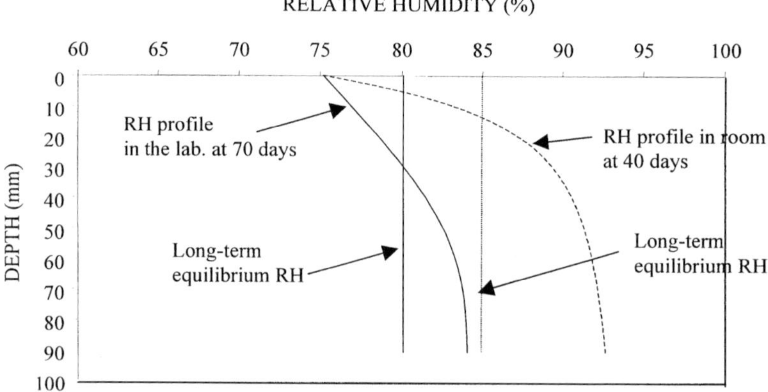

Figure 7 Long-term effect of applying a floor covering in different environments when surface is sealed under RH of 75%

MATHEMATICAL FORMULATION

At present, work is being carried out to mathematically model the transient movement of moisture through the concrete slab taking account of the changing ambient conditions and the non-linear diffusion coefficient [4].

It is anticipated that this model will predict the effect of applying a covering to the surface and the subsequent development of a vapour pressure some time later. Previously [4, 7], prediction methods were used in conjunction with experimental data and by developing empirical equations using various methods. Here, the finite element method (FEM) is being employed, using mathematical principles, to solve the well-known Fick's second law of diffusion [8], given by:

$$\frac{\partial H}{\partial t} = \frac{\partial}{\partial x}(D\frac{\partial}{\partial x}) \tag{1}$$

where H is the relative humidity (%), t is time (days) and D is the diffusion co-efficient (typical units m^2/sec). Following the FE discretisation process, Equation 1 is transformed into a system of ordinary differential equations [9], given as:

$$[C]^{(e)}\left\{\frac{dH(t)}{dt}\right\} + [D]^{(e)}\{H(t)\} = 0 \tag{2}$$

where [C] is the capacity matrix, [D] is the diffusion matrix and H(t) is the humidity as a function of time.

These equations are solved using a time stepping processes [10] with a constant D and boundary conditions. They have been programmed by the authors into a FORTRAN matrices FEMMTC (Finite Element Modelling of Moisture Through Concrete) where the [C] and [D] matrices are assembled. Equation 2 is then solved by FEMMTC using the time-stepping process below:

$$[K_{eff}]\{a\}_n = \{R_{eff}\}\{a\}_{n-1} \tag{3}$$

where:

$$[K_{eff}] = \left[\frac{[C]}{\Delta t_n} + \theta[D]\right] \tag{4}$$

$$\{R_{eff}\} = \left[\frac{[C]}{\Delta t_n} - (1-\theta)[D]\right] \tag{5}$$

and Δt is the time-step, θ is a dimensionless parameter ranging from $0 \leq 1$, $\{a\}_n$ is the current unknown to be solved and $\{a\}_{n-1}$ is the previous solution, where {a} is the pore humidity.

Results from the program are shown in Figure 8, and are compared with the experimental results. In this example, the diffusion coefficient used was constant throughout (1.0×10^{-12} m^2/sec [11]). As shown, there are slight differences in the early stages, signifying that the diffusion coefficient is indeed non-linear early on. However, over time, the accuracy of the model is reduced as D is varying with time (t), depth (x), concentration (H) and temperature (T), or in mathematical terms [12]:

$$D = f(t, x, H, T) \quad (6)$$

The major difficulty in the modelling analysis will be taking account of the non-linear behaviour and work is underway to use non-linear methods to account for the changing diffusion co-efficient. It is anticipated that the model will be able to predict when a slab will be dry enough to apply coverings by taking into account the mix constituents and the ambient conditions. With calibration of the model, the distribution of the residual moisture trapped under the covering will be predicted over time.

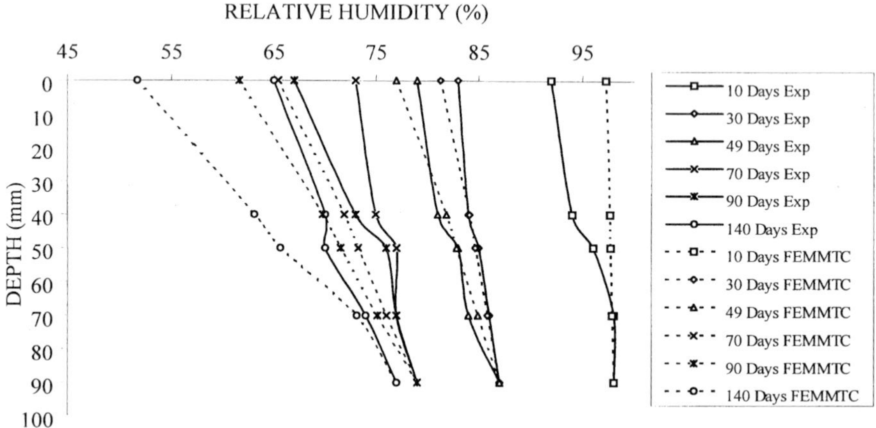

Figure 8 Experimental Results compared to FEMMTC results

CONCLUSIONS

In this paper, the authors present results from experimental work carried out on two sets of identical concrete slabs, one in a laboratory at room temperatures and the other in a control room with an elevated temperature, and the effect the ambient conditions have on the RH profiles within the slab.

The results show, not surprisingly, that the slab in the controlled room dries at a much faster rate than that in the laboratory, particularly at the surface. However, this leads to a significant residue of moisture remaining within the slab, which has been quantified. If a covering were applied at this stage to the slab in the accelerated drying environment, defects may occur leading to expensive repair work. Results from tests carried out over 230 days demonstrate that the normally dried slab had dried sufficiently after 70 days to satisfy the British Standards recommendation that a covering may be applied when the surface reaches a RH of 75%.

However, it was shown that the slab needed to dry for a further 70 days to reach a maximum of 75% RH throughout the thickness the slab. Further, it is predicted that the average residual moisture remaining in the slab, if a coating were applied, in the controlled room would have reached 85%, while the slab in the laboratory would have reached an average RH of 80%.

This former residue of moisture may have developed a significant vapour pressure under a covering, leading to numerous defects later.

Work is currently underway by the authors to develop a numerical model in FORTRAN, employing the finite element method to model the process using equations developed based on a modification of Fick's second law of diffusion. It is anticipated that this model will be able to predict the residual RH profiles through the thickness of the slab for non-linear diffusion coefficients and varying ambient conditions and to estimate the long-term vapour pressure that develops under a floor covering. By doing this, the risk of damage may be reduced and substantial savings may be made in the costly repair work needed.

ACKNOWLEDGEMENTS

The authors would like to acknowledge the technical assistance and advice provided by Mr Alan Rynhart of Tramex Ltd.

REFERENCES

1. O'NEILL M.L., WEST R.P. AND RYNHART A.D., The Measurement of the Moisture Condition of Concrete and Gypsum using the Concrete Moisture Encounter, Proc RILEM Intnl Conf On Non-Destructive Testing and Experimental Stress Analysis of Concrete Structures, Kosice, pp 162-169, 1998.

2. SUPRENANT B.A., Moisture Migration Through Concrete Slabs, Magazine of Concrete Construction, pp 879-885, November 1997.

3. BRITISH STANDARDS INSTITUTION (1996), Code of Practice for installation of resilient floor coverings, BS 8203.

4. BAZANT Z.P. and NAJJAR L.J., Drying of Concrete as a Nonlinear Diffusion Problem, Cement and Concrete Research, Vol 1, No. 5, pp 461-473, 1971.

5. HOLMES N. and WEST R.P., Experimental Investigation of Moisture Migration in Concrete, Colloquium on Concrete Research in Ireland, pp 39-44, September 2001, NUI Galway.

6. VAPOUR EMISSION TEST, Vapour Emission Test for Measurement of Concrete Moisture (test employs anhydrous calcium chloride), The Vaprecision Co, CA, USA, 1994.

7. AKITA H., FUJIWARA T. and OZAKA Y., A Practical Procedure for the Analysis of Moisture Transfer within Concrete due to Drying, Magazine of Concrete Research, Vol 49, No 179, pp 129-137, 1997.

8. CRANK J., The Mathematics of Diffusion, Second Edition, Oxford, 1975.

9. ZIENKIEWICZ O.C. and TAYLOR R.L., The Finite Element Method, Forth Edition, Volume 2, McGraw-Hill, 1989.

10. HUEBNER K.H., THORNTON E.A. and BYROM T.G., The Finite Element Method for Engineers, Third Edition, John Wiley & Sons, 1995.

11. NEVILLE A.M., Properties of Concrete, Fourth Edition, Longman, 1995.

12. RILEM REPORT 12, Performance Criteria for Concrete Durability, Edited by J. Kropp and H.K. Hilsdorf, E & F N Spon, 1995.

THEME TWO:
CONSTRUCTION TECHNIQUES

Keynote Paper

DEVELOPING A GREATER UNDERSTANDING OF THE NATURE AND USABILITY OF CONCRETE IN INDUSTRIAL FLOOR APPLICATIONS

D J J Harvey

Stuarts Industrial Flooring Limited

United Kingdom

ABSTRACT. The increasing use of concrete for the construction of the primary floor slab and wearing surfaces of retail distribution warehouses and production facilities is putting the material under the spotlight. The steady development of methods has kept pace with ever shortening contract schedules and continues to deliver floor surfaces that cope with more and more sensitive high level picking machinery. This has opened the gap between construction practice and design methodology. Concrete is a complex material and needs a great deal of understanding. Contractors have pressed designers, engineers and the industry lead body, The Concrete Society, to upgrade the widely accepted Technical Report TR34 to reflect the changes that have occurred with the introduction of high output techniques and machinery into the United Kingdom. This advance brought fast track just-in-time construction techniques in some conflict with design criteria based on 1920s highway pavement practice. UK research investigations have since delved into the behaviour of slabs in their early life and into the impact of loading joints and corners of slabs. The former has shown conclusively, for instance, that temperature change during construction has a major effect on early concrete slab stresses. The research programmes have been an important part of the determination to bring design criteria up to present practice that is increasingly being honed by specialist flooring contractors.

Keywords: Concrete industrial floors, Substantial areas, Research into early life behaviour, Development of new design and construction codes.

David J J Harvey, is a Director of Stuarts Industrial Flooring Ltd, the UK's largest producer of industrial floors, is a co-founding Member of the Association of Concrete Industrial Flooring Contractors, being elected Chairman 1996 to 1999.

INTRODUCTION

There are two very distinct levels in the knowledge and understanding of how to make the most of the primary material for concrete floors, the concrete itself. We struggle to develop the theories that help us to design concrete slabs. We struggle again when we build with this most versatile of materials, one that is also one of the most variable of building mediums. The construction world itself is filled with classic conditions that have a material effect on concrete in its placement condition. Wind, rain and temperature variations can make design criteria almost laughable when encountering the extremes during the placement process. These variations feature in almost all aspects of this conference's principal topics.

CONCRETE'S STRENGTHS

The Association - ACIFC - has been concentrating on the phenomenon of an industry that has effectively no choice in its lead material. Its Members, in the Association's short life, have increasingly recognised that, despite concrete's variability, there is nevertheless a need for an international code that sets out with clarity and single-mindedness the formula for the use of concrete in industrial floor construction that can be applied around the world. Today there are too many interpretations and confusions in the market place.

Of the many challenges facing specialist-flooring contractors, the biggest client expectation is that floors be constructed effectively to a factory-produced specification of quality and finish and to be maintenance free. That is despite the very real factors on site that can vary from freezing to baking, exposure to wind, rain or sunlight, traffic, bomb scares, concrete mix plant failure to deliver and the continual time pressuring of the now universally adopted "just -in-time" methods. Usage of floors has also changed, with demand for 24hour, 7 day, year after year, as normal requirement. Users of floors and the expectations they have of their mechanical handling suppliers are such that their own factory produced systems are often restricted in operation by the quality of the floor surface simply because it was not factory produced. There are of course expensive surface-course solutions, including self-levelling wearing screeds waiting in the wings.

Concrete nevertheless has proved itself over and over again to have an almost unassailable lead position in application to industrial floor slabs. There is no alternative that provides such reasonable first cost delivery of combined structure and wearing surface with the relative ease of construction and durability to match the demands of all day, every day use throughout the year. But the reality is that the more we think we understand concrete, the greater is the task of obtaining full knowledge of this infinitely versatile material.

THE DRIVE FOR CHANGE

That pursuit of knowledge has driven productive collaborations, where industry and academe have jointly sought routes to greater understanding. This has been increasingly attracting international attention. In the UK the design and construction code for ground supported industrial slabs - Technical Report TR34 [1] - is now being progressed by The Concrete Society to an updated format. The aim is to get closer to the ideal of a performance related code, developed from a guidance document that is already well-respected in the UK and increasingly in France, other European national markets and the Far East.

What is it that is driving UK contractors to press for change? What are the dissatisfactions with present codes that have brought Government funding to match industry's contribution? One difficulty is that the speed of development of the construction methods has left the more prescriptive codes – such as BS8204 [2] – too far behind current practice. The explosive advance of retailing and associated distribution that has hit the UK over the last 20 years has been paralleled by a revolutionary means of placing large area structural concrete slabs. There has been an almost total monopoly of States-side thinking behind so much of this expansion.

Figure 1 Highly mechnised floor slab construction using laser screed laser controlled spreader and alongside application of dry-shake finishing material

However, there is a warning of a clash between cultures and methods. In many respects practical and material, the USA developments encompass the big and economical with acceptance of a more pragmatic approach to floor finish and flatness issues, and a heavy reliance on single layer direct finish concrete. The other side of the coin comes with the Germanic adoption in fast construction of the main structural element of the floor with an overlay that delivers precision, abrasion resistance and "absolute" floor flatness criteria, encompassed within the highly theoretical DIN standards.

In the UK there is a determination to produce a middle way in the new Technical Report TR34 that is aimed to satisfy both camps. It nevertheless assumes the introduction of more Codes of Practice, especially for instance in surveying processes that match guidance on mix design [3] and materials use, to help industry and contractors towards better quality production and management. Constant updating of reference material is essential during this period of change. This is especially so in the relevant parts of BS8204 [4] and BS EN206 [5]. The TR34 development project has not been an easy one and there are differences of view even now as to what should be adopted as acceptable design, construction and surface measuring criteria.

BACKING BY ESSENTIAL RESEARCH

Behind the debate, substantial work has been undertaken by notable UK research teams. For instance, the University College of London's project has been investigating edge and corner cracking of steel fibre reinforced slabs. Supported by ACIFC, University of Loughborough has just completed a thorough investigation of the early life behaviour and design and characteristics of concrete industrial slabs [6] [7] [8], an exercise that has engaged both researchers and practitioners in collaboratively making newly constructed floor slabs available for testing and analysis, flooring contractors and clients in particular allowing access to their buildings. The review by the EPSRC assessment panel of the Loughborough programme was clear. The research, in their view, had been particularly worthwhile in that the project had brought industry and academe together and the research results had advanced knowledge that would help industry to obtain direct benefit, impacting directly on the content of Technical Report 34.

Following on this original work, Loughborough University is involved with an equally valid review and research of load transfer across joints. Again the collaboration between Loughborough and ACIFC contractors is being enlivened for on site measurements and joint load-transfer analysis.

Certain practical issues have taken a while to resolve. Inexperience in the specification and use of admixtures to reduce water content caused contractors to resist their use. The concern was the ability to retain maximum uniformity of appearance of surface over often very large areas. With specialist help however practical guidance was developed [10]. This has overcome contractor resistance and more consistent quality achieved as a result.

Abrasion resistance has been a subject that has not caused much difficulty to UK contractors, due in no small measure to the quality controls applied by the concrete producers and to the workmanship of direct finished surfaces. Nevertheless, Aston University has been reviewing the methods available for abrasion resistance testing as it has become an issue where certain types of topping have been added to enhance the surface of the concrete. The need for testing has shown that there are some deficiencies in the existing equipment available.

Some today consider that the concrete industrial floor within a distribution or warehousing development does not warrant such attention. In the last ten years, however, this sector has seen a significant growth within the construction market and it absorbs concrete at a rate well in excess of 1.5 million cubic metres a year. It is no surprise therefore that, in the UK, this is the largest single recorded use of concrete among concrete suppliers.

The sector has also been innovative, not always beneficially as had been hoped. It has invested heavily in production machinery and plant to deliver structure and surface performance with increasing consistency and, to benefit the main contractor with shortening contract lengths, the placement to cure time has been cut in the last decade by factors approaching five.

The need for more research has not been greater, as exemplified by continuing preference backwards to highway paving design methods of Westergaarde and Meyerhof, dating from the 1920s. The need for clarity in approach to design, especially with the variability of soils making up the foundations, is overdue.

COMMERCIAL DEMANDS

The development of the concrete industrial floor has often been decided by quite contrary aspects. The structural integrity of floors designed for UK use has generally relied on a reasonable factor of safety to cover both loading in use and the effects of minor differential settlement. By using structurally suspended steel reinforced concrete ground slabs, the use of mini piles has also allowed a much-increased use of "brown field" or otherwise unusable sites to be considered.

The structural format of ground supported slabs is well accepted, though there has been a trait among the more competitive to offer a thinner "cheaper" slab, having less of the expensive elements of cement and steel. Proponents from the "longer-life" camp press the case for thicker slabs to overcome the less predictable influences in the life of the slab and much more attention to the bearing capacity of the foundation courses.

Still, however, the user of a warehouse is keen to cut maintenance costs, especially avoidable ones such as the degradation of joints due to the continual pounding from mechanical handling traffic. Current wide bay floors, using steel fibres in concrete to reduce the pre-handling difficulties of steel mesh, are moving increasingly to minimum joint construction. In practice that means the use of day construction joints, with saw cut joints made within hours to deal with heat variations and consequent movement during initial curing. Demand is towards jointless slabs though in practice this does not mean a crack free floor. The reality is that the majority of cracks do not threaten the structure's integrity but they look as if they do - so users object.

Figure 2 Detail of a day or construction joint with armoured steel arris to protect against degradation caused by wheeled handling equipment

One counter to the use of the recently developed fast track construction techniques is where the user needs to install very narrow aisle high level picking systems where floor flatness more critically determines both the stability and the speed of warehouse operations. For this very precise narrow strip construction brings the penalty of a proliferation of joints.

PRIMARY CONSTRUCTION TECHNIQUES

There are three essential techniques of concrete slab construction that are in common use in the UK. These are:

1. Long strip, traditional construction or high tolerance, manually spread and levelled.
2. Flood pour manually or mechanically spread and levelled.
3. Large bay, mechanically compacted and spread by Laser Screed.

Long strip has been the traditional construction technique for concrete slabs, developed in the late 1960s that prevailed until the advent of the laser controlled automated screed. Precisely erected formwork enables teams to deliver to a high standard of finish and floor tolerance. This satisfies up to the Superflat category, as set out in The Concrete Society's TR34 Table 7.1. The quality of surface will satisfy the majority of very narrow aisle (VNA) handling requirements and the structural strength is assisted by steel fibre or bar and mesh steel reinforcement.

The disadvantage is that this type of construction does not fit in with the increasing demand of users for minimal joints to cut maintenance costs. This is because the maximum width of strip is no more than 5m, causing the multiplication of vulnerable joints that can degrade or spall. More precise operation can come only from specialised approaches, such as installing steel rail for the VNA machinery.

Flood pour construction has reduced the number of day joints but at the same time this risks reducing the quality of surface that can be achieved. Nevertheless, by this method, TR34 Category 2 floors can be provided given a skilled experienced workteam. This is an essentially manual operation, although mechanised concreting placement is available for the large volume projects, with concrete pumped to the point of spread, poker vibrated, using vibrating floats and striking by straightedge.

The majority of warehouse and distribution depots however do not generally require the highest qualities of flat floor. This is because the user will not always be using high racking and the likely handling equipment is both stable and able to cope with the flatness variations which will be either long wave, a characteristic derived from the operational characteristics of the laser-controlled spreader and direction of the powerfloating, or of such small dimensions as to have negligible effect.

Large bay construction, using Laser Screed compactor and spreader, is the most advanced, most mechanised of the available methods. This is closest to factory just-in-time production. The Laser Screed is a very high output machine, allowing a quality pavement to Category 2, and with special measures to Category 1, of 150mm slab of in the region of 3000 sq metres to be placed day after day, with a much-reduced attendant work force. To add to this technique of fast track production, special dry-shake toppings [10] increase abrasion resistance or provide colour enhancement to be placed mechanically within minutes of the concrete being placed.

Table 1 Method of determining floor surface and flatness requirements from Table 7.1 for defined areas

Category	Location	Allowable limits (mm)							
		Property I		Property II		Property III			
		300mm I Difference in elevation over 300 mm		300mm 300mm II Difference in slope over 600 mm		Wheel track III Wheeltrack up to 1.5 m		Wheeltrack over 1.5 m	
Superflat (SF)	VNA warehouses with minimum clearance between fixed and moving pallets Maximum throughputs, truck speed and permitted rack height	A 0.75	B 1.00	A 1.00	B 1.50	A 1.50	B 2.50	A 2.00	B 3.00
Category 1	VNA warehouses Racking height between 8 m and 13 m Top guided trucks between 13 m and 20 m	1.50	2.50	2.50	3.50	2.50	3.50	3.00	4.50
Category 2	VNA warehouses Racking height less than 8 m AGV's	2.50	4.00	3.25	5.00	3.50	5.00	4.00	6.00

1. Tolerance of level to datum plane (mm): Categories SF and 1 ±10; Category 2 ±15
2. The floor shall be considered satisfactory when
   a) not more than 5% of the total number of measurements exceed the particular property limit in column A, and
   b) none of the measurements exceed the particular property limit in column B
3. See Section 7.7 with reference to remedial action
4. It is not possible to specify and impose these limits for defined movement unless the precise positions of aisles are known before construction.

Table 2 Means of determining for free movement areas from Table 7.2, from technical report TR34

Floor classification	Location and floor use	Maximum permissible limits				
		Property II		Property IV		
		A	B	C	D	E
	Confidence limits	97%	100%	90%	97%	100%
FM1	• Areas of free movement where very strict flatness and levelness may be required. See note.	2.5 mm	4.0 mm	3.0 mm	4.5 mm	7.0 mm
Note: The FM1 limits are extremely onerous and are only likely to be achieved by specialist contractors using strip construction. FM1 should only be specified where recommended by specialist material handling equipment manufacturers or where narrow aisle racking height could exceed 13 m. See Section 5.						
FM2	• Wide aisle warehousing using reach trucks where potential stacking or racking height is **greater than 8 m.** • Transfer aisles for VNA truck use and AGV areas. Note: End user equipment suppliers may require a higher classification.	3.5 mm	5.5 mm	6.0 mm	8.0 mm	12.0 mm
FM3	• Wide aisle warehousing using reach trucks where potential stacking or racking height **up to 8 m.** • Retail warehouses, cash & carry units • Manufacturing facilities Note: End user equipment suppliers may require a higher classification.	5.0 mm	7.5 mm	8.0 mm	10.0 mm	15.0 mm

Property II

Property IV

Notes to Table 7.2R

1. **Property II**

The floor shall be considered satisfactory when, for a particular classification:

a) Not more than 3% of the total number of measurements exceed the limit in column A for property II (97% limit).

b) None of the measurements exceed the limit in column B, property II (100% limit).

2. **Property IV**

The floor shall be considered satisfactory when, for a particular classification:

a) Not more than 10% of the total number of measurements exceed the limit in column C, property IV (90% limit).

b) Not more than 3% of the total number of measurements exceed the limit in column D, property IV (97% limit).

c) None of the measurements exceed the limit in column E, property IV (100% limit).

3. **Tolerance of level to datum plane**

a) Classification FM1 ± 10 mm

b) Classifications FM2, FM3 ± 15 mm

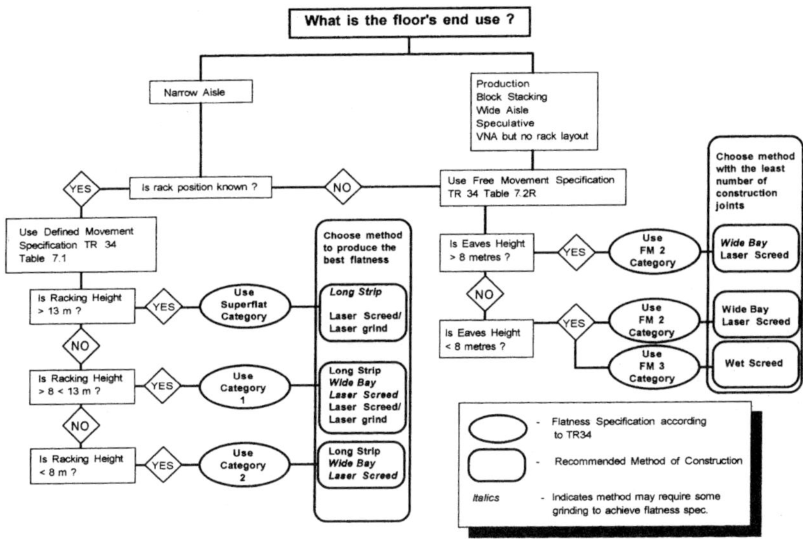

Figure 3 Relationship between floor surface requirements and eaves height of new building – face consultants

ACIFC has access to a useful guidance chart that enables the specifier to match eaves height - the limiting factor of racking systems - to floor tolerances and the main types of slab construction. This shows how direct the relationship is between higher racking and the demand for higher quality, precision surfaces.

MATCHING USERS REQUIREMENTS

Although it is accepted that the ideal floor has no joints if minimal maintenance is to be achieved, there is a cost penalty for this aspiration. Post tensioning for instance provides for specific applications but while others are bar reinforced, the growth is in steel fibre construction and development of jointless floors will continue because it is demand driven.

The last aspect is the "what you see" and the "what you use" after construction. Inevitably the floor surface is seen forever more - and therefore how it is finished has of course attracted all sorts of minds to increase abrasion resistance or to add colours to improve the user environment.

Current construction methods however have tended to cause additional difficulty to the user who is on the limit of the current ability of handling equipment, in particular that of very narrow aisle picking equipment. With these there is little room for variations in the finished surface as they have a direct effect on the stability of the picking unit and crucially the related speed of operation. Floor quality has now reached a plateau in terms of economically obtained quality. It is probably timely for the mechanical handling interests to look to the next stages of suspension development to cope with the new demands.

In the quest for more knowledge in slab construction, the uses of admixtures, steel fibres and the dry-shake topping materials have often caused initial difficulties at the bottom end of the learning curve. Admixtures if incorrectly specified can exacerbate differential concrete set, produce indifferent finishing, and delamination. While with steel fibres, these may stand proud of the finished surface unless, for example, the cement binding of a dry shake topping is worked into the surface of the newly poured concrete. There is one argument that supports a view that the more that is put into concrete the more likely it is that it will bring more problems with them.

Arguably too it is the quality of finishing that puts the value into the floor. While laser controlled screeds can deliver constantly acceptable levels, there is no such aid for the gang sitting on their fleet of counter rotating power trowels. The man on the machine has direct control over quality. And he can cause burnishing or burning by over working the surface. The cause of another potential floor defect, delamination continues to baffle researchers here and in the USA.

Concrete is not forgiving of indifferent workmanship as recent cases have determined. Concrete has an ability to crack, separate, spall, fissure and indeed to wear and dust. It is a measure of the determination of ACIFC Members that it is through its working party structure - and in concert with its supply chain of the concrete producers, suppliers of building products such as admixtures, steel fibres, steel reinforcement, dry-shake toppings, jointing materials, form work, and specialist plant - it can bring attention to recurring construction problems and obtain continuing support to obtain their resolution. This is despite working in a particularly competitive market place.

Figure 4 Sawing joints in a new warehouse floor with 10 m eaves height that will provide 30,000 sq m of storage capacity

The facts nevertheless are these. There is still need to educate those specifying and using concrete industrial floors who have too little appreciation of the floor's contribution to the economics and the life of new development projects.

The answer is to adopt whole life costing. The user must be encouraged to buy a product for life - with all that is implied in long term financing rather than the lowest short term cost that compromises the quality of construction. Until then, mistakes and inaccuracies will continue.

The user must also learn to trust their engineering advice where certain defects are caused by the nature of concrete and will not affect the life or use of the slab. Nevertheless we have the duty and the need to expand our knowledge of both the theory and the practice of how to obtain the best from concrete.

ACKNOWLEDGEMENTS

We acknowledge the assistance of ACIFC members in the preparation of this Paper and particularly Face Consultants for the Table 2 - Floor and Flatness Specification chart. We also acknowledge the permission of The Concrete Society to reproduce from Technical Report TR34 the Table 7.1 and Table 7.2 specification for flatness and tolerances. Photographs were supplied by Stuarts Industrial Flooring Ltd.

REFERENCES

1. THE CONCRETE SOCIETY. Technical Report TR34 Concrete industrial ground floors - a guide to their design and construction. The Concrete Society, Crowthorne, 1994.

2. BRITISH STANDARDS INSTITUTION. BS 8204-1: 1999 Screeds bases and in-situ floorings Part 1: Concrete bases and cement sand leveling screeds to receive floorings – code of practice.

3. ASSOCIATION OF CONCRETE INDUSTRIAL FLOORING CONTRACTORS. Guidance on specification mix design and production of concrete for industrial floors. The Concrete Society. Good Concrete Guide No. 1, 1998.

4. BRITISH STANDARDS INSTITUTION. BS 8204-2: 1999 Screeds bases and in-situ floorings Part 2: Concrete wearing surfaces – Code of practice.

5. BRITISH STANDARDS INSTITUTION. BS EN 206-1: 2000. Concrete. Specification, performance, production and conformity.

6. Behaviour and design of concrete industrial ground floors, Concrete Engineering International, Vol.2, No.8, p.59.

7. AUSTIN, S.A., ROBINS, P.J., BISHOP, J.W., (1998): Behaviour and design of concrete industrial ground floors, Concrete, Vol. 32, No. 6, pp 8-10.

8. AUSTIN, S.A., ROBINS, P.J., BISHOP, J.W., (1999): Techniques for the early-life in-situ monitoring of concrete industrial ground floor slabs. In: Specialist techniques and materials for concrete construction (Eds: Dhir, R.K. and Henderson, N.A, Proceedings of International Conference, Dundee, October 1999) Thomas Telford, London, pp. 317-329.

9. CEMENT ADMIXTURES ASSOCIATION/ASSOCIATION OF CONCRETE INDUSTRIAL FLOORING CONTRACTORS. Specification and use of admixtures for concrete for industrial floors. Concrete Vol. 34, No. 8, September 2000, pp. 35–36. Current Practice Sheet 122.

10. THE CONCRETE SOCIETY. Dry Shake Finishes for Industrial Concrete Floors - An introductory guide. Supplement to Concrete Vol. 35, No 10 November-December 2001.

LARGE AREA POURS FOR SUSPENDED FLOOR CONSTRUCTION

M C Limbachiya

Kingston University

United Kingdom

ABSTRACT. This paper describes a study undertaken to transfer benefits of large area pours (LAP) technology in suspended floor construction. The current perceptions of LAP technology and the obstacles to its increased use within the UK construction industry were identified, through a nation-wide survey and regional meetings with various sectors of the industry. There was general agreement that the use of LAP provides various benefits, leading to an increase in productivity and reduction in construction time. However two key issues, the contractor/ designer relationship and lack of knowledge of benefits, restricted its wider applications in the practice. A case study, containing sections dealing with the structural background, floor construction and a commentary to enable designers and contractors to develop an appreciation of LAP technology rather than apply the technology by rote, was prepared. The implications of adopting LAP techniques in suspended floor construction are also highlighted within the case study. In addition, a simple design approach/flow chart is presented to help designers and contractors to avoid unnecessary restrictions on suspended slab pour areas during concrete placing.

Keywords: Large area pours, Current perceptions, Benefits, Factors affecting pour sizes, Early age thermal cracking.

M C Limbachiya is a Chartered Engineer and Senior Lecturer in the School of Engineering, Faculty of Technology at the Kingston University. His research interests include the recycling and reuse of waste materials, concrete technology and construction, and repair and maintenance.

INTRODUCTION

The floor construction in new industrial or residential developments is one of the most important features from the point of view of its serviceability and appearance. Recently it is widely recognised that the practices of concreting large areas in a single pour, say greater than 200m², can technically improve both productivity and quality of floors. For example, the use of large area pours (LAP) in the construction of suspended slabs can produce superior floors, reduced construction joints, potential for efficient formwork usage and increased productivity of plants, labour and materials [1,2]. Collectively, these can also lead to a reduction in total construction cost. Indeed, current techniques such as pumping, wide-bay finishing, laser screed systems and sprayed curing membranes can allow large areas of concrete to be placed and cured with relative ease. However, restrictions on pour areas still abound, indicating lack of knowledge of benefits and need to encourage greater use. In fact, many specifications limit the pour size either in area or by a maximum linear dimension.

A research commissioned by the Reinforced Concrete Council (RCC) [3] concluded that large pours actually reduce areas of concrete floor subject to potential early thermal cracking. Furthermore, a guide published by the RCC [1] has demonstrated control of early thermal movement by simple reinforcement arrangements and it gives an empirical formula to help engineers to assess risk of cracking. It has shown that the layout of pours, with respect to locations of permanent restraints such as service cores, walls and joints, is of significant importance. The guide shows, through work examples, how to reduce the areas affected by the restraints through selection of fewer large pours rather than a series of conventional small pours. Furthermore, it shows how to reinforce restrained areas to control cracking to an acceptable limit. Hence, any perceived problem with early-age thermal cracking is reduced and the benefits of LAP can be taken in suspended floor construction. The paper reports the finding of a study undertaken, at Dundee University in association with the British Cement Association, to promote technical, quality and economic benefits associated with the use of LAP and to encourage uptake of the technology. Reported work was a part of a technology transfer programme, aimed at examining the factors that inhibit the uptake of most appropriate/ under-utilised technologies and techniques within the UK construction industry.

CURRENT PERCEPTIONS OF LAP

Regional Meetings and Nation-wide Survey

As a starting point a series of meetings, with senior professionals of the main consultants, contractors, architects and concrete suppliers, were held at different venues across the UK to establish industry views and attitude to the use of LAP technology and to discuss reasons for reluctance in using LAP. Discussion from these meetings was then used to prepare a questionnaire for nation-wide survey to determine circumstances when large pours are currently used, reasons for any bias against them as a concept, and how their wider use can be encouraged. In total, 1100 copies of a questionnaire were circulated to targeted consulting engineers, designers, contractors, local authorities and architectures throughout the UK. Of these, over 200 industry representatives from all sectors of the industry responded.

Feedback from the regional meetings and responses to the subsequent questionnaire suggested that the majority of industry is convinced by the benefits of LAP, particularly those involved in design and build type projects. Moreover, there was broad agreement that various benefits can be achieved by using LAP in suspended floor construction. Among these reduction in overall construction time and plant hire duration was considered to be the most important. The ranking of technical, construction and economic benefits are shown plotted in Figure 1. The main criterion for the designers/consulting engineers' acceptance of LAP technique was the demonstrated ability of contractors to carry out such work efficiently. The logistics of the operation in terms of available concrete placing resources, finishing tolerances, changing weather, and site layout were also identified as of significance. It was also quite clear that the consultants' were concerned about the liability issues in many cases and this in turn prevented more specifications allowing construction using LAP technique. The circumstances under which the responding contractors, consultants and local authority representatives would not consider the use of large pours were vary as shown in Figure 2.

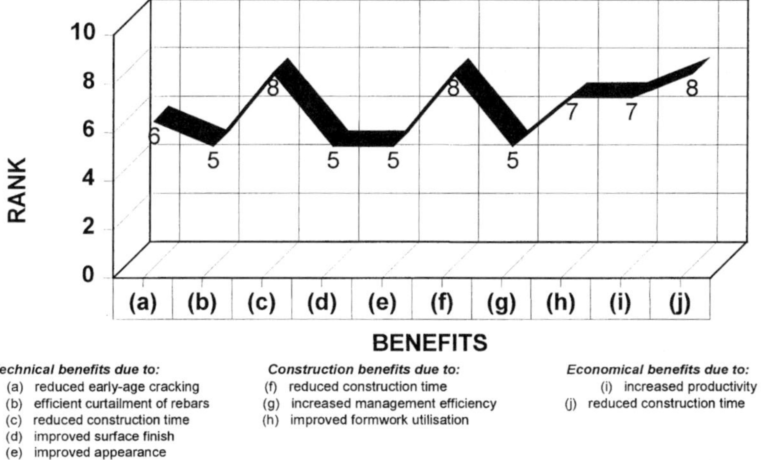

Figure 1 Benefits that can be achieved by using LAP in suspended floor construction

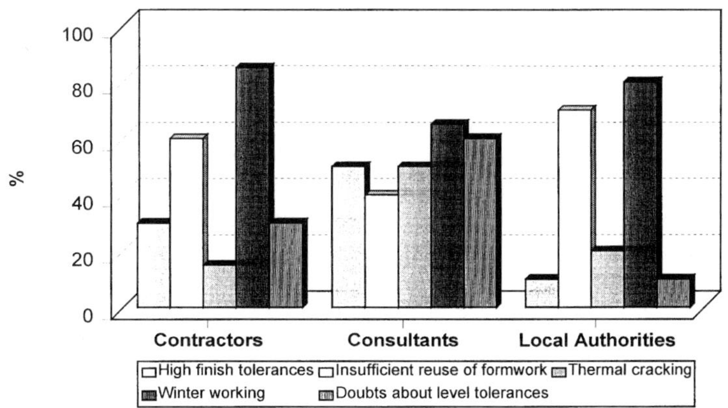

Figure 2 Circumstances under which respondents were reluctant to use LAP

The majority of respondents indicated that an early discussion between the specifiers and contractors, especially at the tender appraisal stage, would encourage LAP usage in suspended slab construction. It was noticed that such discussions could provide solutions to many of the specifier's concerns. It was clear that the authoritative guidance document and case studies are also vital to encourage greater use of LAP technology in practice. The second phase of work therefore proceed to prepare detailed case studies.

CASE STUDY

A case study based on a real project has been presented, as a typical example, for assessing the implications of adopting the LAP technique in suspended floor construction. It contains sections dealing with the structural background, concrete placement and floor construction, and a commentary to enable designers, contractors and clients to develop an appreciation of the LAP technology rather than apply the technique by rote.

The resulting benefits and limitations of the technology were examined and the existing criteria used for determining conventional pour sizes (≤ 100 m^2) and the needs for pour size restriction are questioned. Additional case studies, dealing with a suspended ground floor slabs for retail parks, have been reported in related work elsewhere [4,5].

Structural Background

The chosen structure was the construction of a multi-storey shopping complex with a floor area of about 4800 m^2, Figure 3 shows a typical floor plan of this complex, providing facilities for the high street retail store and various office units to suit client's requirements. The main structure was a reinforced concrete construction and it consists of 4½ two-way suspended slabs directly supported by columns with capital. It had four integral insitu concrete stairs and two lift cores. The support beams were eliminated altogether by selecting suitable proportions of reinforcement.

The slab system included centrally located atrium of size 19.2 x 9.6 m supported by edge beams of dimension 1.0 x 0.45 m, Figure 3. In addition, 2 tower crane openings of dimensions 2.8 m x 2.8 m and three concrete placing boom voids of size of 0.94 x 0.94 m were provided, and the slabs were locally strengthened to accommodate these openings with diagonal bars to control the cracking. On completion of the job, these openings were filled using either pre-cast unit or caging unit on continuity beam and/or pullout bars.

Suspended floor slabs were in-situ concrete on plywood soffit formwork. The flat soffit of the slab also made it possible to use table form falsework system (up to 9.6x2.7m size, Figure 4) to ensure a fast construction programme. In addition, some static formwork was used at edges and corners. The finish surface of the floor required being adequate for general industrial use to give a satisfactory wearing surface.

At an early stage of the project, discussions were held between engineer, designer and frame contractor to decide the structural form and floor construction. The contractor initiated the discussion and others were willing to consider variations required, accommodating large pours. The pour sizes and sequences were decided on the basis of available resources and accessibility on the site.

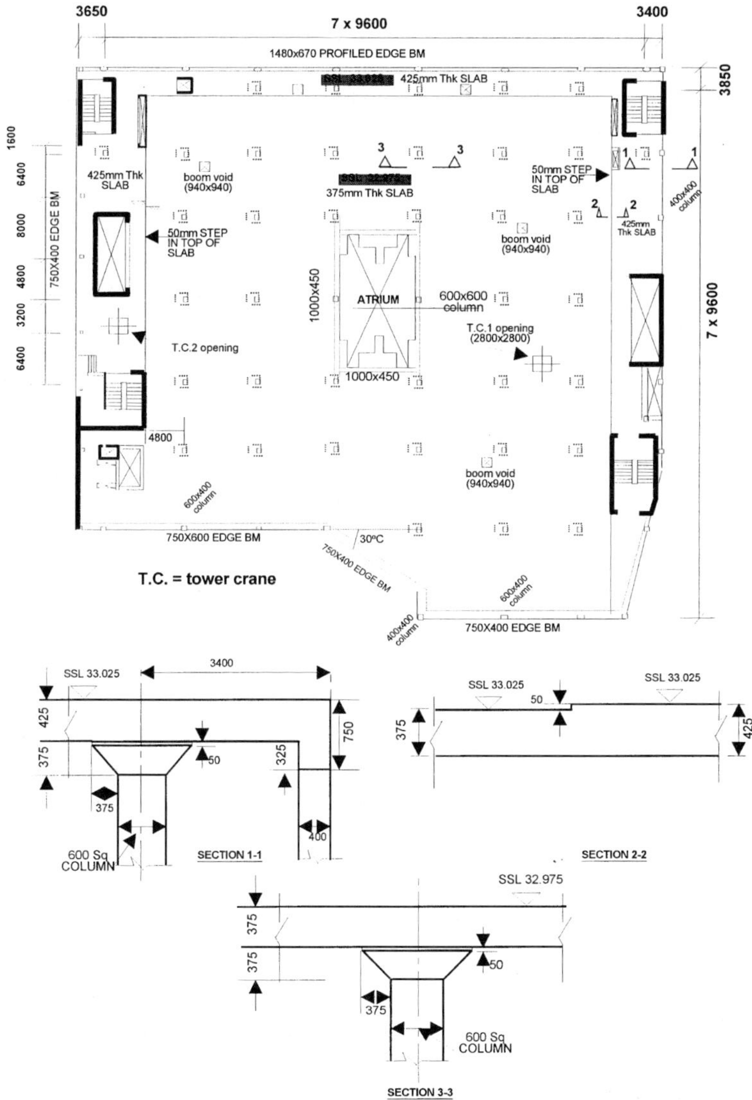

Figure 3 Typical floor plan of a multi-storey shopping complex (≈ 4800 m²)
(based on engineering drawing of O'Rourke Scotland Ltd)

Concrete Placement

The concrete was placed in 6 large pours as shown in Figure 5 with construction joints up to about 39 m long and smallest being 17 m long. The pour sizes/ sequences were designed based on the available of concrete supplier, cranes and other facilities on the site. The pour layouts were selected to minimise two-way stress in the floor. The ready mixed concrete was pumped via placing booms to the point of placing, at a rate of 30-36 m³ per hour. On average, in excess of 250 m³ was placed in a day to meet tight construction programme. Slabs were power floated.

Design Information

- Reinforced Concrete Columns — 0.6 x 0.6 m square columns with capital, 9,6 m centres; 0.6 x 0.4 and 0.4 x0.4 m on edges. Type 2 bars.
- Edge Beams — 0.75 x 0.60 m, 0.75 x 0.40 m, 1.0 x 0.45 m.
- Floor Slab — 375 mm thick reinforced concrete flat slabs with normal weight PC concrete.
- Loading — Typically 4.0 kN/m² plus 1.0 kN/m² for partitions.
- Tolerance — ± 5 mm in 3 m straight edge and ± 10 mm from the datum.
- Concrete — Grade C40, minimum cement content = 325 kg/m³, maximum free water/cement ratio = 0.55 by weight, maximum graded aggregate size = 20 mm, initial drying shrinkage < 0.045%.

Figure 4 Table form falsework system for fast construction process

The mix specification was changed to C40 pump mix with Portland cement (PC) content of 375 kg/m³, free w/c ratio of 0.50 and workability at high range of 75 mm nominal slump. The increase in cement content was required due to the properties of aggregates (gravel) used in concrete production. Water-reducing admixture was used to give additional pumpability. The mix was designed to comply with the requirements of BS5328. Slump tests were carried out on the site before concrete placement and cube samples were prepared daily to check compressive strength. Spraying and covering with polythene sheeting achieved the curing. The striking of formwork was carried out after three days and then cleaned and prepared for further use. Concrete finishing was commenced after 24 hours of the concrete placement in its final state. The cycle time per floor was about five to six weeks.

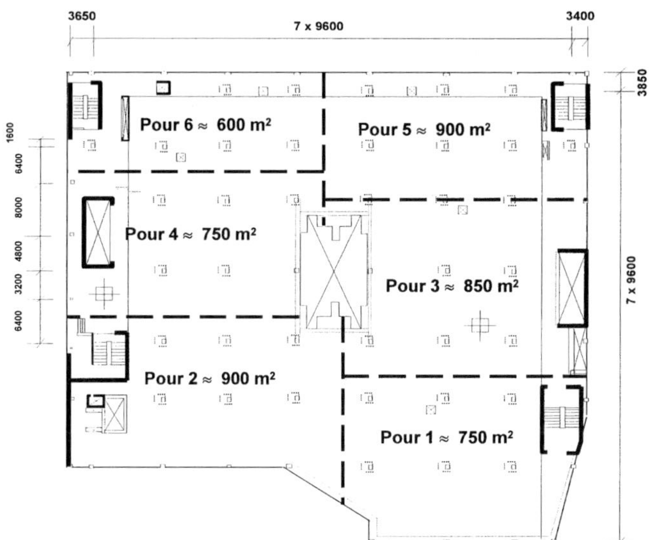

Figure 5 Construction of suspended slabs (375mm thick) in six large pours
(based on engineering drawing of O'Rourke Scotland Ltd)

Comparison Between Conventional Small and Large Area Pours

Figure 6 schematically illustrates how the casting concrete in large pours reduces the area of slab subject to high restraint and as a result potential cracking, compared to the same slab case in 13 conventional small pours of around 120 m². The directions and distributions of restraint are based on the assumption that the area of significant stressed is confined to a 45° triangle, mainly because this gives an adequate representation for the most design purposes [1]. On the whole, assessing the distribution of restraint requires engineering judgement rather than complicated analysis, and the 45°-distribution pattern from an edge restraint has been generally accepted in the practice. If restrained tensile strain induced during the period of cooling from the peak to ambient temperature is greater than the tensile strain capacity of the concrete, then thermal cracking occurs [6], i.e. if :

$$T_1 \ \alpha \ k \ R > \xi_{ult}$$

Where,

- T_1 = the drop between peak temp. after casting and ambient temp.,
- α = coefficient of thermal expansion, per °C,
- k = modification factor,
- R = restraint factor,
- ξ_{ult} = ultimate tensile strain capacity of concrete.

As can be seen from Figure 6, decreasing number of sequential pours leads to a significant reduction in the area of slab subject to high restraint and potential cracking.

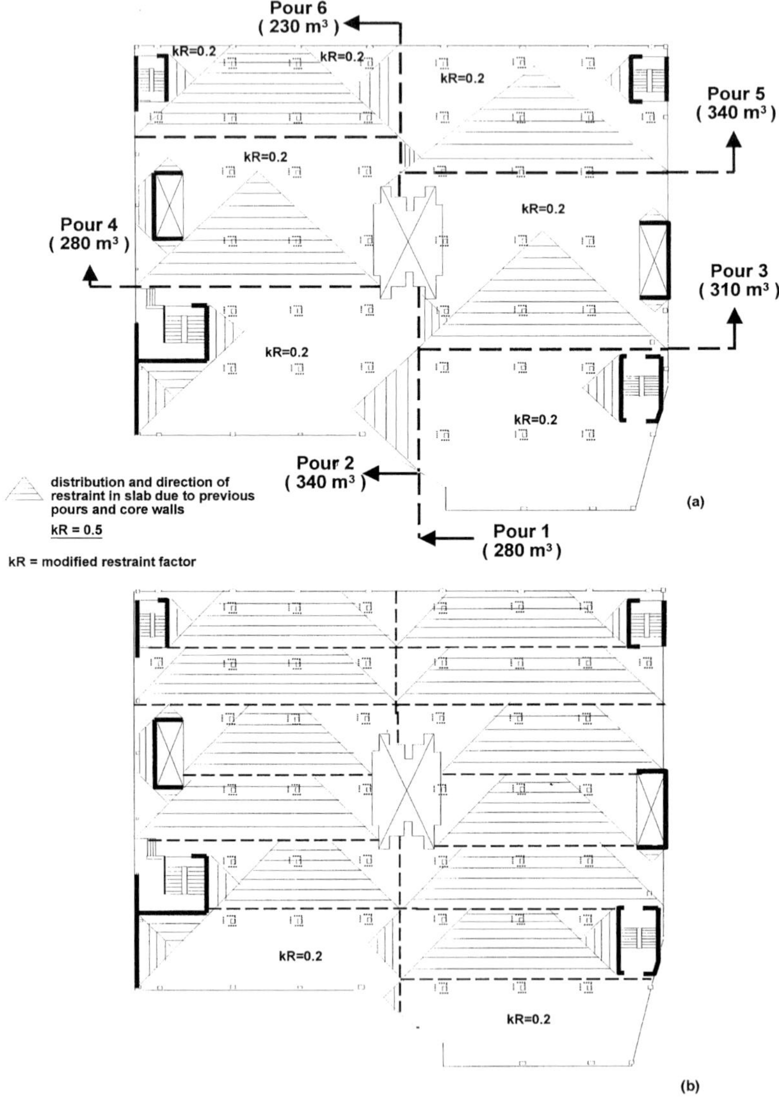

Figure 6 Distribution and direction of restraint in (a) selected large pour sequences, and (b) slabs casted in conventional thirteen small pours

Flow Chart for LAP

The feedback from site meetings and consultation with designers and contractors were then used to develop a simplified approach in adopting LAP for suspended floor construction and is shown as a flow chart in Figure 7.

In the main this approach is divided into 4 different stages. These include (i) initial planning for determination of pour sizes and sequences, (ii) assessment of a feasibility of using LAP technology, (iii) checks on early age thermal cracking and reinforcement details to control cracking, and (iv) implementation of LAP technique. The proposed approach is simple and follows the natural course of events and therefore can be integrated into current practice easily.

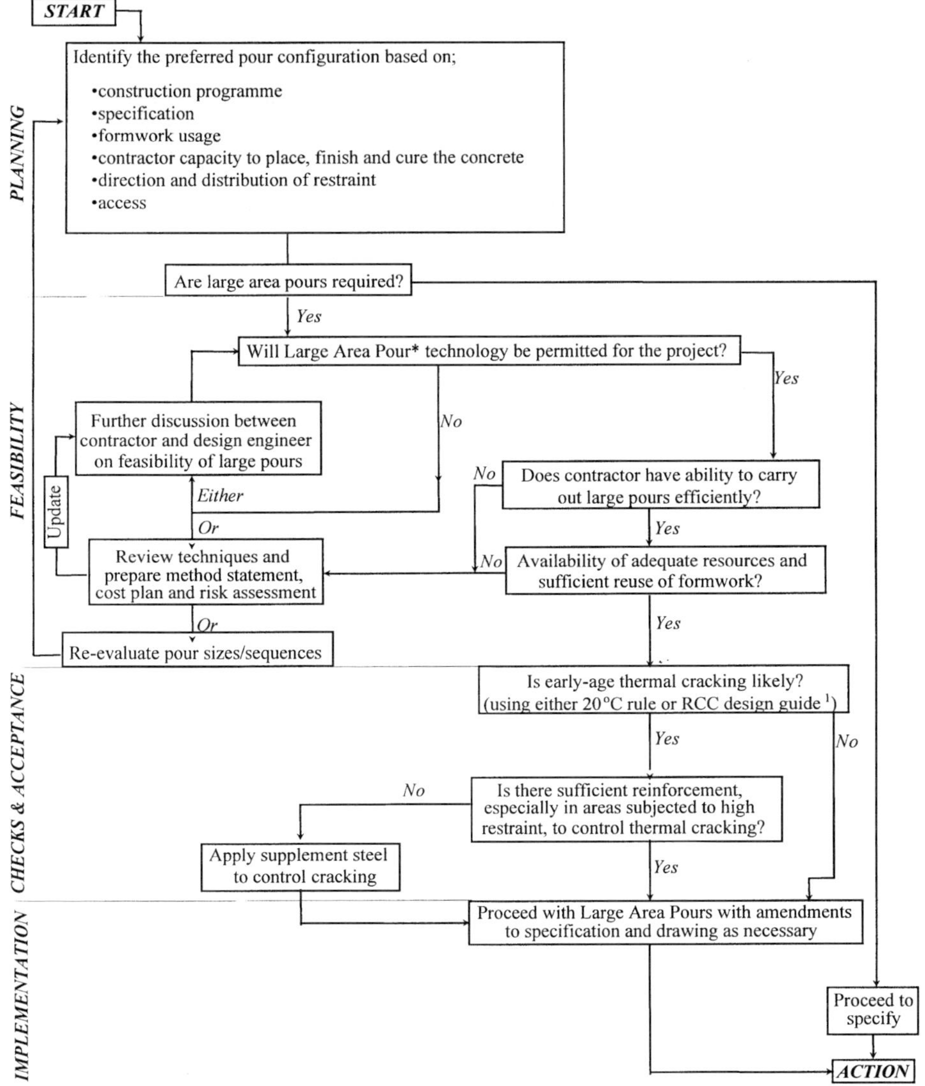

Figure 7 Simplified flow chart for the selection of LAP in the construction of suspended slabs *(* LAP are generally regarded as slab pours in > 200m²)*

CONCLUSIONS

1. The majority of specifiers and contractors had no objection in recommending the use of LAP in suspended floor construction. However, the main criteria for adopting the technique would be proven ability of the contractor to undertake such work efficiently and discussion at the tender appraisal point.
2. Lack of perceived benefits to specifier and ability of contractors to undertake LAP construction effectively are found to be main reasons for the bias against the use of LAP in many cases.
3. On the whole, industry representatives are convinced that LAP can provide improved quality of surface finish, optimum re-use of formwork, reduction in early-age thermal cracking and reduction in potential for construction error, collectively leading to an increase in productivity and a reduction in total cost.
4. The questionnaire response and meetings with industry made it clear that a cost plan, method statement and risk assessment for adopting LAP technique may be required to convince specifiers and end users.

ACKNOWLEDGEMENTS

This work was funded by DETR and British Cement Association and was carried out at the Concrete Technology Unit(CTU), University of Dundee. The CTU Director, Professor R K Dhir is acknowledged for allowing the work to be published. The author would also like to thank Mr J F Reid (BRE Scottish Laboratory), Dr J P Cornish, Dr J E Gibbons and Mr D Stone (Scottish Office-now Scottish Executive) during the project. Special thanks also go to Messrs C Goodchild of Reinforced Concrete Council and S Barber of O'Rourke Scotland Ltd for the advice and providing information on case study respectively.

REFERENCES

1. REINFORCED CONCRETE COUNCIL. Large Area Pours for Suspended Slabs- a Design Guide. Reinforcing Links, Issue 3. Crowthrone, BCA, 1993. pp 8.

2. BUTLER, A. A Specialist Contractor View of Concrete Frames. Concrete. Vol. 30, No.6, November/December 1996. pp 17-18.

3. REINFORCED CONCRETE COUNCIL. Large Area Pours for Suspended Concrete Slabs- C/12. Slough, BCA 1992. pp 72.

4. SPOONER, D C, DHIR, R and LIMBACHIYA, M C. Targeted Technology Transfer. DoE Report BCA/CTU/296, September 1996. pp 53.

5. SPOONER, D C, DHIR, R and LIMBACHIYA, M C. Targeted Technology Transfer. DoE Report BCA/CTU/397, April 1997. pp 65.

6. HARRISON, T A. Early-age Thermal Crack Control in Concrete, CIRIA Report 91, London, 1993

DESIGN OF MECHANICAL-PHYSICAL PROPERTIES OF LAID FLOORS WITH CONSIDERATIONS OF THE NATURE OF INTRODUCED FILLERS AND ADMIXTURES

L B Svatovskaya **V I Kovalev**
N N Shangina **D V Gerchin**
V U Shangin **V Y Solovieva**
L L Maslennikova **P G Komokhov**

V D Martynova
Railway University
Russia

ABSTRACT. This work considers the cement matrix composition cover, the properties of which are controlled both by the introduction of admixtures and fillers and by creation of an inorganic surface by sol-gel technology.

Keywords: Laid floor, Flow, Water retain, Nonmetal fiber, Nature, Hydroxide, Oxide, Portland cement, Sol, Gel, Silicon dioxide.

Professor Larisa B Svatovskaya is the Head of Engineering Chemistry Department, Railway University, St Petersburg, Russia specializing in the chemistry of binders.

Professor Valery I Kovalev is the Chancellor and **Mrs Valentina D Martynova** the Vice Chancellor of the Railway University, St Petersburg, Russia.

Professor Nina N Shangina and **Professor Pavel G Komokhov** are Lecturers in Department of Building Materials, Railway University, St Petersburg, Russia with an interest in binders, concretes and building materials.

Mr Denis V Gerchin is a Post-graduate in the Engineering Chemistry Department, Railway University, St Petersburg, Russia. His areas of scientific interest are dry mixes and concretes.

Dr Vladimir U Shangin, Dr Valentina Y Solovieva and **Dr Ludmila L Maslennikova** are Lecturers in the Engineering Chemistry Department, Railway University, St Petersburg, Russia. They specialize in concrete, dry mixes and clay materials.

INTRODUCTION

In accordance with earlier work at the Railway University of St. Petersburg [1, 2, 3] we consider the creation of flow covers (compositions for the presence of laid floors) based on cement matrix. Various cement compositions, were used, giving the following test parameters:

- ratio between square of material's surface and material's volume;
- viscosity;
- density;

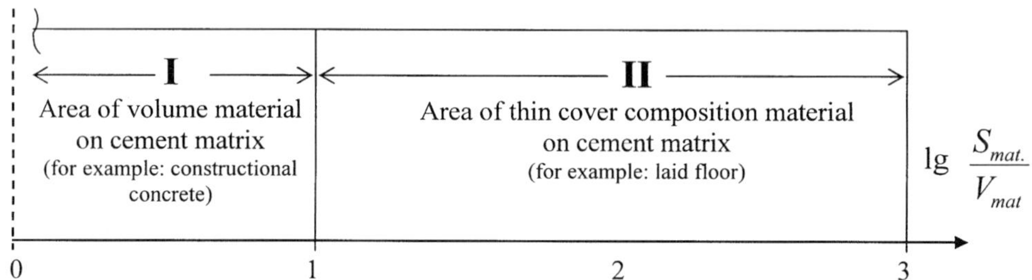

Figure 1 Two groups of the materials on cement matrix ($S_{material}$ – square of all surfase of the material, $V_{material}$ – volume of the material).

Figure 1 shows the relative separation on parts of exploitation and the area II where the parameter lg ($S_{material}/V_{material}$) belongs to the interval [1; 3] – is of interest in this research. The area I – is traditional exploitation of materials on cement matrix in building [4]; area II has been researched very poorly for use of mainly inorganic phases based on mineral materials, except for latex using. Proceeding from that, let's control the mixture's mobility by introduction of fillers. For crack resistance let's introduce fibers by silicate of alumina of another nature. For to increase hardness and abrasive firmness of the single-layer cover let's make the second silica-, alumina- or silicate of alumina layer by the sol-gel technology. But how the mixture's mobility (for material belongs the area II) depends from introduced fillers and admixtures – it has no knowledge in present.

We devote the one part of our work for creation of laid floors and the next one – for creation of super cover on the surface of laid floor.

EXPERIMENTAL DETAILS

Materials

The materials used were: Portland cement M600, M500, M400, sand (fractions 0.14 → 0.35; 0 → 1.25; 0 → 1.6 mm, dust content less than 5%) and a plasticizer admixture MIX-P, created by Department of Engineering Chemistry and Environment Protection, Railway University, St Petersburg, Russia [7, 8, 9].

Methods

The mixture's spread on the flow table was measured by Suttard's viscosimeter. The water-cement ratio is constant for every experiment.

RESULTS AND DISCUSSION

Figures 2 – 6 show the dependence relationship for a mixture's mobility and show that mobility may be controlled by introduction of hard materials of some types of nature. The conclusion is to use in the mixture's structure the cement M 600 and sand fraction 0 → 1.2 mm approximately.

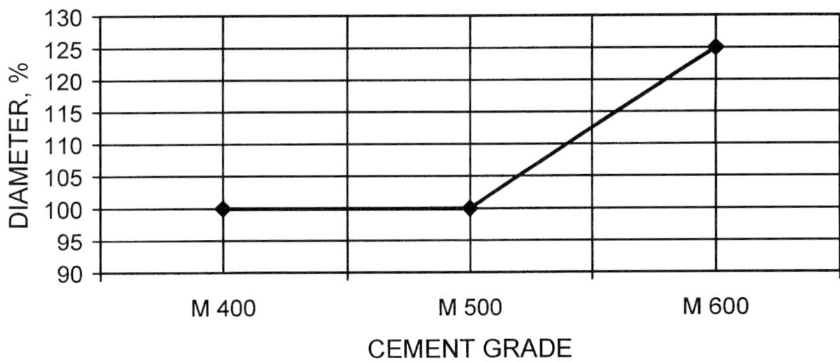

Figure 2 Relationship between diameter of spread of a mixture on a flow table and cement grade

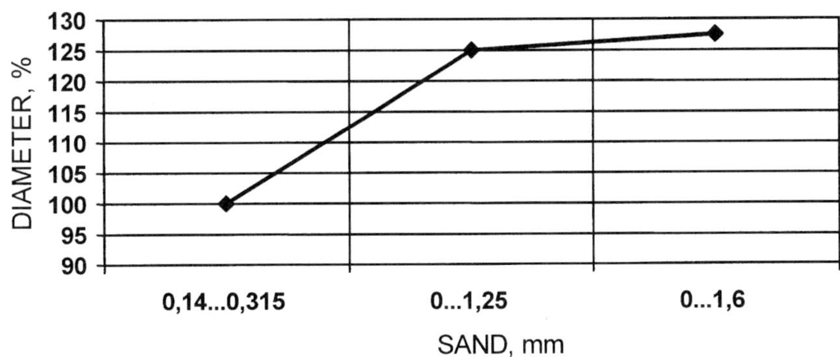

Figure 3 Relationship between diameter of spread of a mixture on a flow table and sands fraction

To use the sand with fraction more than 1.2 mm is not expedient for increasing the diameter of spread of the mixture on a flow table. Using sand fractions less than 1.2 reduces the mixture's mobility.

Here, if use the sand of recommended fraction 0…1.2 mm, we can increase diameter of mixture's spread by 25%.

2% of calcium hydroxide ($Ca(OH)_2$), introduced in a mixture, increased the diameter of the mixture's spread by 20% also (Figure 4) with all other conditions equal.

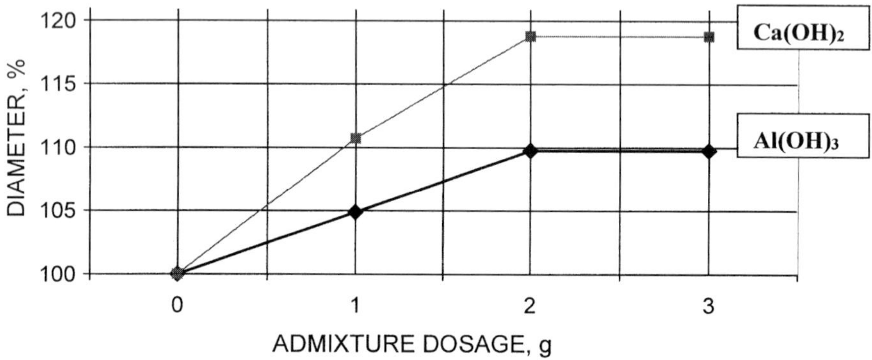

Figure 4 Influence of hydroxides introduced in a mixture of laid floor

Admixtures of oxides and salts are different in their influence on a mixtures mobility (Figure 5) – oxides p- and d-block elements of metals reduce mobility, but salts of s-block elements of metals increase and here we can "find" approximately 20% of mobility increase.

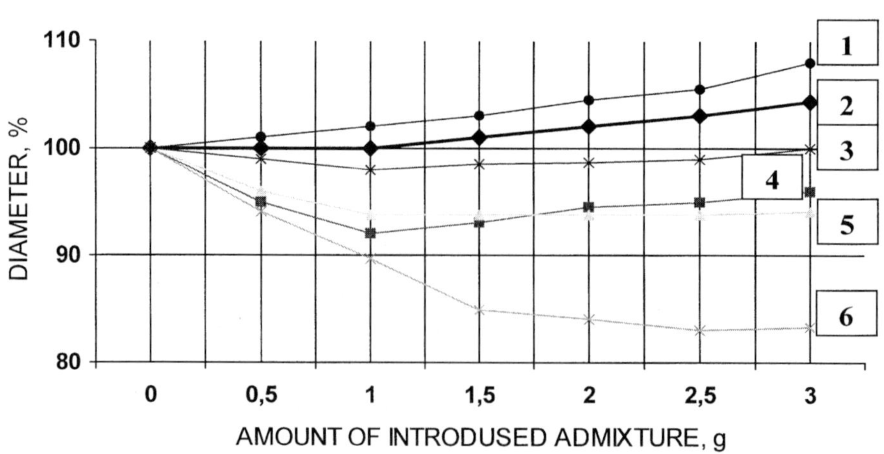

Figure 5 Influence of different admixtures introduced in a mixture of laid floor
1 – $CaCO_3$; 2 - $BaSO_4$; 3 - Pb_2O; 4 - Cu_2O; 5 - CuO; 6 – PbO;

Such influence of salts s-elements may be explained by the construction of electron's cover of ions: s-cover (Ca^{2+}, Ba^{2+}) has an sphere symmetry and has no space orientation and so it has more "degrees of freedom" when compared to p-block elements (PbO, PbO_2, $Al(OH)_3$) and oxides of d-block elements (CuO, Cu_2O), which are oriented in space.

In addition, the energy of p- and d-electron levels is higher than s-electron level, so they hold water more strongly with different oxidation degrees of the same p-block element. It should be noted the influence of low degrees of oxidation. The low degrees of oxidation means a reduce activity of materials, and it influences on a mixture's mobility. The conclusion is confirmed on Figure 6: if the breadth of forbidden zone is high – the mixture's mobility is high too.

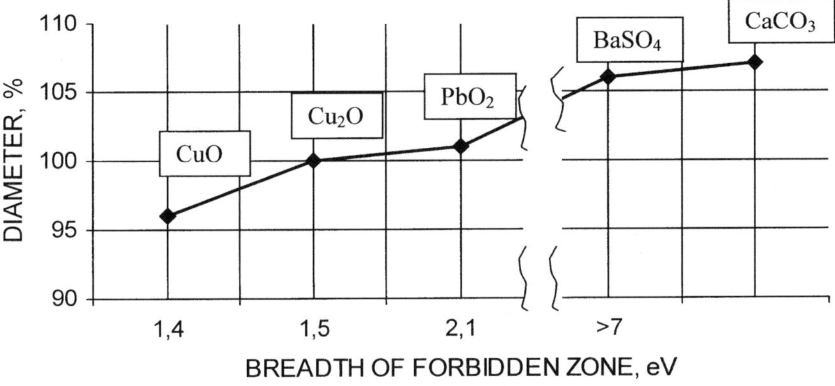

Figure 6 Influence of the breadth of forbidden zone

One of nature's most popular salts – dielectrics is $CaCO_3$ and it is potentially useful for increasing mixture's mobility. Tests show that it increases the mixture's spread on a flow table by approximately 10%.

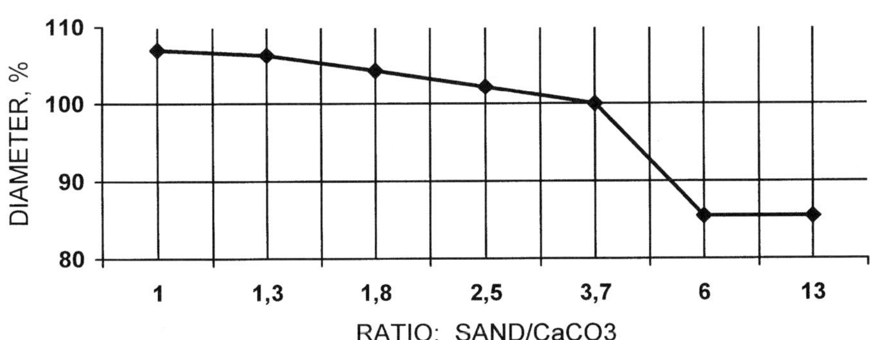

Figure 7 Relationship between diameter of spread of mixture on a flow table and ratio sand/$CaCO_3$

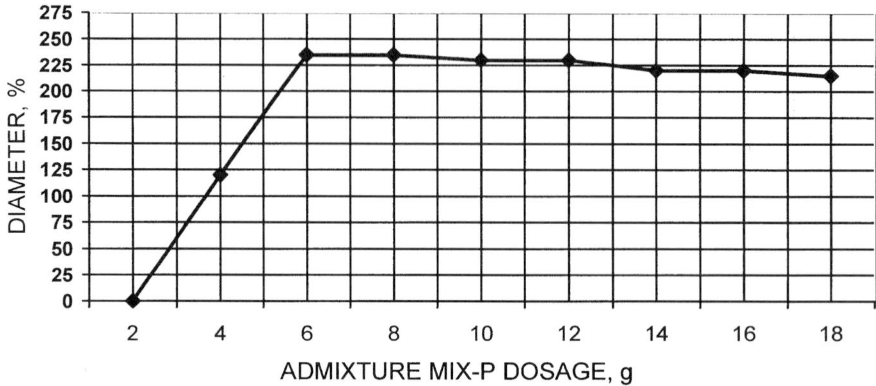

Figure 8 Relationship between diameter of spread of mixture on a flow table and admixture MIX-P

Figure 8 shows the relationship between diameter of spread of mixture on a flow table and amount of introduced admixture MIX-P with polysucroses [9]. This increases the mixture's spread by more than 200%.

A conclusion can be made, that by using plasticizer admixtures, to create a mixture's structure which consists of s-block elements, sand fraction, used cement grade, water-cement ratio, it is possible to increase the mixture's mobility for all other conditions being equal.

In the sum it may be to increase the mixture's mobility 300%. And it is possible to increase the crack resistance of the laid floor, because the system is chemically similar and it creates the system of bonds of cement matrix with introduced dispersion phases, which creates contacts at the phase separation boundary as hydro-carbonate-silicate and aluminates. It is possible to increase the crack resistance of non-polymer inorganic laid floor on a cement matrix by introduction of vollastonit or basalt fibers also.

The exploitation characteristics of material may be forced by introduction of another type of admixtures. It's sol-gel admixtures, which may be introduced, for example, by the technology of super cover of laid floor. It will be double-layer inorganic cover on cement-matrix base in that case. This technology may be used for creation of high hardness and acid resistance. The super cover manufactures from soles SiO_2 or Pl_2O_3 – materials, which have high hardness on Mohs scale in compare to salts hydrosilicates, for example.

That sols are stable when $pH<7$, when $pH>7$ that sols became neutral with creation of gels, which construct a thin super cover, for example – silicon dioxide. It is the way to increase the acid resistance. The hardness of the cover may be increase 40% also.

Inorganic covers are ecologically pure, they does not include polymers. The system has high stability in time. It has enough thermodynamic reserve [5, 6] both by hydration of cement and by creation of inorganic sew at the phase separation boundary. Laid floors on developed mixtures were made and their high physical-mechanical properties have been occurred.

CONCLUSIONS

1. It is shown the influence of s- p- and d-block elements on the characteristics of the inorganic covers based on cement matrix (mixture's mobility, for example). The optimal parameters of dispersion of some introduced fillers were found also.

2. It is shown the way to increase the hardness and acid resistance of laid floors by creation the super cover manufactured by sol-gel technology.

3. Shown covers are inorganic, they does not include polymers. The thermodynamic reserve of system makes better their characteristics in time.

REFERENCES

1. SVATOVSKAYA L B, Development of engineering-chemistry bases of manufacturing and characteristics of binders on Third Century, Modern engineering-chemical bases of building materials, Railway University, St Petersburg, 1999, pp 4-11 (in Russian).

2. SVATOVSKAYA L B, To climbing, Modern natural-scientific bases of building materials and ecology, Railway University, St Petersburg, 2000, pp 3-6 (in Russian).

3. SVATOVSKAYA L B, Properties of engineering chemistry in building materials and ecology, New researching in building materials and ecology, Railway University, St Petersburg, 2001, pp 5-9 (in Russian).

4. NEVILLE A M, Properties of concrete, Moscow, Stroyizdat, 1985 (in Russian).

5. SVATOVSKAYA L B, Hydration properties of cements activated by fluorine anhydrate, Cement and it's application, 1986, №1, pp 10-13 (in Russian).

6. SVATOVSKAYA L B, Thermodynamic and electronic level of strength reserves of cement materials, Bulletin of VUZ (Higher Education School), "Construction" Series, №8, 1998, pp 35-40 (in Russian).

7. SVATOVSKAYA L B, SHANGINA N N, KOMOKHOV P G, SHANGIN V U, GERCHIN D V, Using of domestic admixtures in creation of dry mixes, Modern natural-scientific bases of building materials and ecology, Railway University, St Petersburg, 2000, pp 16-22 (in Russian).

8. SVATOVSKAYA L B, SHANGINA N N, KOMOKHOV P G, SHANGIN V U, GERCHIN D V, Domestic admixtures for dry mixes and buildings mortars, New researches in building materials and ecology, Railway University, St Petersburg, 2001, pp 5-10 (in Russian).

9. SVATOVSKAYA L B, SHANGINA N N, KOMOKHOV P G, SHANGIN V U, GERCHIN D V, Domestic admixtures MIX-types, Railway University, St Petersburg, 2001, p 10 (in Russian).

METHOD OF CONSTRUCTING BURNISH FINISH INDUSTRIAL FLOOR SLABS IN AUSTRALIA

P Ashford

University of Melbourne

Australia

ABSTRACT Client expectations for a high quality, hard-wearing concrete industrial floor slab with long term performance and minimal maintenance costs has become the norm Both designers and contractors require specialised expertise to achieve these results The quality of concrete floors has improved dramatically as a result of technological advances in concrete supply, placement methods, the burnish finishing process, curing and joint sealants. Reduced construction time from larger pour areas with less joints and high quality sealants has improved the life cycle costing of many industrial floor slabs. More stringent control on surface tolerances and finish, together with the requirement to attain a relatively crack free floor have necessitated tighter quality control over preparatory and concrete trowelling works particularly with burnish finishes. This paper reviews the most current practices in Australia and the different floor pouring/placement methods associated with fabric reinforced, steel fibre reinforced and post-tensioned designs. The size of the pour areas and shape thereof often predetermine the choice of both the vibrating screed equipment and trowelling equipment necessary to achieve the slab finish required whilst maintaining surface tolerances. In order to improve the quality of industrial floors it has been necessary to establish working relationships between engineers, developers and particularly subcontractors. This close relationship allows communication between the parties and provides a mutual understanding of all the issues associated with the design and construction methods to be employed.

Keywords: High performance slabs, Burnish finish, Surface tolerances, Vibrating screeds, Scraping straight edges, Bump cutters, Laser screeds, Panfloats, Ride on helicopters, Flatness and Levelness.

P T Ashford is a fellow of the Australian Institute of Building and lecturer in Construction Technology in the Faculty of Architecture, Building and Planning at the University of Melbourne. His research focuses mainly on the design and construction techniques of high-performance industrial floor slabs.

INTRODUCTION

The paper compares the various techniques adopted for the construction of industrial ground slabs. The various pouring methods, levelling techniques, burnish finishes and slab specifications used in Australia are briefly discussed and comparisons made through Six Case Studies. The case studies emphasise the floor laying, levelling techniques and the degree of burnish finish achieved for three pour area types, namely, long strip pours, pours with double width strips with removable rails and large area pours. Methods for alignment and levelling of the formwork and the accuracy required to achieve a very flat surface finish are also mentioned

The case studies represent a variety of pour sizes and shapes and indicate the selection of equipment used by a range of concrete contractors who specialise in a particular type of pour area type. The paper concludes with a "Table of Characteristics and Comparisons" with an indication of future directions, developments and trends.

Predetermined Pouring Methods

The engineer's design has generally been completed well before any contracts have been let for the construction phase, and the documentation of joint types together with the layout has, therefore, often predetermined the pour area and method of construction. This layout and area may not be the most efficient for the successful contractor's time schedules and preparatory works, nor the most efficient for the contractor's workforce and concreting equipment. Pour widths, pour lengths, and the set up of reinforcement once documented and detailed, usually predetermine the concrete placement method. After discussion and negotiations, a re-design of pour sizes and shape may be required for a cost-effective solution for pouring and finishing the slabs.

Typical low slump, low shrinkage concrete requires placement directly from the truck. If the reinforcement has to be chaired up and inspected prior to the pour, direct placement is not possible. The construction method must change so that as the truck moves forward, previously prepared plastic membrane is placed on the subbase and the reinforcement fabric is placed on chairs progressively behind the truck to complete the pour. The final position of reinforcement, cover and lapping in this process can be suspect. Steel fibre reinforced slabs offer the advantage of direct chute concrete placement irrespective of the pour plan shape.

Concrete pumps with placement booms provide an efficient method of placing concrete for both post-tensioned and steel reinforced slabs when the reinforcing method has been set up prior to the pour. With limited roof heights, the horizontal reach of placement booms is limited and this may require the pumps to be set up at several points around the perimeter of the pour so that the placement boom can reach all areas within the pour.

Established Levelling Techniques

The plan shape and size of the pour areas has a major impact on the levelling techniques used during construction. The traditional "Long Strip" layout requires accurate set up and levelling of the edge formwork initially on both long sides of the strip.

Hand screeding and mechanical beam type vibrating screeds supported on the two edge forms can be used to cut and level the concrete surface. Double width long strips with a removable intermediate rail are more efficient since the pour area can be doubled and construction time thereby reduced.

Large area pours with dimensions in excess of 25 m (approx) in each direction can be levelled by taking spot levels (optical or laser) on the concrete surface at regular grids (up to 3.6 m) and then power floated between the spot levels. Floor tolerances expected with this method of construction are perhaps + or – 4 mm in 3.0 m at best. The alternative would be to use a removable intermediate rail which would require the vibrating screeds to span 12.5 m (their upper limit) and floor tolerances would be unlikely to be improved. To achieve higher levels of flatness and levelness a laser guided hydraulic driven mobile cutting and vibrating screed can be used.

Slab Specifications for Finish and Tolerances

Various trowelling techniques are used to produce either a light, medium or heavy burnish finish. Early trowelling with panfloats followed by trowelling with the standard blades and continually working the surface over an extended period, results in a hard wearing abrasive resistant burnish finish, (often referred to as "black concrete").

Tolerances in Australia for industrial and warehouse slabs are usually specified by the architect in conjunction with the requirements of the forklift truck manufacturer. The specifications typically call for a slab to be constructed within + or – 3 mm in 3.0 m with a cross fall not compounding over 10 m and not to exceed 5 mm from RL and a differential across all joints of + or – 1 mm maximum. More recently however, larger buildings have been specified using the "American Standard Specifications for Tolerances for Concrete Construction and Materials, ACI #117R-90 ". This covers the specification and measurement of F_F (floor flatness) and F_L (floor levelness) values for concrete slabs.

CONTRACTOR'S EQUIPMENT

General Overview

The equipment used and the procedures to be adopted for placement, screeding and finishing varies significantly from contractor to contractor in accordance with their available equipment, crew size, quality of surface finish required, their expertise and their specialisation in constructing particular pour area types. Recent improvements in placement, finishing methods and concrete technology have allowed the pour size and joint spacing to be increased without comprising finish, flatness or overall cost. The equipment types and key features of each are briefly described below.

Range of Equipment Used

- Optical levels are generally more accurate but both levelling systems must use a circular bubble to keep the staff vertical. Each pour must be checked using the one set datum point to reduce accumulated error, the setting of new levels cannot be taken from previous pour levels.

- Hand screeding is very labour intensive but still may be suitable for small area pours. The modern power trowels (magic screeds), enable similar work to be done by one operator whilst improving both the surface finish and speed.

- The older mechanical vibrating screeds with solid steel cross sections have been used in the past, but these are being replaced with truss type vibrating screeds with an adjustable precamber to facilitate wider pour strip construction.

- Laser guided hydraulic driven mobile cutting, levelling and vibrating screeds have increased the capacity to pour and level slab areas dramatically. With experienced operators and double passes, surface tolerances of less than + or –2 mm in 3.0 m can be met. Similarly the experienced concreters with two passes of the screed can achieve
- F_F 70+ and F_L 50+.

- Scraping straightedges have replaced the old fashioned bullfloat used by many contractors. They improve the quality of the strike-off and provide immediate improvement to the flatness of the concrete surface. The scraping straightedges vary in widths from 2.4 m - 3.6 m.

- Pan floats (pizza pans) allow both walk-behind and ride-on trowelling machines to commence the trowelling operation earlier, compacting and flattening the top surface of the concrete.

- Walk behind helicopter trowelling machines have improved the surface finish and enable relatively small areas of slab to be burnish finished in a day. Twin blade ride-on power trowels enable larger surface areas to be finished with the same manpower. Triple-blade trowels extend the areas capable of being trowelled. The weight of these ride-on trowels assists and improves the compaction of the concrete surface in order to achieve the abrasive resistance required for a typical burnish finish.

CASE STUDIES

Bullet points for each case study are used to highlight for comparison, the:

- Project description / pour areas
- Specifications / floor finish and tolerances
- Concrete placement method
- Screeding and levelling techniques / equipment
- Trowelling equipment / finishing / duration
- Difficulties experienced

1. " Country Road " Clothing Distribution Warehouse

- The total floor area of 5,200 m^2 was divided into 18 long strip pours each 50 m long by varying widths up to 5.85m. Pouring commenced on completion of the exterior cladding with areas up to 290m^2 per pour. This was the first burnish finish fibre reinforced slab constructed in Victoria.

- A hard-wearing burnish finish was specified for both the racking and order collation areas. Tolerances were specified at + or – 3 mm in 3.0 m with a cross fall no compounding over 10 m and not to exceed 5 mm from RL with + or –1 mm across joints. The constructed slab met these requirements and was checked using an "F meter " at F_F 70 and F_L 40.
- The concrete was placed direct from the truck chute, having reversed down the long strips.
- Truss type vibrating screeds were used and traversed the length of the strip. Edge formwork comprised of timber with a 50 x 4 mm steel flat nailed vertically onto the timber so that the top edge was some 10 mm above the timber. Optical levels to within 2mm were taken at 900 mm centres along the steel edge. A 2.4 m wide bump cutter was used across the strips to remove any surface irregularities.
- Twin blade ride-on power trowels commenced early with panfloats, followed by approximately 4 hrs of helicopter blade trowelling with the blade angle progressively increasing as time passed. At this stage the concrete was quite dark in colour, characteristic of a burnish finish. The machines were equipped with two containers of aliphatic alcohol (a set retarder), which was sprayed onto the surface to adjust non-uniform setting rates. A chlorinated rubber spray on curing compound was applied immediately after the trowelling had ceased. The typical 290 m^2 long strips took on average 9 hrs to complete.
- Some steel fibres were exposed on the surface with occasional fibre tearout at the saw cut joints. The combination of a relatively dry mix design and the fibre reinforcement made the concrete difficult to work.

Figure1 Truss vibrating screed, FRC slab

Figure2 Ride on power trowel, long strip

2. "QP " A Supermarket Chain High Bay Warehouse and Distribution Centre

- The warehouse comprised a 9 storey high bay racking system with a 500 mm thick ground slab to support the superstructure. This slab covered 7,730 m^2 and was poured in 8 stages, covering 46 m x 168 m with each pour approximately 23 m wide x 42 m long (966 m^2). The "distribution centre" slabs were 180 mm thick and post-tensioned. This area covered some 15,200 m^2 in 8 adjacent pours each approximately 38 m x 50 m, (1,900 m^2 each pour).
- The above slabs were specified to have a hard-wearing burnish finish. Tolerances were specified at + or – 3 mm in 3.0 m with all levels not to exceed 5 mm from RL.

- Both slab types were poured using concrete pumps with placement booms. The post-tensioned slab of 1900 m^2 required the use of 2 pumps and 2 vibrating screeds.
- Double and triple width pour strips were necessary to screed these large areas. Truss type vibrating screeds were used in both cases, each utilising removable rails set to the correct height/levels. The rails were removed on completion of the screeding and localised rectification works done. The screeds spanned up to 11.5 m to achieve the pour widths required. Optical levelling of the supporting steel edge piece to within 2 mm was done at 1.2 m centres. Long handle 3.0 m wide scraping straightedges were used in the transverse direction to take out surface irregularities.
- The post-tensioned slab required the use of two, twin blade and one, triple blade ride-on power trowels and several walk behind machines. Early panfloat trowelling was followed by 3 hrs of helicopter blade trowelling which achieved the light burnish finish as specified Each slab pour required a double size crew with an average 14 hrs to complete.
- A continuous supply of concrete for the 500 m^3 "high bay" slab pours, often extended the completion time. Concrete placement and vibration was difficult with closely spaced large diameter reinforcing bars in both the top and bottom of the 500 mm slab. Cold and windy conditions extended the trowelling requirements.

Figure 3 Removable rails, double width strip

Figure 4 Initial trowelling with panfloats

3. "Penguin Books" Warehouse

- The slab was redesigned to a steel fibre reinforced slab 180 mm thick. The total floor area was 21,000 m^2 poured in 21 stages on a 3 wide x 7 long grid pattern. Pours in the 11 m high racking area were 870 m^2 each (28 m x 31 m).
- A hard-wearing burnish finish was specified. Floor tolerances were specified by the supplier of the man-up turret trucks (14 m lift). The specification required a 6 mm maximum difference in elevation of any 2 points 2.4 m apart in the direction of truck travel and a 2 mm maximum difference in elevation of any 2 points across the aisle at the outside of the load wheel path. There was a 1.5 mm maximum difference in elevation of any 2 points 300mm apart in the load and steer wheel path. A F_F 50 slab was considered adequate to meet these requirements. Over the 9 areas poured in the high rack area, the averages achieved were F_F 70 and F_L 60. A comparative optical level grid survey was carried out at 2 m centres with < 2 mm variance.
- The fibre reinforced concrete was placed directly from the chute with two trucks being able to discharge at once.

- A Somero laser screed was used to cut, screed and vibrate the concrete. A uniform head of concrete in front of the screed was no longer required and the screeding time was significantly less, resulting in substantial labour savings and the ability to level/screed much larger areas of concrete. Long handle 2.4 m wide scraping straightedges were used in the transverse direction to take out surface irregularities. Edge formwork was checked with a laser level at 1.5 m centres.
- Two walk behind and two ride-on power trowels were used to finish the slab. Initial trowelling with panfloats commenced 4 hr after the first concrete delivery. At 6 hrs the pans were removed and helicopter blades continued with the slab starting to burnish up at 7 hrs. The slab was completed at 9 hrs and sprayed immediately with a curing compound.
- The concrete was placed with a 28 m wide working face (the full width of the pour) which created some relatively cold faces when there were delays in truck deliveries. This had a direct effect on the trowelling and subsequent work required on the slab surface where fresh and older concrete occur, side by side. Consistent concrete deliveries and uniform setting rates must be maintained. With 6 pours per week, the concreter's progress was restricted by the slower progress of the roof and wall cladding.

Figure 5 Somero laser screed, large area pour

Figure 6 Direct concrete placement

4. " Metabo " Electrical Power Tools Warehouse

- A small warehouse slab of 2,180 m^2 comprising 5 long strip pours, 6.1-6.5 m wide x 61 m long. Each strip runs the full width of the warehouse and covered on average 435 m^2.
- The slab was specified to be a light burnish finish with floor tolerances of + or –4 mm in 3.0 m and not more than 5 mm from datum.
- Concrete was placed direct from truck with access through incomplete edge formwork.
- Preformed metal key sections 2.4 m long were raised some 5 mm above the timber edge formwork and checked at 2.4 m centres using a laser level. The concrete was compacted with immersion vibrators, roughly shovelled level and then spot laser levels taken midway across the strip and at approximately 3 m intervals along the strip. A power float (magic screed) with a 3.6 m blade was then used to screed and float the concrete surface. Permissible floor tolerances made this screeding and levelling technique more economical and faster in contrast to using solid vibrating screeds which were used on two occasions. A long handle 2.4 m wide scraping straightedge was used in the transverse direction.

- Walk behind power trowels with panfloats were followed by a single twin blade ride-on power trowel with helicopter blades. At 6 hrs from commencement of the pour, the slab started to burnish up. With a concrete crew of 8, the final trowelling ceased at 9 hrs with a spray on curing compound applied immediately thereafter.
- The concrete contractors would have preferred the use of set accelerators in the mix in lieu of hot water in the relatively cold conditions (10°C).

Figure 7 Power float, long strip pour

Figure 8 Walk behind helicopter power trowels

5. "Toll Holdings" Transportation Distribution Warehouse

- The total floor area covered 31,680 m^2 and was poured in 20 stages on a 2 wide x 10 long grid pattern. Each pour measured approximately 36 m x 44 m (1,584 m^2). The slab was 160 mm thick post-tensioned with ducts at 2.1 m centres each way. On completion of the stressing, the 900 mm wide edge strip between the slab and external precast walls was poured.
- The specification was extremely brief and called for a standard burnish finish. Floor tolerances were specified at minimum F_F 50 and F_L 40. After each pour, an "F" meter was run over the floor in lines around the perimete. Average readings for the slabs completed were F_F 62 and F_L 50.
- The slabs were poured using a concrete pump with a placement boom. The hydraulic arm could not reach beyond 27 m, so 9-10 m of flexible rubber pump hose supported on timber blocks had to be attached to the end of the pump arm in order to reach the 36 m wide pour.
- Screeding and levelling was done using a Somero laser screed. Heavy duty aluminium removable ladder rail tracks were set up on stools above the tendon ducts, these extended the full width of the pour and included a ramp section to pass over the edge formwork. A long handle 3.6 m wide scraping straightedge was used to smoothen out minor surface irregularities. The laser screed used two passes to achieve the tolerances required Timber edge formwork had a 30 x 5 steel edge plate nail fixed at 450 mm centres and checked for level at 1.2 m centres using a laser level with circular bubble on the staff.
- Three, twin blade ride-on power trowels were used to finish the slab. Initial trowelling with panfloats commenced 4.5 hrs after the first concrete delivery. At 6.5 hrs, helicopter blades were used with the slab starting to burnish up at 8 hrs. The slab was completed at 13 hrs with a crew of 12 "F" meter readings were taken prior to the slab being sprayed with a curing compound.

- Additional immersion vibrators were required to ensure adequate compaction of the concrete under the tendon ducts (normally the laser screed is sufficient). Setting up the 10 m of rubber hose extension to the pump slowed down the concrete placement process.

Figure 9 Laser screed on post-tensioned slab

Figure 10 Ladder tracks & concrete pump

6. "Glaxo" External Carpark Paving Slab

- This fabric reinforced concrete slab covered an area of 3,500 m^2 and was poured in 3 stages. The pours were 50-70 m long x 20 m wide with a constant cross fall of 1:100 in the 20 m direction.
- Tolerances on the uniform cross fall were specified to be within + or – 5 mm for external paving works with a stipple, non-slip surface finish. Perimeter timber formwork was set to the required falls or levelled with laser spot levels at 2.4 m centres and to within 3 mm.
- Placement was by concrete pump with a hydraulic boom arm and required two set up positions to cover the area. Two trucks fed the concrete pump.
- With a uniform cross fall, spot levels were obtained by measuring down from a taught string line raised on 150 mm high spacer rods located on each side of the pour. The spot levels were taken at 3.6 m centres under the string line both across and along the pour area. This enabled power floats (magic screeds) with a 3.6m blade to be used between the spot levels, creating the cross fall in one direction and level slab in the other direction. The concrete had been previously compacted with immersion vibrators and roughly screeded with shovels. A long handle 2.4 m wide scraping straightedge was also used.
- Walk behind and ride-on power trowels with panfloats only were used for 2-3 passes to achieve the non-slip surface finish required.
- Contractors would have preferred the use of a set accelerator in the concrete mix in lieu of hot water only to enable the trowelling to commence earlier.

Figure 11 Power float with spot levels

Figure12 Power float & scraping straight edge

Table 1 Characteristics and comparison of case studies

CASE STUDY NO.	TOTAL FLOOR AREA SQ.M.	CONC. POUR AREA SQ.M.	POUR AREA TYPE/ METHOD	SLAB DESIGN	PLACE-MENT METHOD	TOLER-ANCES	SURF-ACE FINISH	LEVELS & SCREED TYPE	TROWEL METHOD & FINISHING
1	5,200	290	Long strip	Steel fibre	Direct from truck	3mm/3m 5mm RL 1mm joint	Hard burnish	Optical Truss vibrating screed	Ride-on panfloats Ride-on blades
2	7,730	966	Double strip	Steel rein.	Concrete pump	3mm/3m 5mm RL	Hard burnish	Optical 2 Truss vibrating screeds	Ride-on panfloats Ride-on blades
3	21,000	870	Large area	Steel fibre	Direct from truck	FF 50 FL 40	Hard burnish	Laser Laser vibrating screed	Ride-on panfloats Ride-on blades
4	2,180	435	Long strip	Steel rein.	Direct from truck	4mm/3m 5mm RL	Light burnish	Laser Power trowel	Walk behind panfloats & blades
5	31,680	1,584	Large area	Post-tension	Concrete pump	FF 50 FL 40	Standard burnish	Laser Laser vibrating screed	Ride-on panfloats Ride-on blades
6	3,500	980	Large area	Steel rein.	Concrete pump	5mm/3m Cross falls	Non-slip	Laser Power trowel	Walk behind panfloats

SUMMARY & CONCLUSIONS

The concrete contractors in Australia, have in recent years, imported a wide range of equipment best suited to meet their own specialised construction method. The variety of equipment is chosen to enable the contractors to meet the surface finish and surface tolerance specifications particular to their market sector. There is significant market segmentation within the industrial ground slab concrete industry. This is particularly noticeable with those contractors who have imported expensive laser screed equipment and now only construct large area pours to F_F and F_L requirements.The six case studies represent a range of slab pour types, levelling and finishing techniques adopted in Australia, but it is still primarily the engineer who dictates the pour area/layout. Pour areas around or in excess of 1000 m^2 are becoming the norm with the traditional large concrete crews being replaced by the use of laser guided screed equipment.

ACKNOWLEDGEMENTS

To all the Engineers, Contractors and Concrete Contractors for all six case study projects.

FLOW APPLIED SCREEDS

N Beningfield

RMC Readymix Limited

United Kingdom

ABSTRACT. The paper briefly outlines the recent development of flow applied screeds, primarily within Europe. It examines the current need and requirements for these materials within the UK and continental Europe and how these may be met. The use of different cement types is considered. Mix design, material specification and test procedures are also considered. Protective measures for screeds that still have to lose moisture are briefly introduced. Key British and European standardisation in this field is briefly reviewed.

Keywords: Floor screed, Flow, Calcium sulfate, Cement, Aggregate, Testing, Placing, Curing, Drying, Water resistance, Shrinkage, Curling, Cracking.

Neil Beningfield, is Divisional Co-ordinator and Product Development Manager with RMC Readymix Limited. He specialises in mortars and screeds and has lectured and published widely in the field. He is Chairman of the British Standard Mortar Committee, B/519/2, principal UK expert on the European Mortar Committee, CEN TC125/WG2 and a member of various other British and European Standard Committees in these fields. He is Chairman of the Mortar Industry Association and immediate Past President of EMO, the European Mortar Industry Organisation.

INTRODUCTION

Floor screeds are arguably one of the most problematic elements of concrete floor construction. Whilst in some countries they are used little, if at all, the UK, France, Germany, Holland and some other European countries design a high proportion of floors to incorporate floor screeds, as opposed to the alternatives of eg directly finished concrete or dry, laminar systems.

Because screeds are often regarded as something of an afterthought, perhaps even as a necessary evil, they are often little considered in the design process.

This may result in several problems. Screeds are often selected by sub contractors or flooring companies, rather than being specified in an engineering sense. This commonly leads to loose or non-existent engineering, testing or compliance criteria being required of the materials.

So called "semi-dry" screeds fail not infrequently, with commonly observed problems relating to mixing homogeneity, cracking, honeycombing, curling, lack of compaction and excessive dryness. Problems may also result as a function of the bonding of the screeds to the sub-base.

Correctly engineered flow applied screeds, generally based on calcium sulfate, Portland cement or mixed binder offer a way of addressing these issues.

Production of material at appropriate workability for flow application ensures a partial solution to many of the problems. The use of an appropriate flow test, both at the production plant, and particularly on site, to ensure correct workability provides a further technical measure. As the placing process needs no compactive energy, issues of poor compaction are resolved. The use of formulations with virtually no drying shrinkage also overcomes the shrinkage and related problems.

These issues of specification, and in particular of formulation and testing are discussed in this paper, which describes and quantifies the relevant material properties affecting actual performance on the construction site.

THE INFLUENCE OF PLACING ON THE FINAL SCREED PROPERTIES

It is self evident that placing, compaction and curing profoundly influence the final quality of a screeded floor, (and indeed of all cementitious mortars and screeds). In the case of semi-dry screeds however, these influences may be disproportionately high, leading to a concomitantly high incidence in both the number and severity of problems.

Mechanical compaction is rarely, if ever used in the application of semi-dry screeds and operatives rely on the use of their screed batten to compact the material whilst they spread it. Although in theory they could compact the screed during the spreading process by tamping conscientiously, in reality this does not happen. The material is almost always just spread across the floor and given a closed surface finish, which gives it the appearance of compaction, whilst in reality only the top few millimetres are compacted as a result of the finishing process.

This is hardly ever adequate, a problem that is frequently exacerbated by excessive dryness. Excess dryness makes compaction more difficult and as the excess dryness increases, so the energy required for proper compaction also increases to a level greater than that attainable even by the most conscientious operative.

This situation is further compounded by the lack of curing that is commonplace in the area of screeds. For many building sites, drying the screed is a preoccupation, curing is not a consideration.

The use of the appropriate flow applied screed completely overcomes these issues of compaction, simply as a function of their workability; they are applied at a DIN flow table value of between 240 and 260mm, as further defined and discussed later in this paper.

At this high flow value, the materials are exceptionally fluid so that no voids or fissures are present. Moreover, the high moisture content and rapid hydration of some types means that curing is completely unnecessary. There is a sufficient excess of water to ensure hydration without the need for the slightest period of curing. This point is also addressed later in the paper.

GENERAL CHARACTERISTICS OF FLOW APPLIED SCREEDS

Flow applied screeds are placed by pump at high flow values and as a function of their very high workability require virtually no compaction. In this respect, they are analogous to self-compacting concrete in that the mix design prevents segregation, even at the high placing workabilities used.

A further property of those based on calcium sulfate or some specialist mixed binders is that they possess virtually no drying shrinkage, a function of their hydration reaction. This means that non-structural curling and cracking, commonplace in semi-dry screeds, and arising as a function of shrinkage is absent.

An additional characteristic of a flow applied screed, when compared to a semi-dry screed, is that for similar compressive strengths the latter will have a flexural strength of only about half that of the flow applied material. This is due to the virtual absence of the relatively large voids associated with imperfect compaction in the latter.

This major reduction in voids also greatly reduces the acoustic transmission properties of the screed, an important issue, particularly in structures that are designed for multi-occupancy or as public buildings. The thermal conductivity is enhanced, also as a function of the absence of voids, which is of interest for underfloor heating applications.

PORTLAND CEMENTS FORMULATIONS FOR FLOW APPLIED SCREEDS

Attempts have been made to produce flow applied screeds using many cement types, including Portland, calcium aluminate and calcium sulfate but not all have been successful. Early production of thick section Portland cement screeds, that is those for placing at a thickness of between about 30 and 70mm, produced initially in Germany in the early nineteen eighties, behaved satisfactorily during placing. At early age they also appeared to be satisfactory but they then frequently failed in the longer term, at ages of from several months old to two or three years old. These issues were so severe that most systems were withdrawn from the market place.

The cause of the problems is clearly the shrinkage associated with longer term differential water loss throughout the floor section from the hydration compounds associated with the normal Portland cement hydrates.

As is well known [1], these hydrates possess the fundamental characteristic of moisture movement in response to changes in the inter molecular water associated with their gel pore structures. As this relatively weakly held moisture is gradually lost in the first few months and even years of the life of a structure, the top becomes dryer than the bottom thus resulting in a moisture gradient gradually developing throughout the depth of the floor section. Differential shrinkage throughout the section results, with the top drying and shrinking more and thus producing a bending moment which causes the screed to curl upwards. This is particularly pronounced at the corners where there is both less water loss and less restraint. Attempts have been made to overcome these problems by the implementation of several strategies, used either separately or in combination.

Firstly, the water cement ratio may be reduced to as close as is possible to a level at which virtually all the water is combined chemically to form hydrates. Although these hydrates themselves may still possess some inherent potential for shrinkage, as discussed earlier, this will be much reduced as the amount of water itself reduces down towards the level necessary to just achieve hydration. As is known [2], this approximates to a water cement ratio of about 0.25. However, it is clear that this approach of water reduction is technically incomplete because even at a very low water cement ratio, indeed one that approaches or reaches the optimum of 0.25, there is still free water. It is not the case that all the water is available or used for hydration, some is still "free", to act as a cause of shrinkage.

Moreover, at these low water cement ratios, there is insufficient workability for flow to take place and so the approach has to be modified by the addition of powerful plasticisers.

Secondly, close control of grading, particle shape and aggregate cement ratio, often coupled to the use of additions like p.f.a., is used to produce the lowest possible water cement ratio, but this approach alone cannot solve the problem of shrinkage, although it will help.

The third approach is to use one of the newly developed shrinkage reducing admixtures. These work by coating the unstable/accessible pores with a thin layer that inhibits access of water molecules, thus tending to reduce at source the root causes of the shrinkage problems. None of these approaches is entirely effective alone but there are indications that the use of a combination may be effective.

The problem with the practical application of this philosophy is that production control has to be excellent, as a change in any one of many parameters can give rise to substantial shrinkage with the concomitant probability of failure due to curling.

At the time of writing, (2002), current practice is to place areas of no greater than about $80m^2$ as one bay, separated from adjacent bays by movement joints. This compares with a size of about $2000m^2$, more in some favourable circumstances where bays are approximately rectangular, for screeds based on other non-shrinking cementitious systems. However, it is believed that future development will increase bay size maxima for the Portland cement based systems.

Another approach with Portland cement screeds is to completely seal in all of the moisture at a very early stage, before any differential shrinkage has had a chance to develop. This is usually carried out relatively soon after laying, perhaps within four to seven days, by applying a coating of virtually impermeable epoxy or similar resin.

Any free water present then establishes equilibrium throughout the depth of the slab, thus avoiding differential shrinkage due to the presence of a dryer surface and a resultant moisture gradient. It is probable that this concept works additionally by trapping the water within the screed where it remains available for hydration. This process then further depletes the available free water, whilst improving the degree of hydration in the screed and thereby further improving its strength.

BLENDED CEMENT SYSTEMS

There are many blended cementitious systems, using blends of a great variety of cement types. Some of these systems use two main cement types but others contain three or more, are specially developed and rely on patented blends or formulations.

The precise nature and relative proportions of the hydration products in some of these may be very difficult to classify accurately. Some rely on conventional hydration species but others consist of a mixture of final hydrates.

Some of these may not be stable in the presence of moisture, slowly degrading, whilst others may exhibit considerable destructive expansion in wet conditions, making their use entirely inappropriate where these may occur. Some of the systems fail because they are formulated to use the ettringite form of calcium sulfoaluminate as a principal hydration species. These formulations may however contain further unreacted sulfate, often deriving from the use of calcium sulfate in the original formulation.

Calcium aluminate cements, sometimes referred to as high alumina cements are also used in some blended systems and such materials usually show rapid strength development.

There are a large number of proprietary systems based on blended cements, with widely differing properties, and in some cases produced in quite small amounts. For these reasons, they are not considered further in this paper.

CALCIUM SULFATE SCREEDS

The basic chemistry of their hydration reactions mean that the shrinkage of calcium sulfate screeds is not an issue.

Stable hydrates form at a much earlier age than with Portland cement and some types achieve fifty percent of their ultimate strength at an age of one day. The hydration reaction is not accompanied by shrinkage associated with early moisture loss and the hydrates do not have the same water sensitive gel pore structure as Portland cement. This means that there is virtually no linear shrinkage and a complete absence of differential shrinkage and related curling. Table 1 below (due to Dickerhoff), shows the relative pore size distribution for Portland cement and calcium sulfate hydrates.

Table 1 A comparison of pore sizes for different cement types

PORE SIZE UM	PORE SIZE DISTRIBUTION %	
	Calcium sulfate screed	Cementitious screed
0.001-0.01	5	54
0.01-0.1	4	10
0.1-1	21	19
1-10	67	7
10-100	3	10

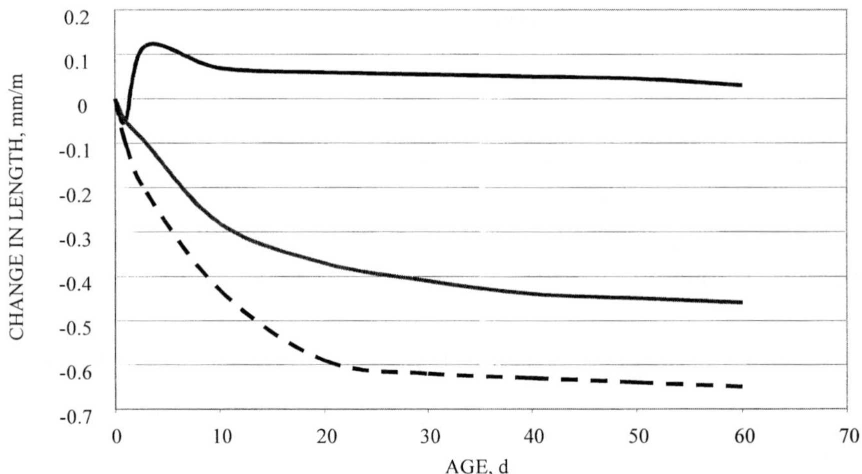

Figure 1 The relative expansion and shrinkage of different screed types

Figure 1 (also due to Dickerhoff), shows how these differing pore size distributions affect the shrinkage or expansion of the final floor screed. The hydration reaction of calcium sulfates is simple, and follows one of two general models,
depending on reactant type as shown below.

$Ca\ SO_4 + 2H_2O \rightarrow Ca\ SO_4{\cdot}2H_2O$ (for the anhydrite form)
$CaSO_4{\cdot}1/_2H2O + 1^1/_2H_2O \rightarrow CaSO_4{\cdot}2H_2O$ (for the hemihydrate form)

It should be noted in respect of the second equation above that the hemihydrate is probably a member of a zeolitic system.

The form of hemihydrate used is usually the alpha type but this may sometimes be blended with the beta form, to enhance the early strength development even more. Whatever the hemihydrate form used, it is formulated with several constituents before final use. Plasticisers and other working property enhancers are used, and most final products show some thixotropy. In addition, retarders are needed in order to control initial set, in a way that is perhaps analogous to the inclusion of gypsum in Portland cements.

In contrast, the anhydrite forms require an initiator, as opposed to a retarder.

Different total amounts of hydration take place according to whether the material is anhydrite or hemihydrate, with about 60% of anhydrite hydrating in the final material but 85-90% reacting in the case of hemihydrate.

AGGREGATES FOR FLOW APPLIED SCREEDS

The requirements for the aggregate for flow applied screed are not complex and generally, good conventional screed sands will be adequate. The only special need is for the virtual complete absence of lighter species, as these will readily float to the surface. This clearly means that lignite, peat and similar materials must not be present.

Maximum aggregate size may vary from 8 or 10mm, down to 4 or even 3mm. For usual application thicknesses, the larger the size the better from the engineering point of view, but local custom and practice often quite subjectively constrain the stated maxima, with the result that 5 or 6mm are generally used in the UK as the practical maximum sizes, although in continental Europe the larger sizes are often used.

The use of either siliceous or calcareous aggregate is quite acceptable and particle shape is not critical, with angular as well as rounded shapes performing adequately.

THE PROPERTIES OF FLOW APPLIED CALCIUM SULFATE SCREEDS

Using calcium sulfate contents of from about 600kg/m^3 to about 800kg/m^3 produces compressive strengths of from c.20N/mm^2 to c.30N/mm^2. These are in line with the requirements of the draft European and British standards which are based on existing DIN standards.

As earlier discussed, the flexural strength for a given compressive strength is higher than for conventional Portland cement screeds, probably due to the relative absence of voids.

The empirical relationship between compressive and flexural strength, often taken as about 1:7 for conventional cementitious species [2], is found to be nearer to 1:4 for these materials. The thermal conductivity of flow applied calcium sulfates is better than that of conventional screeds, with a value of from 1.8 to 2.2W/mK, compared with 1.4W/mK for the latter. This property finds application in the field of heated floor screeds where encapsulating the pipes with these materials results in good and even heat flow throughout the system.

Acoustic testing on completed structures shows enhanced performance when compared to conventional screeds, almost certainly as a function of the lack of voids.

Impact resistance, determined using the BRE impact tester is very good, with values of from 0.05 to 1.5mm being typical, thus easily complying with the most onerous requirements of BS8204 [3].

The durability of calcium sulfate screeds differs greatly to that of screeds based on Portland cements. Unlike the latter, calcium sulfates resist mildly acid conditions, although the inclusion of minor amounts of calcium hydroxide or even of Portland cement to many formulations, added to neutralise the acidity of the commonly used by-product materials, means that acid tolerance is lower than might be expected.

In one aspect of durability, water resistance, calcium sulphates are clearly inferior to conventional screeds. They are unstable in the continued presence of water and should only be used in dry service conditions, although occasional wetting, even saturation, is unlikely to cause problems.

Because there is less free alkali, calcium sulphates do not at first sight appear to passivate embedded metal as well as Portland cement types. However the picture is complex. Because of the minor additions of lime or cement referred to earlier, they do generally produce an alkaline pH. Moreover, because of their density and lack of voids they resist ingress of atmosphere which leads to carbonation in conventional materials, thus reducing pH and degrading corrosion protection to embedded metal in the latter.

Resistance to elevated temperatures is poor for calcium sulfates, as the combined water is only weakly held and the hydration species begin to dissociate at the relatively low temperature of c. 45^0k. However, in practice this mechanism is rarely a problem, and electrical heating elements are often successfully incorporated within these screeds.

Contact with other parts of the structure containing Portland cements should be avoided if there is any likelihood of the area becoming wet, as in the presence of water the sulfate is leached from the system and is available to potentially react with the Portland cement, thus having the potential to cause sulfate attack.

PLACING, CURING AND DRYING OF FLOW APPLIED SCREEDS

The placing of these materials is almost always by pump, with volumes of up to about $12m^3$ per hour achieved using conventional plant, usually worm and stator although sometimes piston. This produces a potential laid area of some 300-350$m/^2$, about ten times the amount achieved with semi-dry materials.

Areas of up to between 2000 and 2500m^2 may be placed as a single bay, with no need for movement joints or division into separate bays, so long as plan aspect ratios are relatively square. As aspect ratios become elongated, simple joints are needed. The screeds may be bonded or unbonded.

In section, abrupt changes in depth should be avoid as these may induce cracking although this problem is rare, probably due to the good tensile and flexural strength and is much less likely to cause an issue than is the case with cementitious flowing screed.

Conventional curing is unnecessary, as the strength development is so rapid, with some types achieving nearly half of their ultimate strength at 24 hours. However, it is necessary to protect the surface from draughts for that time as excessive early moisture loss can lead to plastic shrinkage, in a manner analogous to that seen in some concretes.

After this time, it is beneficial to allow as much airflow as possible, in order to hasten drying of the screed. Prolonged wetness, as may occur in conditions of low airflow or high humidity, is undesirable and may further delay final hardening and drying. Additionally, with some types of formulation a carbonation induced laitence appears within a few days, comprised of a layer of calcium carbonate. This is often worse if drying is prolonged. Moreover, if left in place, it has the further effect of stopping, or at the least greatly slowing, the drying time. It is general practice to remove it using a light sanding with a floor polishing/grinding machine in order to ensure that the drying process proceeds unimpeded.

Prior to application of the final floor covering, the screed is required to have reached a relative humidity of not greater than 75%, if moisture sensitive covering, eg vinyl is to be used. Alternatively, a moisture content of 0.5% may be used, or 1.0% in the case of non moisture sensitive floor coverings. Failure to permit the screed to dry properly to these moisture content levels may well cause problems later.

In situations where the screed has to be subjected to traffic before it has dried sufficiently to be covered with the final floor covering, and a protection is required that still allows it to dry, it may be covered with a layer of specially selected polymer.

Polymers have been developed, some based on styrene acrylates, that are moisture vapour permeable but resistant to the passage of liquid water. This means that they allow the screed to dry out but protect it against abrasion and site damage, as well as against saturation in the event that the surface is subjected to water spillage or similar wetting.

These polymers also act as surface primers for the screed, prior to the application of adhesives for the fixing of the vinyl or other applied surface finish. Without priming, adhesive bond may not be maximised.

THE TESTING OF FLOW APPLIED SCREEDS

The workability, is determined using the simple Haegermann flow table, Figure 2.

Figure 2 The Haegermann flow table test

This test is simple and reliable, with good reproducibility and repeatability and is robust and easy enough to be used on a building site. For the hardened properties, the compressive strength and flexural strength are determined using prism specimens, with the flexural strength being determined first and the compressive strength then tested for using equivalent cubes taken from each of the broken prism halves.

The material must be air cured, not water cured.

Impact resistance is determined using the BRE impact tester, on the actual site floor. This means that compaction, curing and strength development in-situ are all taken account of, in contrast to eg laboratory testing of small samples, even if these are sampled from the site.

CONCLUSIONS

1. Flow applied screeds may be produced from a large number of cement types.
2. Screeds based on calcium sulfates are characterised by their very low shrinkage, which leads to a lack of curling problems on site. They may be laid without joints in much larger areas than conventional screeds.
3. Calcium sulfate screeds are not moisture resistant.
4. They possess improved thermal conductivity and better acoustic performance.

REFERENCES

1. TAYLOR, H. F. W., Cement Chemistry, Thomas Telford, 2nd edn, 1997.
2. NEVILLE, A. M., Properties of Concrete, Longman, 4th edn, 1995.
3. BRITISH STANDARD BS8204-1:1999, Concrete bases and cement sand levelling screeds to receive flooring - Code of Practice, British Standards Institution London.

THEME THREE:
SPECIALIST REQUIREMENTS

Keynote Paper

PERFORMANCE EVALUATION OF FLOORS FOR STATIC CHARGE ON HUMAN BODY – PROPOSAL OF A NEW METHOD

H Watanabe

Takenaka Corporation

H Ono

Tohoku Institute of Technology

H Kaizu

Tajima Corporation

Japan

ABSTRACT Human body electrification can be a major cause of electrostatic disasters. Typical conventional evaluation methods measure electrical resistance of installed floors. Despite their handiness, they do not provide a possible electrostatic potential, which is vital to assess a damage. This lead the authors to develop a quantitative and practical evaluation method for anti-static floors. A concept to simulate human body electrification on floors was proposed, followed by the development of an anti-static tester. The tester measures the maximum electrostatic potential (Vm) on the measured floor and decay time of the potential (Dt). Meanwhile, a scale for anti-static performance of floors ("U-scale") was established by standardizing judgement of experts of floor finish. Finally, an equation was proposed to estimate U with Vm and Dt.

Keywords: Human body electrification, Electrostatic disasters, Electrostatic potential, Anti-static floors, Anti-static tester.

H Watanabe is an engineer in charge of research and development in construction technology, working at Production Headquarters at Takenaka Corporation.

H Ono has been working on the performance evaluation of floors. He is currently a Professor at Department of Architecture, Faculty of Engineering in Tohoku Institute of Technology (Doctor of Eng).

H Kaizu chairs the technical committee of the Association of Interior Floor Industry and has been active in standardization of interior finishing materials.

INTRODUCTION

Human body electrification on building floors can be a major cause of electrostatic disaster. On the other hand, a variety of anti-static floor coverings are developed, designed, and installed. To judge their performance, however, is difficult because of a lack of quantitative and objective evaluation method for their anti-static effect.

An evaluation method for the anti-static performance of floors has been a concern of "the Society for the Study of Specific Performances of Building Floors", to which the authors belong. The present paper presents the results of the study by the group since 1984 from a viewpoint of architectural engineers. Detailed versions of the present paper including description of the anti-static tester were previously published in Japanese [1] and English [2]. The method is to be standardized as Japanese Industrial Standard (JIS) [3].

EVALUATION METHOD FOR ANTI-STATIC FLOORS

Human body electrification is expressed by the following equation:

Electrification = (Generated static electricity) – (Leaking electricity)

A variety of evaluation methods have been used as shown in Table 1. They can be categorized according to one of the two indexes measured for evaluation. The most frequently employed are the test methods measuring leak resistance. Their advantage is that the complex factors (ex. human body) do not affect the evaluation. However, they are not satisfactory as a method to decide the degree of human body electrification. The test methods measuring generated static electricity are well represented by the stroll method. Since they employ human body potential as a direct measure, they are considered most reliable. For a discussion on the direct causality of electrostatic disaster troubles, however, further study is required because of dispersion in measurements due to human body factors and the absence of the dissipation effects after the movement. Thus, a new evaluation technique with handiness as well as quantitative measurement is required.

Table 1 Typical evaluation methods for anti-static floors

STANDARD	TITLE	OBJECTS OF MEASUREMENT	
		GENERATED ELECTRICITY	LEAKING ELECTRICITY
ASTN F 150	Standard Test Method for Electrical Resistance of Conductive Resilient Flooring	---------	Leak resistance
UL 779	Standard for Electrical Conductive Floorings	---------	Leak resistance
ANSI/NFPA 99	Standard for Health Care Facilities	---------	Leak resistance
IEC 61340-4-1	ELECTROSTATICS PART 4: Standard Test Methods for Specific Applications. Section 1:Electrostatic Behavior of Floor Coverings and Installed Floors.	---------	Leak resistance
JIS L 1023	Test methods for several characteristics of textile floor coverings.	Potential of Walker on Carpet	

Aim and Methodology of the Study

The authors' approach to develop a new test method was to establish an evaluation index based on judgments made by experienced research engineers on anti-static performance of floors. To assure the reliability of judgement, it should be a collection of a number of individuals. The other approach of the authors was to develop an anti-static tester to reflect the judgement.

Thus, the process of the present study are described as follows:

1. To propose a concept to simulate human body electrification on floors.
2. To develop an anti-static testing apparatus (referred as "the anti-static tester" or "the tester", hereafter) which realizes the above concept.
3. To establish an objective and quantitative scale for anti-static performance of floors.
4. To investigate and present evaluation method based on the measurements by the anti-static tester and the established scale.

ANTI-STATIC TESTER

Human body electrification on building floors is contributed to rubbing and separating of a footwear and floor while walking. Figure 1 presents the factors involved in human body electrification. The authors created an electric and mechanic model whose electrification correlates with that of a human body.

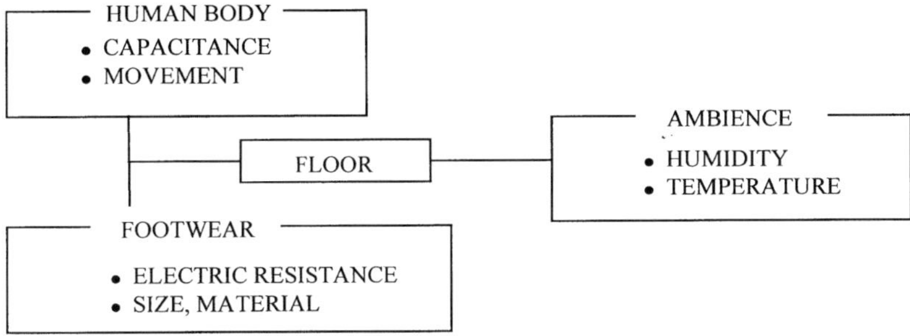

Figure 1 Factors affecting human body electrification on floors

Constitution of the tester

Figure 1 also shows the essential components that reflect factors on human body electrification, ie, rubber rollers and driving component. The rubber rollers replace footwear, one of the major factors on human body electrification. The authors decided that the tester should simulate electrification with anti-static shoes specified in a Japanese standard [2] or its equivalence (the electronic resistance of the sole ranging from 10^5 to 10^8 ohms). The reason is that the anti-static performance of a floor alone is usually not sufficient to avoid electrostatic disaster when the resistance of the footwear is very high.

Though evaluation involving footwear with higher resistance than anti-static shoes needs further study, the above condition does not deteriorate the significance of the present study since it is a common practice to wear shoes to avoid electrostatic disaster.

The driving component rotates the rubber rollers on the specimen floor. The pressure is given to the rollers to provide slip resistance comparable to that of anti-static shoes. As shown on Figure 3, it consists of a shaft, which is driven by a motor accumulating static electricity, and two rubber rollers at the ends of the arms from the shaft.

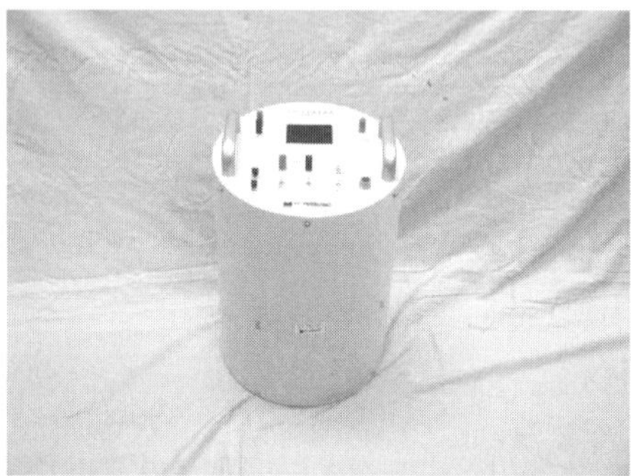

Figure 2 Anti-static tester

Figure 3 Driving component (the tester shown upside down)

Measurements

The tester is designed to measure two values; static potential and decay time. (A decay time is defined here as the time elapsed before the voltage reduces to a half of the initial value). The reason for measuring static potential is that the potential is relevant to the occurrence of electrostatic disaster. The reason for measuring decay time is the authors' assumption that dissipation of the human body potential after movements can be an index value relevant to anti-static performance.

A sensor measures the maximum potential of the rollers while they are forced to rotate for two minutes. This is followed by decay time measurement where a plunger eliminates the potential by earthing the shaft and charges a standard voltage of 50 V. The reason for adopting the standard, as low as 50 V, is that the static damage to semiconductors occurs at a voltage lower than 100 V and that decay time under low voltage was, thus, considered an effective index value.

Specifications

To materialize the concept of the tester, the authors decided the specifications of the major parts as shown in Figure 4.

Preliminary experiments using the tester indicated that the accuracy and repeatability in measurement was satisfactory.

FACTORS INVOLVING HUMAN BODY ELECTRIFICATION		SPECIFICATIONS OF PARTS IN THE TESTER
CAPACITANCE OF HUMAN BODY -Generally 50 to 200 pF	⇨	CAPACITANCE OF TESTER -100 to 150 pF
MOVEMENT IN WALKING -Speed : approx. 100 steps/min -Area of contact: 24 cm^2 /step 2,400 cm^2 /min -Pressure at contact: 0.14 MPa	⇨	RUBBING AND SEPARATION BY RUBBER ROLLERS -Area of contact: 0.6 cm^2 (standing still) 2400 cm^2 /roller/min (rotating) -Pressure at contact: 0.13 MPa
FOOTWEAR -Resistance: 10^5 to 10^8 ohms	⇨	RUBBER ROLLER -Resistance: 10^4 ohms (Lower reststance was employed to avoid influence of the resistance of the rollers.)

Figure 4 Replacement of factors involving human body electrification

EVALUATION SCALE

Although a variety of evaluation scales for anti-static performance have been presented, none of them are related quantitatively with actual electrostatic disaster. This is because the following reasons make analyzing the mechanism of electrostatic disaster difficult;

1. Information on damage to electronics devices tends to be confidential;

2. Many factors on electrostatic makes it impossible to analyze their direct and precise effects.

On the other hand, experienced research engineers are able to judge anti-static performance of floors by the data on their electric property. Such background had lead the authors to build an evaluation scale depending on the judgement of engineers.

Measurement of Panelists' Judgement

Experienced engineers on anti-static performance usually make judgements based in the measured human body potential and leak resistance. In the present study, human body potential by the stroll method specified in the JIS L 1023 and leak resistance by NFPA 99 (see Table 1) were measured for the specimen floors of various materials. The specimen floors used are shown in Table 2. They were installed on an earthed aluminum plate. The size of the plate was 85cm x 85 cm x 0.2cm. The ambient temperature and humidity for the measurements were 21°C to 23°C and 20 %RH to 27 %RH, respectively.

The results (shown on Table 3) were presented to 13 panelists, who are experienced research engineer of floor finish. The panelists' judgements were expressed by selecting one of the categories described in Table 4. The assumption for evaluation was that the ambient humidity should be less than 30% RH and that the footwear should be anti-static shoes specified.

Table 2 Specimen Floors

TYPE		NUMBER OF SPECIMENS
PVC tile	Anti-static	2
	General use	1
PVC sheet	Conductive	1
	Anti-static	2
	General use	1
Tile carpet	Anti-static	2
	General use	2
Carpet	General use	2
Floor coating	Conductive	2
	Anti-static	3
	General use	3
Total		21

Table 3 Data presented to panelists

SPECIMEN FLOOR	Vh(V) : HUMAN BODYPOTETIAL BY STROLL METHOD	LEAK RESISTANCE(Ω)	
		R_{50} CHARGE VOLTAGE =50 V	R_{500} CHARGE VOLTAGE = 500 V
#1	62	5.0×10^{10}	3.0×10^{10}
#2	1311	1.5×10^{13}	5.0×10^{7}
#3	28	1.8×10^{6}	1.0×10^{6}
#4	2170	2.0×10^{12}	1.5×10^{12}
#5	176	1.2×10^{11}	7.0×10^{10}
#6	888	1.0×10^{11}	1.8×10^{12}
#7	70	3.0×10^{10}	6.0×10^{9}
#8	420	4.0×10^{11}	1.1×10^{11}
#9	74	5.0×10^{8}	1.7×10^{7}
#10	16	3.0×10^{7}	2.5×10^{7}
#11	34	2.5×10^{6}	2.5×10^{6}
#12	2	7.0×10^{9}	3.5×10^{8}
#13	1	1.0×10^{7}	3.5×10^{6}
#14	35	2.0×10^{10}	1.1×10^{10}
#15	8	1.5×10^{8}	1.0×10^{6}
#16	502	1.5×10^{13}	3.0×10^{7}
#17	12	1.0×10^{8}	5.0×10^{7}
#18	6	1.0×10^{5}	8.0×10^{4}
#19	781	1.5×10^{12}	6.0×10^{11}
#20	1470	2.5×10^{11}	1.5×10^{11}
#21	5	3.0×10^{10}	4.5×10^{9}

Table 4 Categories for evaluation of anti-static performance of floors

CATEGORY	DEFINITION
7	Combustibles (ex Hydrogen or acetylene) are handled safely IC devices are manufactured without trouble
6	(In middle of categories 7 and 5)
5	Liquid or solid hydrocarbons are handled safely Products with semiconductors are assembled without trouble
4	(In middle of categories 5 and 3
3	Frequency of static-electrification is low and electric shocks are avoided
2	(In middle of categories 3 and 1)
1	No anti-static performance is expected

The mean values of the categories selected for each specimen floor are presented at the column of "U value" in Table 6.

An analysis of variance made on the judgements by the panelists lead to the results shown in Table 5. The ratio of variance (Fo) and contribution (c) is very large for the floor specimen among other factors. This proves a significant difference among the specimen floors in view of their anti-static performance and, furthermore, high consistency of the judgements by the panelist. It is concluded that the evaluation scale obtained has a highly significance and is able to represent the panelists' evaluation. A method of succèssive categories was, then, employed to construct a scale, "Users' evaluation scale for reducing human body electrification," or "U-scale", hereafter. Table 6 gives the U values of the specimen floors.

Table 5 Results of analysis of variance

FACTOR	S	f	V	Fo	c(%)
Floor specimens	1117.44	20	55.8722	79.0171**	82.1
Panelists	56.85	12	4.7373	6.6998 **	3.6
Error	169.70	240	0.7071	-	-
Total	1343.99	-	-	-	-

S: Standard deviation
f: Degree of freedom
V: Variance

Fo: Ratio of variance
c: Contribution
* *: significant at a risk of 1%

Table 6 U Value, maximum potential (Vm) and decay time (Dt) of specimen floors

SPECIMEN FLOOR	U VALUE	Vm (V)	Dt (ms)
#1	2.33	1187	2367
#2	0.90	556	9999*
#3	5.61	20	17
#4	0.27	1371	9999*
#5	2.08	139	9999*
#6	0.65	1503	9999*
#7	2.77	361	9999*
#8	1.12	421	9999*
#9	4.09	47	282
#10	5.18	6	15
#11	5.23	13	11
#12	5.29	34	166
#13	6.00	5	88
#14	2.72	100	959
#15	5.18	19	267
#16	1.69	329	9999*
#17	4.93	14	121
#18	6.16	14	3
#19	0.56	881	9999*
#20	0.55	668	9999*
#21	2.66	159	827

Some discussion on the validity of U-scale is as follows.

1. As to the objectivity of the U-scale, it is argued by the high consistency among the well-experienced panelist. The U-scale has a sufficient objectivity under present conditions as mentioned on the conventional evaluation methods.
2. The U-scale indicates an anti-static performance under a low relative humidity, which enhances human electrification.

ESTIMATION OF "U-VALUE"

Relation with Conventional Index Values

In this section, the conventional indexes for anti-static property are discussed in relation with the U-scale. Figure 5 illustrates the relation between the U values and human body potential with a stroll method (Vh). Although the two values have a significant correlation, the figure also indicates some disorder in their relation, which implies existence of some other electric factors. Dissipation factor, which reduces the human body potential after a one-minute stroll, was considered to be included to attain a better correlation with the U values. Meanwhile, dissipation factor is assumed to be dependent on leak resistance (R). That leads to the following estimation:

$$U_1 = a_1 * \log|Vh| + b_1 * \log R + c_1 \quad (1)$$

U_1: estimate of U value

a_1, b_1, c_1: constants

A multivariate regression analysis was applied to decide a, b, and c using Vh, R_{50} (Table 3), and U (Table 6).

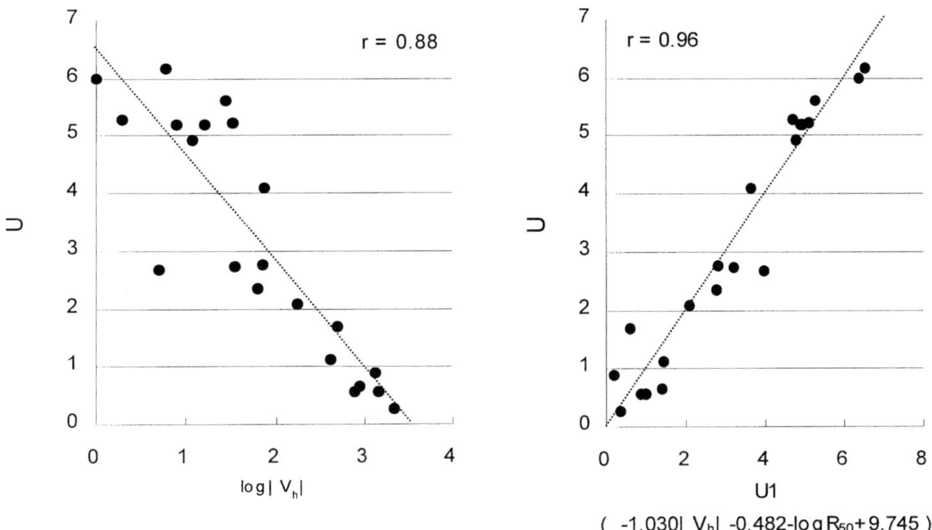

Figure 5 U vs Vh (Human Potential)

Figure 6 U vs U_1 (U-value Estimated with Vh and R_{50})

They both show that the estimation was sufficiently significant. The findings above proved that anti-static performance could be explained as a performance that includes not only human body potential dissipation as well as the potential under body movement, which has not yet been clearly indicated yet. Furthermore, under the condition that difference in human body and its movement and difficulty of measurement are negligible, the relations illustrated in Figures 5 and 6 are satisfactory as new indexes using conventional index values. Although this conclusion may seem to be a matter of course, it should be regarded as new findings: it succeeded replacing some index values based on experts' judgements with the quantitative index values based on measurements.

Estimation of U-value by the tester

The maximum potential (Vm), and decay time (Dt) of the specimen floors were measured by the tester under the same ambient temperature and humidity as the measurements of Vh and R. The results are shown on Table 6. The results of a regression analysis indicated that correlation of logVm vs. log|Vh| as well as logDt vs. log R_{50} were both significant. That lead us to Equation (2), where log|Vh| and log R_{50} in Equation (1) are replaced by logVm and logDt, respectively.

$$U_2 = a_2 * \log|Vm| + b_2 * \log Dt + c_2 \quad (2)$$

U_2: estimate of U

a_2, b_2, c_2: constants

A multivariate regression determined the constants as follows.

$$U_2 = -1.382\log|Vm| - 0.774\log Dt + 8.168 \quad (3)$$

Calculated values of U_2 for the specimen floors are illustrated in Figure 7 in relation with their U values; the figure indicates that their correlation is satisfactory. Thus, it is concluded that Equation (3) can be used as an index for quantitative and objective evaluation.

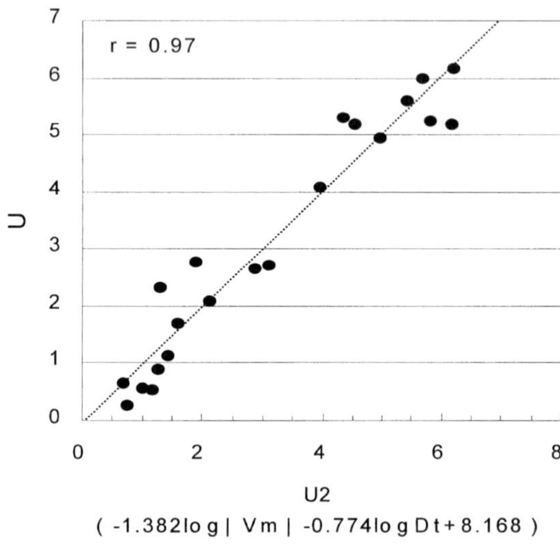

Figure 7 U vs U2 (U-value estimated with Vm and Dt)

Proposal of the Test Method of Anti-static Floors

The authors propose the following evaluation method:

"Use the tester to measure the maximum potential (Vm) and decay time (Dt) of the floors to be tested. Refer to Equation (3) or Figure 7 to evaluate its anti-static performance quantitatively. Although a calculated U value shall be, strictly speaking, a relative index, it has an absolute meaning using the categories in Table 4 under an ambient condition for possible electrostatic disaster."

CONCLUSIONS

A floor's performance for reducing human body electrification was discussed. The conclusions are summarized as follows;

1. The authors developed a tester for the performance for reducing human body electrification. The tester measures two index values, the maximum potential and decay time.
2. An objective and quantitative scale for anti-static performance of floors (U-Scale) was established by standardizing experts' judgements.
3. Proposed evaluation method is to use the tester to measure the two index values of a floor and to estimate its U-value.

ACKNOWLEDGEMENTS

The authors would like to acknowledge the support of member companies to the activities of "the Group for the Study of Specific Performances of Building Floors" to make a standard test method.

REFERENCES

1. ONO, H et al., Study on Evaluation Method of Anti-static Floor for Reducing Static Charge on Human Body, the Journal of Construction Engineering, Architectural Institute of Japan, No501, 1997, pp 25-31.

2. ONO, H, WATANABE, H., TABATA, Y., NAGAHASHI, N., NAGATA H., Evaluation Method of Anti-static Floor for Reducing Static Charge on Human Body, Industrial Floors '99, Munich, 1999.

3. JIS A (Number yet to be decided) Anti-static effect of floor coverings and installed floors - Methods of measurement and evaluation, Japanese Standards Association.

4. JIS T 8103-83 Anti-electrostatic footwear with/without safety toes, Japanese Standards Association, 1983.

A SUMMARY OF THE DEVELOPMENTS IN THE ABRASION RESISTANCE TESTING OF INDUSTRIAL CONCRETE FLOORS

V Vassou

R J Kettle

M Sadegzadeh

Aston University

United Kingdom

ABSTRACT. Extensive research on the abrasion resistance of concrete floors has been carried out at Aston University. A standardised portable test apparatus and method have been developed for measuring abrasion resistance. The initial work revealed the importance of mix design, finishing techniques, curing regimes and surface treatments. Further research work was concerned with the effects of cement replacements. It was demonstrated that cement replacement materials in the concrete mix could lead to better long-term performance in floor slabs. An additional experimental study assessed the influence of the aggregate quality and hardness. It was concluded that low–grade (strength) coarse aggregates could be used to provide industrial floors for medium industrial environments although, when used as a fine aggregate, the performance is impaired and such concretes may not be appropriate for any industrial floors. The current work investigates the potential use of fibres in ground floor slabs, and gives the opportunity to explore the effects of the different fibre types and their properties on the abrasion resistance of concrete. This paper aims to present the findings of both the previous and the current research projects.

Keywords: Abrasion resistance, Cement replacements, Concrete floors, Coarse aggregate hardness, Curing regimes, Fibres, Finishing techniques, Surface treatments.

V Vassou is currently a Research Assistant and Consulting Engineer in the Civil Engineering & Logistics Subject Group in the School of Engineering & Applied Science at Aston University. Her most recent research concentrates on the abrasion resistance of fibre reinforced concrete floors.

R J Kettle is a Chartered Civil Engineer and the Head of the Civil Engineering & Logistics Subject Group in the School of Engineering & Applied Science at Aston University. His research interests focus on the abrasion resistance of concrete floors and has a long list of publications in his specialist topic.

M Sadegzadeh is a Chartered Civil Engineer and Research Fellow in the Civil Engineering & Logistics Subject Group in the School of Engineering & Applied Science at Aston University. His research focuses mainly on the efficiency of cementitious materials and concrete durability.

INTRODUCTION

Over the past 20 years at Aston University, extensive research work has been carried out on the abrasion resistance of concrete floors. A standardised portable apparatus and test method have been developed for measuring the abrasion resistance of concrete [1]. This has been used both for laboratory and in situ testing. This early work revealed the importance of mix design, finishing techniques, curing regimes and surface treatments on the abrasion resistance of concrete. It was demonstrated that the surface layer of the concrete is of primary importance as far as abrasion resistance is concerned. A field study lead to the development of abrasion criteria for concrete slabs in a medium industrial environment and these have been incorporated in the floor classifications in BS 8204 [2] and TR –34 [3]. Further research work [4] was concerned with the effects of cement replacements on abrasion resistance and concluded that appropriate curing was critical to the abrasion performance of mixes containing these materials. A more recent experimental study [5] assessed the use of low–grade coarse aggregates in concrete slabs. It was concluded that low–grade (strength) coarse aggregates could be used to provide industrial floors for medium industrial environments although, when used as a fine aggregate, the performance was significantly impaired.

During recent years, floor construction methods involving the addition of steel, polypropylene or glass fibres into a concrete mix have become a commonplace in the UK. However, only a handful of workers [6, 7, 8, 9] have actually validated their remarks through experimental work but many [10, 11, 12] only claim that the introduction of fibres into concrete results in a greater surface abrasion resistance as compared to that of conventional concrete. This lack of experimental evidence was considered to be a significant gap in the literature. The current project [13] at Aston University is investigating the potential use of fibres in ground floor slabs, and gives the opportunity to explore the effects of the different fibre types and their properties on the abrasion resistance of concrete.

DEVELOPMENT OF THE ABRASION APPARATUS

It is generally accepted that adequate abrasion resistance is an essential requirement for floor performance. Until the early 1980s, however, there has been an absence of a simple, portable apparatus for assessing the abrasion resistance of concrete in the UK and so it was common to indirectly predict the abrasion resistance through more easily assessed factors such as compressive strength and/or cement content [1, 4, 5]. Unfortunately, while these indirect factors can indicate the potential performance of the concrete, they are not reliable predictors of abrasion performance [1, 5]. In particular, problems of inadequate abrasion may still occur because the concrete surface layer is very dependent on:

- Construction procedures – i.e. finishing techniques, curing regimes and surface treatments [1].
- Its constituent materials – cement replacements, low–grade aggregates and fibre reinforcement [4, 5, 13].

In order to study the influence of these factors on abrasion resistance there was a clear need to develop a test apparatus to measure the abrasion resistance. A collaborative research programme involving both laboratory and field testing was carried out by Aston University and the British Cement Association (BCA).

The main conclusions [1] are summarised in the following sections, but the most significant outcome was a simple test for assessing the in-situ abrasion resistance of concrete. This test has now been formally included in the latest edition of BS 8204: Part 2: 1999 [3].

Test methods for assessing the abrasion resistance of concrete

Specific research on the abrasion resistance of concrete can be traced back over a century. Nevertheless, there was no simple apparatus that would simulate all the action to which a concrete floor could be exposed during service [1, 5]. The techniques have involved impact with steel balls, water and steel shot, rubbing with steel plates and rolling wheels [1, 5].

Some of these methods were only able to simulate the forces associated with one type of industrial environment, whereas at least three industrial environments need to be considered, namely light, medium and heavy. This implies that either several test apparatuses need to be used to simulate the forces present in the different industrial environments, or that one apparatus must be developed so that its mode of action may altered to suit the action it is required to simulate. It was decided that the development of a versatile apparatus would be more advantageous than the use of three different sets of apparatus, this latter approach having been followed by the ASTM (C779). The main emphasis was placed on the rolling wheels system, as this was believed to be the appropriate for most environments and would also enable inter – laboratory studies with the then existing BCA machine. It was also essential to produce a flexible system, which should be portable, so that tests could be undertaken both in the laboratory and on site

Description of the Aston abrasion apparatus

The Aston design is identical to the BCA accelerated wear machine [1, 14] and consists of three 75 mm diameter hardened steel wheels, mounted tangentially on a circulate plate and able to rotate freely on their individual fixed axles.

Figure 1 Three accelerated abrasion testers available in the UK
(from left Aston, BCA and Commercial testers)

Figure 1 shows the Aston tester, the original BCA tester and the commercial version available from Wexham Developments. In all three machines the rotating plate is connected by a vertical shaft to a single-phase electric motor, suitably geared to drive the plate at 180±5rpm. The shaft, plate and motor are retained on a fixed top plate, which can move vertically relative to a restraining frame. The frame bares directly onto the floor surface and is located by pins inserted through two diagonally opposite legs into holes drilled into the floor. In this way the apparatus is prevented from moving away from the test position but, because the frame is not rigidly fixed down, the wheels are free to follow the surface profile thus the contact between the wheels and the floor is maximised. The loading on the wheels imposed by the self-weight of the plates, shaft and motor is increased to 65 kg by the addition of a lead collar. The wheels are mounted on the rotating plate such that they wear an annular groove with a mean diameter of 225 mm and a width of approximately 20 mm, the width of each rolling wheel being 20 mm. The duration of the test is approximately 15 minutes, ie 2850 revolutions. The depth of wear is measured, using a depth gauge, at eight locations fixed around the circular abrasion path and the mean value is used to express the abrasion resistance of the tested surface. Although the rotating plate is able to accept three different abrasion heads (i.e. revolving plates, rolling and dressing wheels) the standard tests are undertaken with the rolling wheels. A new design with modified dressing wheels [15], is currently being evaluated as a possible alternative test system for very demanding surfaces such as dry–shakes.

The initial study [1] concluded that the test apparatus fulfilled the five necessary conditions for the assessment of abrasion resistance, specifically:

- The test method must subject the test surface to a treatment that is expected in service.
- The test method must be sensitive to variations in surface conditions.
- The test results must be repeatable.
- The test apparatus must be portable, so that in situ tests may be carried out.
- The test method must be easy to follow, and the cost and time of testing should not be excessive.

It was therefore decided to use the rolling wheel machine to investigate various key factors that affect abrasion resistance and the results [1, 4, 5, 13] are summarised in the following section.

FACTORS INFLUENCING ABRASION RESISTANCE OF CONCRETE FLOORS

The experimental programmes undertaken to date have two main themes, being formulated to study the effects of construction factors (ie finishing techniques, curing techniques and surface treatments) and mix constituents (ie cement replacements, low–grade aggregates and fibre reinforcement) on the abrasion resistance of concrete.

Construction factors

In order to closely simulate site practices, a mould size of 2.0 x 1.5 x 0.1 m was selected because it enabled full use to be made of conventional power plant in the production of test specimens. The construction factors [1] that were investigated are listed in Table 1, together with an appropriate key. The same process was also adopted for more recent studies [4, 5, 13] on the role of mix constituents on abrasion performance.

A large number of tests were performed and the typical results from this early work [1] are summarised in Figure 2. These only represent a small proportion of the total data produced during the investigation but they illustrate the influence of the various factors listed in Table 2.

Table 1 Key to laboratory investigation

CATEGORY			CODE
Mix			
Cement content (kg/m^3)	Free water/cement ratio	Compressive strength (N/mm^2)	
365	0.44	65	1
345	0.52	47	2
300	0.65	29	3
Finishing Technique			
Hand float and trowel			H
Power float and 1 initial pass of power trowel			PF
Vacuum dewatering followed by PF			VD
Power float and repeated passes of power trowel over a period of several hours			RPT
Curing			
Air			AC
Wet hessian			WH
Polythene sheet			PS
Resin based sprayed on membrane – 90 % efficiency			SM
Liquid surface treatments			
Sodium silicate			Si
Magnesium fluorosilicate			
In surface seals – 3 types			ISS
Control specimen – no hardening			O
Dry–shake surface treatments			
Metallic			M
Natural			N
Cement			C

It was concluded from this investigation that the finishing procedure is more critical than the initial mix proportions, and that the compressive strength is not necessarily an appropriate means for specifying the abrasion resistance of a concrete floor. Repeated power finishing and vacuum dewatering produced highly significant improvements in the abrasion resistance of all the concrete mixes. Curing was also found to be important in controlling abrasion resistance, in particular, it is of prime importance with the higher water/cement ratio mixes.

The abrasion resistance was greatly increased by the application of a resin-based, sprayed curing membrane.

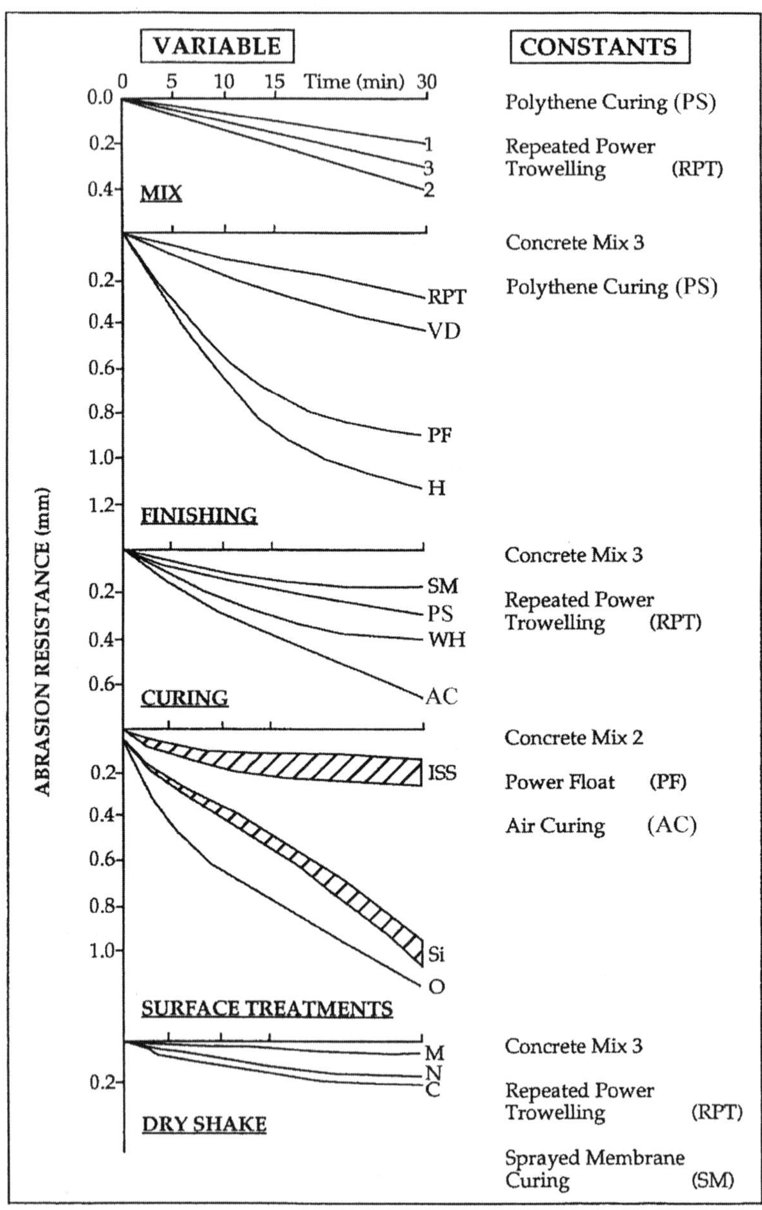

Figure 2 Construction factors affecting abrasion resistance [1]

The application of concrete surface hardeners increased the abrasion resistance of concrete only very slightly. They were more effective on lower water/cement ratio mixes than on the higher ones. Adequate curing appeared to be more effective in increasing abrasion resistance, than the application of surface hardener types of surface treatment. However, in – surface seals based on polymers in aromatic solvents, were found to produce significant increases in abrasion resistance. The metallic aggregate dry–shake improved the abrasion resistance although both the natural aggregate and cement dry–shakes did not produce significant improvements in the abrasion resistance in comparison to the control slabs subjected to repeated power finishing. This may be due to critical influence on this particular finishing technique and was not directly related to the compressive strength.

Cement replacements

While considerable work has been undertaken on the effects of cement replacements, particularly pulverised fuel ash (pfa) and ground granulated blast–furnace slag (ggbs), on many of the properties of concrete, only limited information has been reported on their influence on abrasion resistance. As a result further work has been undertaken [4] and additional data obtained from the continuing experimental investigation. Many of the factors previously identified [1] as having a major influence on the abrasion resistance of concrete had a similar impact on the abrasion resistance of concrete containing these cement replacements, but certain factors had a crucial influence. In particular, the need to provide a positive curing regime to these slabs immediately after completing the finishing procedures was clearly demonstrated. Selective data from this study [4] are given in Table 2 and it is clear that, while the mixes with pfa appear to be slightly more sensitive to poor (air) curing than the control mix, the mixes containing ggbs are very severely influenced. Thus great care will be required if ggbs is to be used in concrete to provide abrasion resistant surfaces.

Table 2 Abrasion data for mixes containing cement replacement materials subjected to different curing regimes [4]

MIX DETAILS		COMPRESSIVE STRENGTH	ABRASION DEPTH (mm)	
CEMENT REPLACEMENT			POLYTHENE CURING	AIR CURING
TYPE	AMOUNT (%)	(N/mm^2)		
0	0	42.0	0.30	0.47
pfa	20	36.5	0.33	0.65
pfa	40	41.0	0.39	0.79
ggbs	40	43.5	0.38	2.17

Further work with these concretes demonstrated that, for replacement levels up to 40%, the level of cement replacement was not critical providing that the mix designs were adjusted to achieve a constant strength level. This trend is illustrated in Table 3, although there is some scatter in the results, which is attributed to local variations in ambient humidity and temperature within the laboratory. There was some limited evidence [4] that when concrete containing pfa was subjected to extended curing beyond 28 days, through the use of curing membranes, these showed benefits from the pozzolanic reactions.

Their 3 and 6 month abrasion depths were lower than those at the 28 days, i.e. an improved abrasion resistance which could lead to a better long-term performance when used in floor slabs.

Table 3 Abrasion data for mixes containing different replacement levels of pfa [4]

REPLACEMENT LEVEL (%)	W/C	COMPRESSIVE STRENGTH (N/mm^2)	ABRASION DEPTH (mm)
0	0.56	42.0	0.30
10	0.54	40.5	0.51
20	0.51	37.5	0.35
30	0.46	36.5	0.45
40	0.44	41.0	0.39
30	0.65	25.0	0.90
30	0.37	53.0	0.24

Low–grade aggregates

A more recent investigation [5] on low–grade aggregates has demonstrated the vital role played by the sand. This work involved three crushed rocks - carboniferous sandstone, magnesium limestone and dolomitic limestone and involved using these materials as:

- The coarse aggregate in mixes containing Trent Valley (TV) sand as the fine aggregate.
- The fine aggregate in mixes containing TV crushed gravel as the coarse aggregate.
- Both the coarse and fine aggregates in particular mixes.

Table 4 Abrasion data for low strength aggregates [5]

AGGREGATE		COMPRESSIVE STRENGTH (N/mm^2)	ABRASION DEPTH (mm)	
FINE	COARSE		NORMAL TEST	WET TEST
TV	TV	51.5	0.21	0.46
TV	SST	27.5	0.32	2.06 *
TV	ML	38.5	0.19	1.74 *
TV	JL	33.0	0.14	0.92
SST	TV	35.5	0.39	0.76 *
ML	TV	47.0	0.40	0.75 *
JL	TV	49.0	0.56	0.77 *

Key: TV = Trent Valley; JL = Jurassic Limestone
SST = Carboniferous Sandstone; ML = Magnesium Limestone
* = Values after 10 minutes as abrasion apparatus became unstable after 15 minutes

The results are summarised in Table 4, and as expected, replacing the TV gravel with various low strength coarse aggregates produced a significant reduction in strength, which can be attributed to the premature failure of the low strength particles. In contrast, when crushed material from the low strength rocks is used as the fine aggregate the effect on the strength is much less marked. This can be attributed to related effects, namely:

- As crushing the coarser particles produced the fine aggregate, many of the weak zones in these particles are eliminated as a consequence of the particle breakdown during crushing.
- Any remaining weak zones in the fine particles will be more spatially limited within a given test specimen and so have a much reduced probability of producing premature failure.

The abrasion results suggest that, although the coarse aggregate mixes had significantly lower cube strengths, they also achieved abrasion depths (0.14 to 0.30 mm) which indicate they have appropriate abrasion resistances for certain industrial premises. However, the abrasion resistances of similar mixes containing the inferior fine aggregate, are significantly lower – the abrasion depths typically exceeding 0.40 mm. This suggests that it is the nature of the fine aggregate that predominantly controls the abrasion resistance. Previous studies [1, 4] of abrasion resistance have demonstrated that the porosity and hardness of the top 1 mm of the slab are the key factors with regard to abrasion resistance. So fine particles from porous rocks are expected to increase the total porosity and so reduced the hardness leading to the lower abrasion resistance shown in Table 4.

Various observations on in–service floors have raised reasonable concerns that wetting the floors – such as through controlled washing in an abattoir – can reduce the resistance to abrasion. To assess this phenomenon, a number of test specimens were subjected to both the standard and test and to similar tests in which the surface is kept wet through the test period.

These data are also given in Table 5 and show for all the mixes that wetting significantly increased the rate of wear. In particular the effects of wetting appeared to be very dependent on aggregate type, while the wear of the control slab increased by some 50 %, the low–grade slabs showed increases in excess of 100 %. All the values in wet state exceed 0.40 mm and so it is clear that concrete slabs to be used in such an environment would require special surfacings such as dry–shakes to provide suitable long-term performance.

Fibre reinforcement

The current investigation [13, 16] is exploring the role of fibre reinforcement on the abrasion resistance of concrete floors.

Table 5 suggests that there is a significant improvement in the abrasion resistance of concrete due to fibre inclusion. It was observed that fibre concrete achieved a range of improvements from 8 to 79 % when compared to the equivalent plain concrete. This is dependent upon the mix design and the curing regime. The best performance was achieved by mix A2 (s/c, 0.51%). The results in Table 5 obtained from the specimens cured with a curing compound are very similar to those obtained from the specimens cured in a polythene sheet. It must be noted that the thin film created by the application of the curing compound had been removed with white spirit prior to abrasion testing. Further work is currently undertaken to study the long-term consequences of using curing compound [16].

Table 5 Abrasion resistance and compressive strength of plain and fibre reinforced concrete [13]

MIX ID	W/C	COMPRESSIVE STRENGTH (N/mm^2)	ABRASION DEPTH (mm)		
			POLYTHENE CURING	AIR CURING	CURING COMPOUND
A1, s/c, 0.51 % - 45 mm	0.44	64.50	0.22	0.32	0.17
B4	0.44	60.00	0.29	0.73	0.32
A2, s/c, 0.51 % - 45 mm	0.52	55.50	0.11	0.17	0.15
B5	0.52	53.50	0.44	0.79	0.46
A3, s/c, 0.51 % - 45 mm	0.65	45.50	0.50	0.78	0.59
B6	0.65	42.00	0.61	0.95	0.64

Key:
A = Fibre reinforced concrete mix
B = Plain concrete mix
s/c = Steel crimped
% = Percentage of fibre volume
mm = Fibre length

The current work [13, 16] has also investigated the effects that different types of fibres, such as polypropylene, glass and a blend of steel and polypropylene fibres, have on the abrasion resistance of concrete, detailed results will be published elsewhere [16]. It was however observed that inclusion of polypropylene fibres at 0.1% by volume improved the abrasion wear by 86% when compared to the equivalent plain concrete. Similarly, the inclusion of the blend of steel and polypropylene fibres into the concrete mix reduced the abrasion wear by 73%. Glass fibres also improve abrasion resistance but to a less marked extent, ie 43% reduction in abrasion wear. The ongoing work is also examining the role of the length of steel fibres on the abrasion resistance, with the initial data suggesting that the longer fibres lead to higher abrasion depths [16].

FIELD ASSESSMENT

Although the laboratory work clearly demonstrated the importance of a number of factors, it was considered important to try to establish a link between the abrasion resistance measured in the test and the field performance of concrete, this enabling the test to be used for the in-situ testing to assess the quality of concrete floor surfaces. It is not possible to provide full details of this investigation [1, 17, 18] in this paper, but the principal factors can be summarised. The programme involved testing slabs that had service lives of between 5 and 15 years. Initially the main objective was to assess the relationship between the abrasion resistance, as measured by the accelerated abrasion test, and the actual service performance.

To achieve this in a quantitative manner requires periodic measurements of the actual depth of wear by reference to a series of datum points on the actual floor slab. It is also necessary to know the traffic loadings to which the floor slab has been subjected. This would have to include type of traffic, frequency, wheel loads, type of wheel, etc. Furthermore, the wear on the section of floor slab will be more severe where the traffic changes direction. These suggest that the quantitative determination of the relationship between accelerated wear and actual service wear is extremely difficult and the accuracy of such an analysis will be open to question. In this investigation, a quantitative assessment was made of the actual wear, which has taken place since each floor slab was constructed, together with the user rating.

In the various premises, there was a vast range of fork-lift trucks, including differences both in manufacture and type. The industrial slabs were classified in accordance with the type and frequency of traffic operating on the floor slab. Broadly, those floor slabs on which the trucks operated with pneumatic or solid rubber tyres were classified as light industrial use, and the accelerated abrasion test was conducted with the revolving pads type of head. When trucks with polyurethane tyre were being frequently used to handle moderately heavy loads (up to 3000 kg), the slabs were considered to be exposed to medium industrial use and the rolling wheels type of head was used in the accelerated abrasion test. Those slabs, which were subjected to heavily loaded trucks, operating on steel wheels, were classified as receiving heavy industrial use and both the rolling wheel and dressing wheels system were used for the accelerated test.

Whilst the large majority of the floors were classified as being subjected to medium industrial use, there were sufficient data to also cover the light and heavy environments – these latter floors having generally been treated with metallic – based dry–shakes [1, 17]. However, his paper concentrates on the data for the typical medium environment.

The procedure adopted for the assessment of the actual service wear followed a standard format. Initially, the floor slabs were visually inspected and the locations of the maximum wear were determined. These were usually at places either where the traffic changed direction or in the loading areas. Attempts were made to measure the depth of wear at these locations with a straightedge feeler gauge. A minimum of 50 readings was recorded at these locations, and the average value was found for each floor slab, which was referred to as the mean estimated depth of wear.

In addition each floor slab was assessed subjectively through a combination of a visual comparison with other floor slabs in the study, together with the opinion of the user regarding the performance of the floor slab during its service life, solely in terms of wear resistance and dusting. The floor slabs were classified with a rating system, in which the rating was related to the combination of these findings. Three scales were used; these were "Good", "Normal" and "Poor".

Accelerated abrasion tests were performed at the same section of the concrete floor slab. These tests were performed with the appropriate type(s) of head. In most cases one test was performed adjacent to the location where the actual depth of wear had been assessed. These chosen areas had not been exposed to any type of traffic, eg under shelving track, etc. A total of three abrasion tests was conducted on each floor slab, the exception being the heavy industrial floor, where six tests were performed, three with the rolling wheels and three with the dressing wheels head.

DEVELOPMENT OF PERFORMANCE CRITERIA

Typical results from floor slabs exposed to medium industrial traffic are summarised in Table 6, additional data for the light and heavy environments are available elsewhere [1, 17].

Table 6 Summary of the results for floor slabs in medium industrial environment [17]

GRADE OF CONCRETE (N/mm^2)	TIME IN SERVICE (YEARS)	ABRASION DEPTH (mm)		RATING
		MEAN ABRASION DEPTH	MEAN ESTIMATED WEAR DEPTH	
30	6	0.18	0.84	Good
30	11	0.41	3.35	Poor
30	11	0.45	3.12	Poor
20	8	0.55	3.10	Poor
20	6	0.65	2.15	Poor
20	8	0.33	1.89	Normal
30	12	0.15	1.37	Good
30	12	0.12	1.18	Good
30	10	0.35	2.67	Normal
30	5	0.30	1.21	Normal
30	5	0.27	1.05	Normal
30	6	0.05	0.55	Good
25	12	0.47	2.96	Poor

From these results it is observed that the mean estimated depth of wear is not uniquely related to the accelerated abrasion depth. It is suggested that the main reason for this scatter in the results is due to differential traffic intensity between the various slabs. These values have been included only as a general guide, but more confidence is attached to the designated rating system, as this includes allowance for such factors as age, intensity of traffic, and the overall performance of the floor slab.

Three main categories have been used to judge the performance of the floor slabs under study, these being “Good”, “Normal” and “Poor”. There is a relatively large number of floor slabs, subjected to medium industrial traffic, so that these categories may be related to the abrasion depth determined with the accelerated abrasion test. It is clear that depths between 0.05 to 0.18 mm, 0.27 to 0.35 mm and 0.41 to 0.65 mm are associated respectively with the “Good”, “Normal” and “Poor” categories.

Using these results, it was possible to propose specific limits for the accelerated abrasion depth of concrete slabs in the medium industrial environment; this is presented in Table 7. This has been supported by subsequent site data [18].

Table 7 Classification of concrete floor slabs in a medium industrial environment [17]

QUALITY OF SLAB	ABRASION DEPTH
Good	< 0.20 mm
Normal	0.20 – 0.40 mm
Poor	> 0.40 mm

Although the data from the heavy industrial environment was not so extensive, it is clear that a good performance in such a severe environment is associated with depths of wear of less than 0.05 mm being achieved with the rolling wheels. Indeed, with these floors, it is apparent that the severe form of the test, with dressing wheels, produced more extensive wear – compatible with wear achieved in service, and so demonstrates the necessity of simulating the condition to which the slab will be exposed during its service life. The original classification derived form the findings of the field investigation and subsequently extended [18] has been included in the latest edition of BS 8204 – Part 2: 1999 [3] together with the specification of the commercial apparatus and test procedure.

CURRENT DEVELOPMENTS OF THE EXISTING ABRASION APPARATUS

The classification in BS 8204: Part 2: 1999 [3] has generally been developed for floor in medium duty environment. However, with concrete floors being required to perform satisfactorily in ever more demanding circumstances (heavy duty), the use of surface treatments has increased and recent collaborative work with the Concrete Society has been centred on assessing these floors. Very often floors that have undergone a surface treatment, produce abrasion depths of the order of 0.00 – 0.05 mm when subjected to the standard test and are therefore classed as "special" floors. Once the special class has been determined, the existing apparatus is not sufficiently sensitive to determine the differences between the several types of dry–shake toppings and certain high grade concretes. A potential problem is foreseen especially with the increasing availability of the dry–shake toppings while standard materials, with the benefit of repeated power flooring and good curing, can produce adequate results for a floor to be classed as "special". In order to overcome this problem it was suggested that once the initial class of the floor is determined and it satisfies the "special" class then further tests should be carried out in order to determine the quality of the surface matrix of the floor. It was proposed that several more aggressive heads replace the standard head for a secondary test. It is not possible to present results from the current study in this paper as it is still ongoing but the interim data published [15] confirmed the above concerns.

SUMMARY

This paper has drawn attention to a number of points concerning both the assessments of abrasion resistance and factors affecting abrasion resistance. These are summarised below:

- A portable accelerated abrasion test has been developed to assess the abrasion resistance of both laboratory and in-situ floors slabs.

- Through in-service testing it has been possible to produce performance criteria to rate the abrasion resistance of in service slabs, which have now been incorporated in BS 8204: Part2: 1999.

- Attention to a number of key factors can maximise the abrasion resistance of a given concrete mix, particularly:

  - Positive finishing procedure
  - Provision of a good curing regime
  - Use of high quality surface treatments to satisfy demanding performance requirements.

- Cement replacements can be used in floor slabs without being detrimental to performance providing that adjustments are made to the mix design to achieve the specified strength.

- The characteristics of the sand have a major influence on the abrasion resistance of the concrete so care is needed over its selection.

- The presence of water on the wearing surface significantly reduces the resistance to abrasion; this effect being particularly marked with lower grade aggregates.
- The inclusion of fibres produced an improvement to the abrasion resistance of the concrete. The shape of the steel fibres was shown not to be a factor that affected the abrasion resistance. However, the type of fibre is significant, with the largest improvement obtained with polypropylene fibres.

- With concrete floors being required to perform satisfactorily in ever more demanding circumstances, it was necessary to develop special, hardened test wheels although the basic machine and the test procedure are not compromised.

REFERENCES

1. SADEGZADEH, M, 1985, Abrasion resistance of concrete, PhD thesis, Aston University, Birmingham.

2. CONCRETE SOCIETY, 1994, Concrete Industrial ground floors – A guide to their design and construction, Technical Report No 34, The concrete society, Slough.

3. BRITISH STANDARDS INSTITUTION, 1999, BS 8204: Part 2 Screeds, bases and in-situ flooring – Concrete wearing surfaces, BSI, London.

4. Phitides, M, 1991, The abrasion resistance of concrete with pulverised fuel ash, MPhil thesis, Aston University, Birmingham.

5. Webb, M, 1996, The effect of low–grade aggregates on the abrasion resistance of concrete, PhD thesis, Aston University, Birmingham.

6. LIU T C, Maintenance and preservation of concrete structures EM DASH 3. Abrasion-erosion resistance of concrete; Technical Report - US Army Engineer Waterways Experiment Station, Jul 1980, No C-78-4, p 129.

7. NANNI, A, Abrasion resistance of Roller-compacted concrete, ACI materials Journal, Vol 86, No 53, 1989, pp 559 – 565.

8. ALEXANDERSON, J, Polymer cement concrete for industrial floors; Polymers in Concrete, International Congress on Polymers in Concrete, Vol 1, 1982, pp 360-373.

9. SUSTERSIC, J, MALI, E, URBANCIC, S, Erosion – abrasion resistance of steel fibre reinforced concrete, in Durability of Concrete; Second International Conference, Montreal, Canada; Ed by MALHOTRA V M; American Concrete Institute, Detroit, MI, Vol 2, (ACI SP-126), 1991, pp 729 – 743.

10. PARAMESWARAN V S, Research and applications of FRC - Indian scenario; Indian Concrete Journal, Vol 70, No 10, Oct 1996, pp 553 – 557.

11. CARR, K, Making good concrete better, Concrete, November/December 1998, pp 16 – 17.

12. KNAPTON, J, 1999, Single pour industrial floors, Thomas Telford, London.

13. VASSOU, V., KETTLE, R J, Abrasion resistance of fibre reinforced concrete floors, Proceedings of Materials Week, International Congress on Advanced Materials, 25-28 September 2000, ICM – International Congress Centre, Munich (may be downloaded from www.materialsweek.org/proceedings).

14. KETTLE, R J, VASSOU, V, SADEGZADEH, M, Comparative investigations using three accelerated abrasion testers, Proceedings of the 10th BCA Annual conference on higher education and the concrete industry, Concrete Communication Conference, 29 – 30 June 2000, British Cement Association, Berkshire, UK, pp 383–395.

15. VASSOU, V, KETTLE, R J, SADEGZADEH, M, The concrete abrasion test apparatus in accordance with the BS 8204 (Part 2: 1999). The reality!; Proceedings of the 9th International Conference on Structural Faults & Repair – 2001, ed Forde MC, CD – ROM: ISBN 0-947644-47-4, Engineering Technics Press, Edinburgh.

16. VASSOU, V, The abrasion resistance of fibre reinforced concrete floors, PhD Thesis, Under preparation, to be released October 2002, Aston University, Birmingham.

17. KETTLE, R J, SADEGZADEH, M, Field investigation of abrasion resistance, Materials and Structures, Vol 20, No 116, 1987, pp 96 – 102.

18. CHAPLIN, R G, Site testing for abrasion resistance of concrete floors, Proceedings of the Second International Colloquium on Industrial Floors, Technische Akademie Esslingen, Stuttgart, Germany, 1991, pp 629 – 639.

ABRASION RESISTANCE OF POWER FLOATED CONCRETE INDUSTRIAL FLOORS – A STATE OF THE ART REVIEW

T Hulett

Concrete Society

United Kingdom

ABSTRACT. This paper summarises the key findings of a review of the factors affecting abrasion resistance in power finished industrial floors. This work formed part of the state of the art review of all aspects of the design and construction of industrial floors for the new edition of The Concrete Society's third edition of Technical Report TR34, Concrete Industrial Ground Floors – *A guide to their Design and Construction.* The surface zone of the floor is quite different in composition to the remainder of the concrete beneath. The importance of the mechanised finishing technique and curing is recognised. Cement contents should be optimised and not maximised. There is no direct relationship between compressive strength and abrasion resistance. Water content should be limited. The potential performance of topping materials should be analysed in the same way as the bulk concrete of the slab.

Keywords: Abrasion resistance, Abrasion mechanisms, Surface structure, Power-trowelling, Curing, Aggregates, Cement, Water/cement ratio, Fibres.

T Hulett is a principal engineer with the Concrete Society specialising in industrial floors. He was project manager and lead author for the third edition of Technical Report TR34, Concrete Industrial Ground Floors – *A guide to their Design and Construction*

INTRODUCTION

In 2002, The Concrete Society published the third edition of its Technical Report TR34, Concrete *Industrial Ground Floors – A guide to their Design and Construction*[1]. The new edition provides state of the art advice on the aspects of materials and construction methods that can affect abrasion resistance in power-finished floors.

The report is concerned primarily with the top-surface zone that has been found to be in the order of one to two millimetres in depth. Floors with coarse aggregate exposed, as a result of abrasion, would generally be considered unfit for use in most warehousing or production environments.

Power finishing has two stages. Firstly, the concrete surface is power-floated. This levels the surface and depresses coarser aggregate particles beneath an open textured layer of finer aggregate and cement paste. The power-finishing machine is usually fitted with circular steel dishes, known as pans, for this stage. When the concrete has stiffened sufficiently, the process of power-trowelling begins. This compacts and closes the surface to create a flat, polished finish. For this stage, the power-finishing machine is fitted with steel trowels.

ABRASION MECHANISMS AND SURFACE STRUCTURE

Abrasion resistance [2] has been defined as the ability of the concrete surface to resist wear caused by rubbing, rolling, sliding, cutting and impact forces; and wear, which is the removal of surface material, as a process of displacement and detachment of particles or fragments from the surface. Wear can be a result of true abrasion or fracture.

True abrasive wear is the result of a rough surface sliding across another surface to remove material. This process is analogous to the action of glass paper on a surface[3].

Fracture wear is a process of cracking of material in a surface as a result of sliding, rolling or impact motion.

Observation of floors in use suggests that cutting, scraping and fracturing by impact are the most significant factors in warehousing and industrial premises. A simplified model of the mechanisms of cutting etc is shown in Figure 1. There are two interacting processes. Firstly, asperities, the cause of roughness, are removed by scraping or cutting motion and/or crushing. Secondly, the picking out of and/or crushing of particles causes indentations. Indentations create adjoining asperities and so the two processes are inter-linked and progressively remove the surface.

The top surface zone of a power-trowelled floor consists of a mortar of fine aggregate and cementitious binder and by observation can be considered to be in the order of 2mm thick. The importance of this layer is confirmed by Sadegzadeh *et al*[4]. Determination of micro-hardness gradients and pore size gradients demonstrated the effects of repeated power trowelling on the surface zone to a depth of 2 to 3mm.

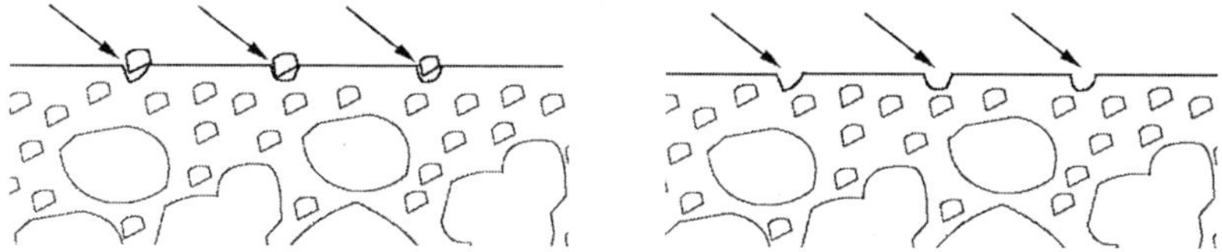

Figure 1 Simplified model of wear

The proposed model of wear mechanisms suggests that, to enhance abrasion resistance, a concrete surface needs to be smooth and of a very tight or closed structure. It should consist of materials that are not likely to fracture or abrade.

Abrasion resistance is tested using the BCA machine described in BS 8204-2: 1999[5]. The machine simulates a wearing mechanism by the use of three hardened steel wheels mounted on a revolving plate spun at 180 rpm for a total of 2850 revolutions or nominally 15 minutes. The resultant annular depression in the surface is measured and the average depth calculated and reported to the nearest 0.01mm. The results may be used to classify a floor in accordance with Table 4 of BS 8204.

FACTORS AFFECTING ABRASION RESISTANCE

Finishing Technique

Finishing technique is considered to be the most significant factor affecting abrasion resistance. Chaplin[6] reported that the traditional finishing techniques associated with plain and granolithic floors involved manual trowelling to flatten and compact the surface mortar, thereby reducing the effective w/c ratio at the surface to produce hardwearing concrete. Fentress[7] demonstrated that wear rates could be reduced in the order of 20% by the use of hard steel trowelling when compared to a normal float finish. Manual trowelling has now been replaced by power trowelling. A typical power-trowel is shown in Figure 2. Most of these machines are "ride on" types that weigh about 300kg.

Figure 2 Power-trowel

Power-trowelling changes the composition of the surface zone by removing moisture as a result of vibration and compression. This reduces the size of pores and the total pore volume. Fine aggregate particles are forced into a more intimate bond. As the vibration is delayed beyond initial setting, the effect is that of revibration. This process is known[8] to increase bond strength probably by the closing of plastic shrinkage cracks formed around aggregate particles and the removal of voids beneath aggregate particles. Sadegzadeh[2] suggested that the composition of the surface zone is changed by up to three repeated passes of the power-trowel. Sufficient time is needed between each pass allow any surface water generated by the previous pass to evaporate.

The result of repeated power trowelling is a smooth dense surface zone that is characterised by its dark colour. A number of authors[9,10,11] have reported on the relationship between surface permeability and abrasion resistance. Figure 3 demonstrates the close and statistically significant relationship between the initial surface absorption as measured by ISAT and abrasion depths, confirming that pore structure is an important factor.

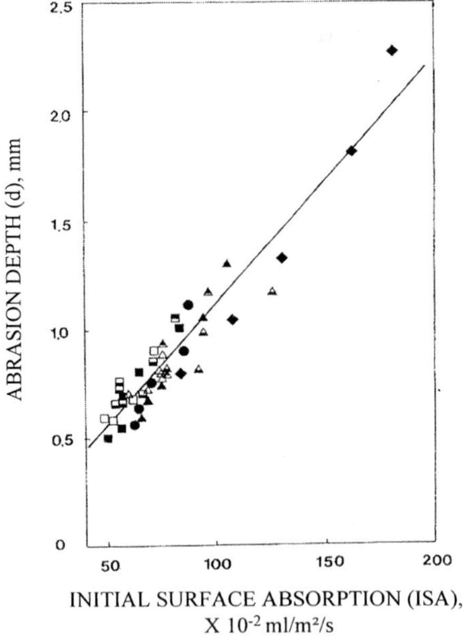

Figure 3 Relationship Between Initial Surface Absorption and Abrasion Resistance. (From Reference 9)

Toppings, which are blends of dry fine aggregate and cement reduce the w/c ratio at the surface in its early plastic state. This may advance the process of water reduction in the surface zone to an extent whereby the combined effects of a lower starting w/c ratio and water reduction by repeated trowelling are greater than that of repeated trowelling alone.

Curing

Kettle *et al*[12] and Plimmer[13], demonstrated that curing is a major factor in the development of abrasion resistance. Curing with polythene sheeting or resin-based sprayed membranes were shown to give best results. Table 1 shows the abrasion depths from test results for air-cured and resin-cured samples for a range of w/c ratios.

Chaplin[14] demonstrated the importance of adequate curing with particular reference to the use of pulverised fuel ash (pfa) and ground granulated blast furnace slag (ggbs). Both cement types require more attention to curing, with ggbs giving particularly poor results when air cured. Both cements gave performance equal to portland cement (PC) when adequately cured.

It has been recognised that resin-based sprayed-on curing compounds have the potential to generate apparently high abrasion resistance results. Although this is acknowledged in BS 8204-2:1999[5] and ASTM C 779 – 95[15], tests are commonly carried out with curing compounds present. To provide validation, large slab samples were cast at The University of Aston using typical flooring grade concrete. The samples were cured with curing compounds or with polythene. When tested, those with curing compound had negligible wear of 0.02mm whereas those cured in polythene had wear of 0.28mm.

The degradation processes of curing compounds and their potential for providing long-term resistance to abrasion are not established. Therefore, a programme of long-term tests for the effect of ultraviolet light and other reactions has begun but cannot be reported here.

Compressive Strength of Concrete

In the UK, abrasion resistance has traditionally been related to concrete compressive strength. However, it is now established[16] that compressive strength and other mechanical performance factors are primarily a function of w/c ratio. Furthermore, the effective w/c ratio of the surface zone is significantly influenced by repeated power trowelling and is only indirectly related to the w/c ratio of the original bulk concrete.

Sadegzadeh[2] and Rushing[17] have confirmed that the compressive strength of the bulk concrete is not a good indicator of abrasion resistance in power-trowelled floors.

Table 1 The Effect Of Curing On Abrasion Resistance. (From Reference 12)

FREE W/C RATIO	ABRASION DEPTH mm/CURING REGIME	
	Air cured	Resin cured
0.44	0.83	0.57
0.52	1.10	0.61
0.65	1.68	0.74

CONSTITUENTS OF CONCRETE

Coarse Aggregate

Although, coarse aggregate is not present in the surface zone, overall aggregate gradings and particle shape should be such as to limit water demand, excessive bleeding and associated settlement. Coarse aggregate should not contain soft materials or impurities that would affect the stability of the layers underlying the surface zone.

Webb *et al*[18] showed that the use of poor quality Magnesian limestone coarse aggregate had no significant effect on the abrasion resistance of samples. The limestone was characterised by high absorbance and by high rates of loss measured by the Los Angeles abrasion test[19]. However, the use of crushed Magnesian limestone as a fine aggregate had a significant effect as discussed later.

Existing guidance (TR34 1994[1], BS 8204-2: 1999[5], BS 882: 1992[20]) suggests that both the ability of the coarse aggregate to resist abrasion and its crushing value are important. However, the literature indicates that these factors are not primarily relevant, as the coarse aggregate is not normally exposed to abrasion.

It may be appropriate to infer qualities of fine aggregates from the coarse materials from which they are produced by crushing. Hard brittle materials, commonly used in concrete, may have relatively high Los Angeles coefficients but still provide satisfactory abrasion resistance. This can be explained by the fact that, as coarser particles are crushed, weak zones in the particles are eliminated leaving sound fine materials[21]. Softer materials will also have high Los Angeles coefficients and may produce poor fine aggregate, although some friable aggregates such as sandstones can break down to mechanically sound fine aggregate.

Softer materials such as soft sandstones or soft limestones are likely to produce fine aggregates with similar characteristics that will in turn lead to poor concrete abrasion performance. These materials also have higher absorption characteristics.

This perhaps suggests that the porosity of the coarse aggregate is a better indicator of the concrete's abrasion resistance than its mechanical soundness as indicated by the Los Angeles test.

Fine Aggregate

Wear mechanisms have earlier been characterised as cutting, picking and crushing. The ability of the fine aggregate and the surrounding cement paste to resist these actions is dependent on particle packing, particle shape and bond, and on the toughness or hardness of the fine aggregate.

It can be appreciated that very soft materials will provide poor abrasion resistance. It has been shown that soft aggregates such as Magnesian limestone[18] perform poorly. In practice, fine aggregates in normal use in concrete would seem to give adequate performance[22] and only known soft materials should be excluded.

The fine aggregate can be a constituent of a dry shake topping. Toppings are produced in factory conditions using dried aggregates. They can be constituted from a range of sizes and gradings can be closely controlled. Theoretically particle packing in the surface zone can be enhanced.

Some product literature[23] suggests that the hardness of aggregates in toppings is a factor in abrasion resistance. However, researchers have not found a relationship between hardness of aggregate and abrasion resistance. They have, however, reported on the effect of using metallic aggregates.

Scripture *et al*[24] reported the following conclusions from an investigation to evaluate the influence of aggregates on wear resistance of concrete:

- There is no relation between the hardness of the aggregate and resistance to abrasion
- All brittle aggregates, including slags, give the same resistance to abrasion
- Malleable-iron aggregate has an abrasion resistance entirely different from that of brittle aggregates. Iron aggregates, although relatively soft, show 400% greater resistance to abrasion than brittle aggregates.

Sadegzadeh[2] concluded that dry shakes containing minerals from natural sources, produced no improvement over a control base concrete, whereas, metallic aggregates did. Metals used in dry shake toppings can be both hard and ductile. This ductility allows metal aggregate particles to deform under load without shattering or losing strength.

Cement

Cements in common use in the United Kingdom are suitable for industrial floors. Current published guidance is that abrasion resistance increases with cement content.

Thinking has changed and more emphasis is now placed on the w/c ratio and its relationship with mechanical and durability factors including abrasion resistance.

Neville[8] states that rich mixes are undesirable for abrasion resistance, a cement content of 350kg/m^3 being probably a maximum. Another early study[17] confirmed this, suggesting an upper limit of about 335kg/m^3 beyond which any further increase in cement content had only a limited effect on abrasion resistance. Harrison in his preliminary research[25] for the review of BS 5328[26] found no data to substantiate a relationship between increased cement content and increased abrasion resistance.

Water/Cement Ratio

Researchers have established[12] that the resultant effective w/c ratio at the surface is lower than in the bulk concrete. There may be a limit on the extent to which the power-trowelling process can reduce initial high w/c ratios in the surface zone. In Sadegzadeh's[2] work, mixes with w/c ratios of 0.44 and 0.52 achieved nominally the same abrasion resistance after repeated power-trowelling whereas the mix with a w/c ratio of 0.65 had less resistance. The research suggests that abrasion resistance is maximised both by reducing the w/c ratio of the bulk concrete and by the further reduction in the surface zone by repeated power-trowelling.

Dry shake toppings reduce the w/c ratio in the surface zone in its early plastic state. They are likely to be particularly beneficial if the water content at the surface is high, for example where the bulk concrete has excessive bleed. Such concrete is also likely to be deficient in smaller sizes in the fine aggregate grading, a factor which in itself could lead to an open surface texture. Such concrete is likely to be less resistant to abrasion and therefore likely to benefit from the addition of a topping.

Fibres

Vassou *et al*[27] investigated the effects of polypropylene and steel fibres. Both materials were effective in reducing wear rates at addition rates of 0.9kg/m^3 of polypropylene and 40kg/m^3 of steel fibres. The results are shown in Table 2.

The effect of further increases in fibre addition was to increase wear rates. There were no tests on samples with fibre contents lower than those shown in the table.

The researchers concluded that the improvement in abrasion resistance is a result of reduced bleed in mixes with both fibre types and consequent reduction of the w/c ratio in the surface zone. The mixes with polypropylene fibres were observed qualitatively to have less bleed than those with steel fibres. The steel fibres were not noticeable at the surface and were judged not to have directly influenced the abrasion test.

Table 2 Effect of fibre addition on abrasion resistance (From Reference 27)

FIBRE TYPE AND ADDITION RATE	DEPTH OF WEAR (mm)
Control – plain concrete 345kg/m^3-w/c 0.52	0.44
Polypropylene @ 0.9 kg/m^3	0.06
Steel @ 40kg/m^3	0.12

SUMMARY

Most industrial buildings used for warehousing, distribution and manufacturing have power-trowelled floors. A floor can be considered durable if the surface layer or zone of approximately one to two millimetres thickness has not been penetrated or removed during its design lifetime.

The abrasion resistance of the surface zone is predominately a function of the repeated power-trowelling process and curing and to a lesser extent it is a function of the fine aggregate used in the surface of the concrete. Fine aggregate in the surface zone is either that present in the bulk concrete used for the floor or it can be a constituent of a dry shake topping applied to the surface.

Aggregates for concrete in normal use are satisfactory. Fine aggregates should have continuous gradings. Fine aggregates including soft materials or having high contents of very fine materials should be avoided. Coarse aggregates have no direct effect on abrasion resistance.

Toppings can be beneficial either as a result of their contribution to lowering the w/c ratio in the surface zone and/or because of the aggregates they contain. Aggregates in toppings will give better performance than the fine aggregate in the base concrete if they provide improvements to particle packing, or because metallic aggregates are used. The hardness of the fine aggregate in a topping is not a good indicator of abrasion resistance.

Increasing cement contents beyond the range of 335 - 350kg/m^3 does not increase abrasion resistance. It is important to avoid high w/c ratios, although abrasion resistance is not sensitive to w/c ratios in the range 0.45 to 0.50 as excess water in the surface zone can be removed by the process of repeated power-trowelling.

There is a test method for assessing the potential abrasion resistance of floors, however, true assessment is difficult because the commonly used resin-based curing compounds render the test method ineffective. The long-term effects of these compounds are not well established and therefore tests should be taken on samples or floor areas that have been cured in or under polythene instead of a curing compound.

REFERENCES

1. CONCRETE SOCIETY TECHNICAL REPORT 34, CONCRETE INDUSTRIAL GROUND FLOORS, A guide to their design and construction, Second Edition, The Concrete Society, 1994.

2. SADEGZADEH, M., "Abrasion Resistance of Concrete". PhD Thesis, Aston University, Birmingham, UK, 1985.

3. MOORE, M.A., "A review of Two Body Abrasive Wear", Wear, Vol.27, 1974, pp.1-17

4. SADEGZADEH, M., PAGE, C.L.& KETTLE, R.J. "Surface Microstructure and Abrasion Resistance of Concrete", Cement and Concrete Research Vol. 17, 1987, pp. 581-590.

5. BRITISH STANDARDS INSTITUTION, BS 8204-2: 1999, Screeds, bases and in-situ floorings, Part 2. Concrete wearing services - Code of practice, 1999.

6. CHAPLIN R.G. "Abrasion Resistance Concrete Floors", Advances in Concrete Slab Technology, 1980 pp. 532-543.

7. FENTRESS, B., A.C.I. Journal 1973 pp. 486-491.

8. NEVILLE, A.M., Properties of Concrete, Fourth Edition 1995.

9. DHIR, R.K., HEWLETT, P.C., & CHAN, Y.N., "Near-surface characteristics of concrete: abrasion resistance", Materials and Structures, Vol. 24, 1991 pp. 122-128.

10. TOMSETT, H.N., "What is the Value of NDT in Structural Appraisal", Proceedings International Conference on Structural Faults, Vol. 1, Edinburgh, Engineering Technics Press, 1989, pp.3-9.

11. CONNELL, M.D., "A Study of an Abrasion Test for Concrete and some of the factors affecting Abrasion Resistance" ACT Project, 1985, pp. 72.

12. KETTLE, R.J., & SADEGZADEH, M.,. "Development in Concrete Testing for Durability", Aston University, 1984, pp10.

13. PLIMMER, J. M., "An assessment of the Schmidt-Rebound Hammer as a method of determining the potential surface wearing characteristics of industrial concrete floors." Report on a City and Guilds Advances Concrete Technology project. 1977. (Unpublished), pp. 38.

14. CHAPLIN, R.G., "The influence of ggbs and pfa additions and other factors on the abrasion resistance of industrial concrete floors", BCA, 1990 pp. 64.

15. AMERICAN SOCIETY FOR TESTING AND MATERIALS, C 779-95, Standard Test Method for Abrasion Resistance of Horizontal Concrete Surfaces, 1995.

16. DHIR, R.K., TITLE, P.A.J., & McCARTHY, M.J., "Role of Cement Content in the Specification for Durability of Concrete, University of Dundee, 2001, pp. 301.

17. RUSHING, H.B., "Concrete Wear Study", Report PB. 193410 Louisiana Dept. of Highway, June 1968, pp. 28.

18. WEBB, M., KETTLE, R.J., & COLLINS, R.J., "Abrasion Resistance of Concrete Containing Lower Grade Aggregates", Concrete for Environment Enhancement and Protection, E & FN Spon, 1996, pp. 12.

19. BRITISH STANDARDS INSTITUTION, EN 1097-2:1998, Tests for mechanical and physical properties of aggregates.

20. BRITISH STANDARDS INSTITUTION, BS 882: 1992, Specification for Aggregates from natural sources for concrete, 1992.

21. KETTLE, R.J., & SADEGZADEH, M.,. "Field Investigations of Abrasion Resistance", Materials and Structures, Vol. 20, 1987, pp. 96-102.

22. BRITISH STANDARDS INSTITUTION, prEN 12620, Aggregates for Concrete, 2001.

23. PRODUCT LITERATURE: Armorex Ltd, Don Construction Products Ltd, MBT Feb Ltd, Roy Hatfield Ltd, Permaban Ltd.

24. SCRIPTURE, E.W., et al. "Floor Aggregates", Proceedings, Am. Conc. Inst., Vol. 50, No. 4, 1953, pp. 305-316.

25. WORKING PARTY of THE CONCRETE SOCIETY "The Influence of Cement Content on the Performance of Concrete". The Concrete Society, 1999 pp. 47.

26. BRITISH STANDARDS INSTITUTION, BS 5328: 1997, Concrete Part 1: Guide to specifying concrete.

27. VASSOU, V., & KETTLE, R.J., "Abrasion resistance of fibre reinforced concrete floors", Paper No. 349, Materials Week International Congress, Munich, 2000 pp. 9.

MONITORING OF FLAT SLAB FLOOR PANELS AT BRE CARDINGTON

R H Scott

University of Durham

S G Gilbert

Queen's University Belfast

United Kingdom

ABSTRACT. A programme of long term service load tests was undertaken on the sixth floor of the full-scale, seven storey, reinforced concrete building at the Large Building Test Facility of the Building Research Establishment at Cardington. By using internally strain gauged reinforcing bars cast into an internal and external floor bay during the construction process it was possible to gain a very detailed record of slab strains resulting from the application of two intensities of test loads. Three site visits were made over a two year period to record results from the strain gauged bars. Results from this monitoring exercise are presented in this paper.

Keywords: Flat slab, Strain gauged reinforcement, Service load testing, Reinforcement strain distributions

R H Scott is a Reader in the School of Engineering at the University of Durham. He now has twenty years experience in the manufacture of strain gauged bars for use in reinforced concrete research and in the necessary instrumentation and software for the interpretation of results. Several hundred bars have been manufactured during this period involving the installation of around 10 000 electric resistance strain gauges. In addition to research projects at Durham, collaborative programmes have been undertaken with organisations both at home and abroad.

S G Gilbert is a Senior Lecturer in the School of Civil Engineering at Queen's University, Belfast. He has nearly twenty years research experience on reinforced concrete slabs and is the author of over twenty technical papers. He has a particular interest in the behaviour of slab-column connections and has been involved with several research programmes in this field.

INTRODUCTION

Flat slab floors are popular and offer advantages over other types of in-situ concrete floors such as beam and slab construction. The flat soffit leads to simpler formwork which, in turn, leads to a faster construction time. In addition, the shallow slab depth minimises the overall floor depth and provides excellent flexibility for services. However, difficulties can arise in the design of flat slabs due to the occurrence of high shear stresses in the column regions and the development of large deflections in the slab, particularly with the current trend for spans up to 9 m being required.

Opportunities to correlate design procedures for flat slabs with results from full-scale testing are rare and so construction by the concrete industry of the full-scale, seven storey, reinforced concrete building at the Large Building Test Facility of the Building Research Establishment (BRE) at Cardington [1] provided an ideal and unique opportunity to continue the study of the behaviour of these slabs at full-scale. This building was one of three full-scale buildings constructed at Cardington for experimental purposes, the other two being an eight storey steel framed building and a six storey timber building. Design of the floors in the in-situ concrete building used a variety of analytical procedures and reinforcement layouts.

Previous laboratory tests by the authors on slab-column connections had included internally strain gauged reinforcing bars to provide detailed information regarding strain and bond stress distributions. In collaboration with BRE seven of these internally gauged bars were installed in April 1998 as part of the reinforcement schedule in the floor slab at sixth floor level of the full-scale test building.

It was hoped that the slab bays which incorporated these bars would, eventually, be load tested to ultimate. However, it was considered that, without affecting the integrity of the structural frame, useful data and experience could be gained by conducting a programme of service load tests with the required loads being applied by loading the floor area with large sand bags. Consequently, the authors have been involved with two such test programmes, the first of which, in December 1998, applied six arrangements of short-term loading to the floor bays containing the gauged bars. Shortly afterwards, all floors in the building were permanently loaded by BRE to 3.0 kN/m^2, again using sand bags, and this provided the opportunity to monitor the behaviour of the slabs over a longer period of time.

This paper describes the methodology and results for the long term testing

INSTRUMENTATION

The experimental building has a 4 x 3 bay configuration and a typical floor plan is shown diagrammatically in Figure 1. Internal columns are 400 x 400 mm and external columns 400 x 250 mm (orientated as indicated in Figure 1). The floors are of flat slab construction and are 250 mm thick. Figure 1 also indicates the extent of the loaded area used for the long term tests.

The seven gauged bars (see Table 1) were all T16's and comprised two U-bars (Bars 1 and 2), three long straight bars (Bars 6, 7 and 8) and two short straight bars (Bars 11 and 12). Bar manufacture followed the procedure which has been successfully developed at Durham University whereby electric resistance strain gauges are installed in a central longitudinal duct [2]. The gauges had a 3 mm gauge length and the duct cross-section was 4 x 4 mm. Altogether 163 gauges were installed, divided between the seven bars as shown in Table 1. Bars 6, 7 and 8, at 3800 mm long, are believed to be the longest internally strain gauged bars yet produced.

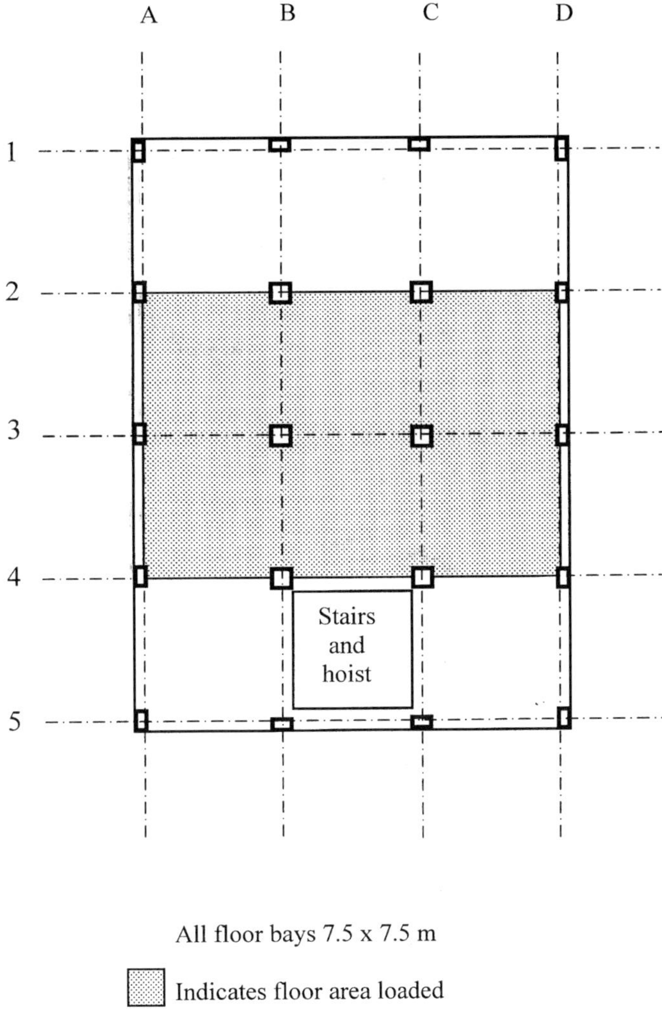

Figure 1 Floor layout showing area load tested

A three wire system was used for the strain gauge wiring with the wires being routed along the duct and out of the ends of the bars. These wires were very fine (0.2 mm diameter) and considerable care had to be taken to ensure that they were adequately protected where they emerged from the bar ends.

This was achieved by sealing the wires in metal boxes connected to the bar ends with short lengths of plastic tubing. All connections were carefully sealed with mastic so that the whole system was watertight and able to resist the rigours of site installation and concreting.

Table 1 Details of Gauged Bars

	BARS 1 AND 2 (U-BARS)	BARS 6, 7, 8 (STRAIGHT)	BARS 11 AND 12 (STRAIGHT)
Bar length (mm):	1630 mm (top leg)	3800 mm	1900 mm
No of gauges per bar:	13 (all in top leg)	41	7
Bar position:	Top of slab (top leg)	Top of slab	Bottom of slab

Due to time constraints (the building was already under construction when funding for bar manufacture was confirmed), the instrumented bars could only be installed in one of the upper floors of the building (ie the sixth floor). Here the slab reinforcement comprised a blanket cover of two-way mats having T12 bars at 150 mm centres in both directions, top and bottom, with additional loose bars in regions of high bending moments adjacent to the columns and at mid-span regions. The gauged bars replaced some of the loose bars. Bars 1 and 2 were at or adjacent to edge column D3, Bars 6, 7, and 8 were at or adjacent to internal column C3 whilst Bar 11 was at mid-span on Grid 3 and Bar 12 was at mid-span midway between Grids 2 and 3, all as shown in Figure 2.

Strain gauged data were logged using hardware brought to site from Durham for each visit.

TEST PROGRAMME

Three site visits were made to monitor long term strains - June 1999, July 2000 and April 2001. All but five of the 163 gauges were in full working order. This was the same quality of performance as is normally achieved under laboratory conditions when working with these gauged bars. Drift over the monitoring period was very small.

The strain gauge data logging hardware was located well clear of the areas being loaded. Specially fabricated wiring harnesses (15 m in length) were used to connect the very fine wires from the gauged bars to the logger input cards. The strain gauge autobalance values for zero load (obtained during the first site visit in December 1998) were input into the data logging hardware at the start of each visit to initialise the logging system prior to new data being recorded. At the completion of each site visit the fine wires from the gauged bars were sealed inside their metal boxes to keep them in good condition for the next site visit.

Trial weighings indicated that sand bags each weighed approximately 1 tonne. They had been at the Cardington facility for some time and thus it was assumed that the humidity of the sand, and hence the weight of each sand bag, had stabilised.

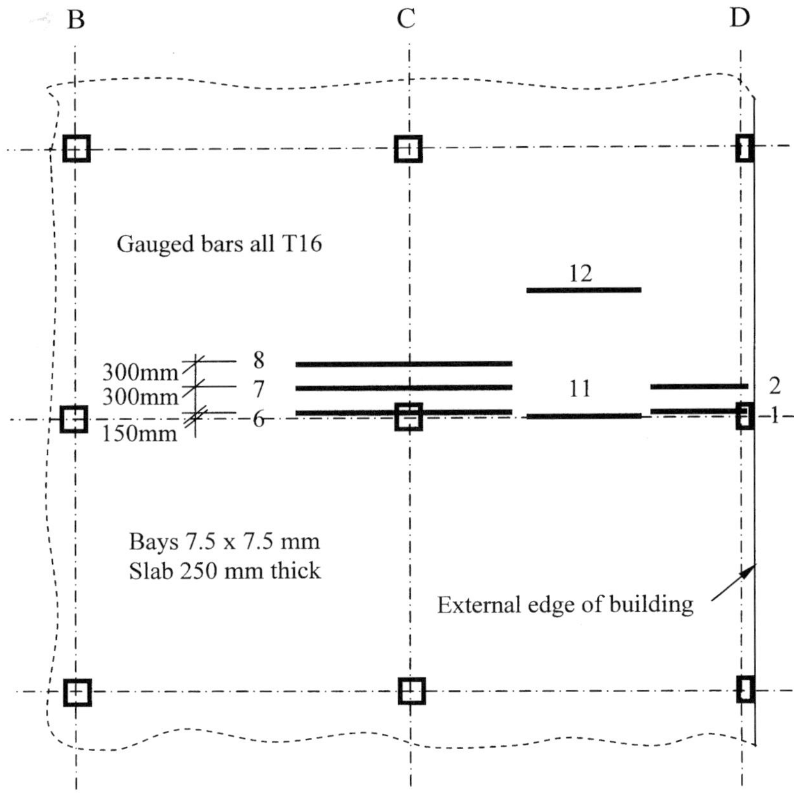

Gauged bars replace "normal" bars already scheduled

Bars 1 & 2 : Top steel (U-bars)
Bars 6, 7 & 8 : Top steel
Bars 11 & 12: Bottom steel

Figure 2 Diagrammatic Layout of Gauged Bars

In June 1999 data was recorded for the 3.0 kN/m^2 loading, known as Test F. This was the load applied by BRE and achieved by placing 96 sand bags over the six floor bays indicated in Figure 1 (i.e. 16 bags per bay in a 4 x 4 arrangement on each bay). In July 2000 data was again recorded for Test F, after which the authors increased the load on all six bays of the sixth floor to 4.7 kN/m^2 by placing additional sand bags to bring the total up to 150 (ie 25 bags per bay in a 5 x 5 arrangement on each bay). Data was then recorded for this new loading, known as Test H (Test G was an intermediate load stage). Finally, in April 2001, data was again recorded for Test H. This loading was left in place for future visits, which, unfortunately, now appear unlikely due to the closure of the Cardington facility.

RESULTS

Good quality strain data were recorded throughout all stages of the test programme. The results presented here are for the effects of the test loads only as the tight construction programme did not allow time for recording strains due to slab self weight. Figure 3 gives an overview of the results for both load intensities by showing strain distributions along Grid 3 for Bars 6, 11 and 1 (see Figure 2). The sign convention is tension positive.

The overall pattern of the strain distributions was as expected with peaks occurring at positions of maximum hogging and sagging moments. The additional load added to Test F to produce Test H caused an overall increase in strains, again as expected. However, and interestingly, time dependent changes can be seen to have been generally small for both tests, for which there are two possible explanations.

Firstly, when load was first applied in early 1999 to produce Test F, the slab was already nine months old. Thus the concrete was relatively mature which would have limited the potential for creep in compression. Also, the reinforcement arrangement of two-way mats top and bottom over the entire floor area rendered the slab very stiff, further severely limiting any tendency for the concrete to creep.

Another, and perhaps less obvious, factor with potential to affect long term strains was loss of tension stiffening in the tensile zones of the slab cross-section. This is recognised in Part 2 of BS8110 [3] by prescribing concrete stresses at the level of the tension steel of 1MPa for short term loads and 0.55MPa for long term loads. This gives a reduction in slab stiffness, which leads to higher reinforcement strains in the long term, but BS8110 gives no timescale at which "long term" can be expected to commence. However, current research at the Universities of Durham and Leeds (also reported at this congress [4]) suggests that this reduction in tension stiffening will have occurred within 2-3 weeks after the slab was first loaded i.e rather earlier than hitherto assumed. Thus the slab would already have reached a largely steady state, so far as creep and tension stiffening were concerned, by the time the load for Test F was first applied. The results reported here confirm this steady state condition.

Early in 1999, the authors were told that the slab had cracked in the column regions at the construction stage due to premature removal of the props supporting the formwork. Consequently, the slab stiffness at these locations would have been reduced relative to that in the uncracked span regions and this would account for peak strains rising by around 40% in Bar 6 at column C3 and by over 50% in Bar 1 at column D3 when the additional load to produce Test H was applied.

There were some detailed changes adjacent to column C3. These are indicated in Figures 4-6 which show strain distributions for Bars 6, 7 and 8 respectively. Considering Test F first, the results for Bar 6 indicate that, local to the column, the two slab bays either side acted remarkably independently. Strains peaked on both sides of the column, but were very low within the column width suggesting that load transfer across the column width was small. Unlike Bar 6, which passed through column C3, Bars 7 and 8 were entirely in the slab at increasing distances from the column (Figure 2). Accordingly, for Test F, strains in Bars 7 and 8 were lower than those in Bar 6 (see Figures 4-6) and the distributions were flatter where these bars crossed Grid C, particularly so with Bar 8. Thus, although there was very little transfer of bending moment between the internal bay and edge bay across the column itself, away from the column bending moments were able to "flow" through the slab from one bay to the other.

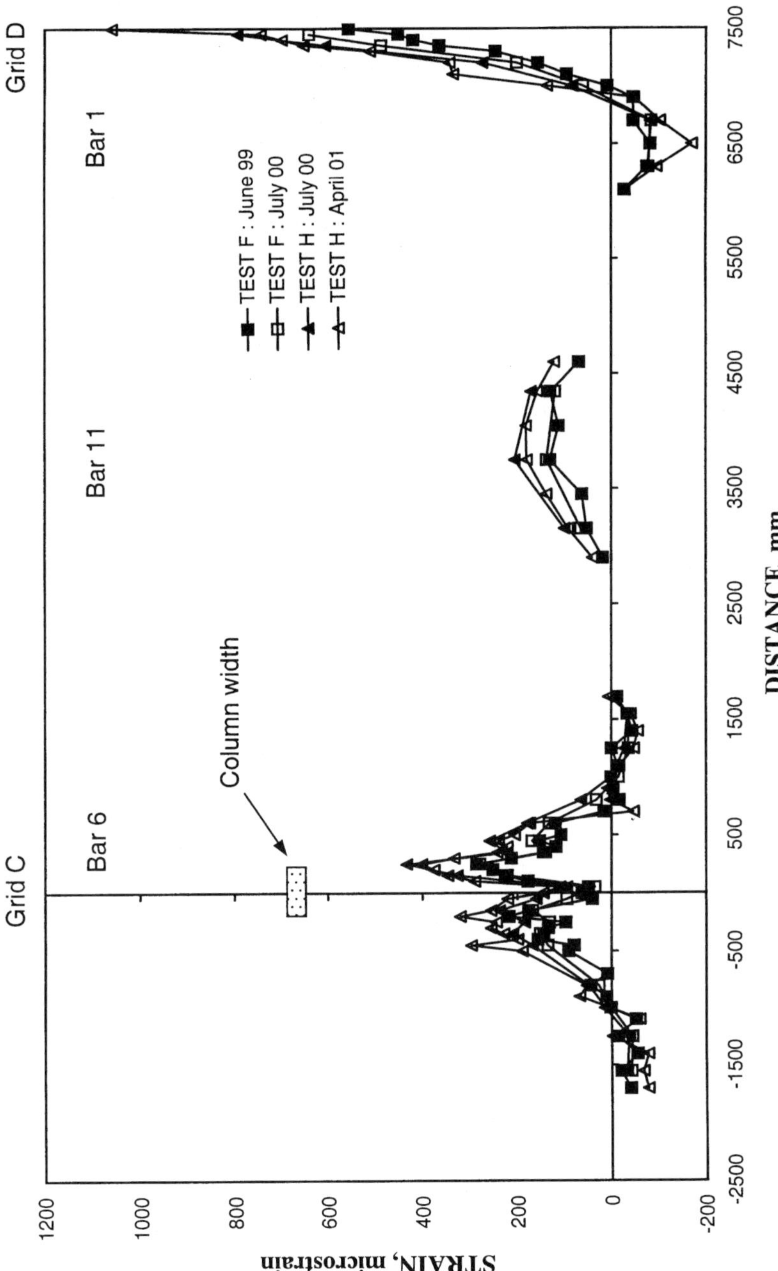

Figure 3 Strain distribution along Grid 3

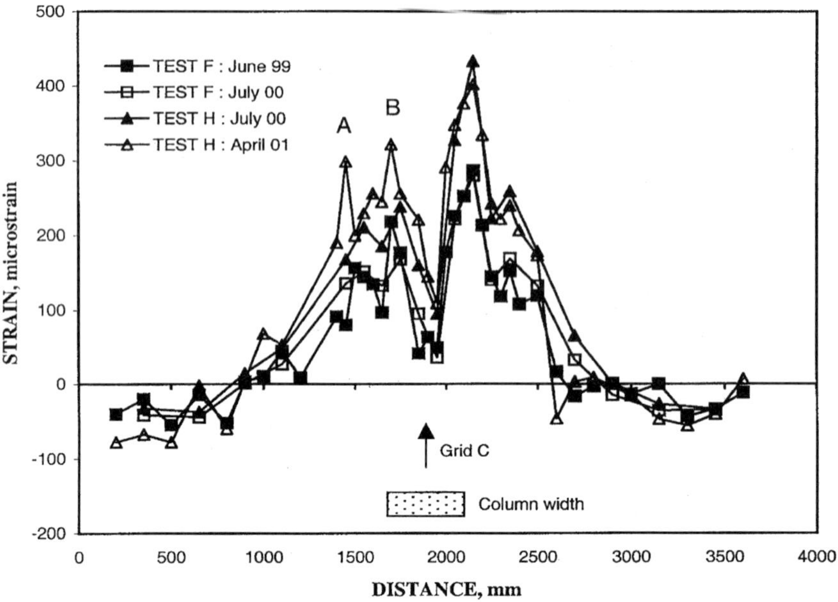

Figure 4 Strain distributions for Bar 6

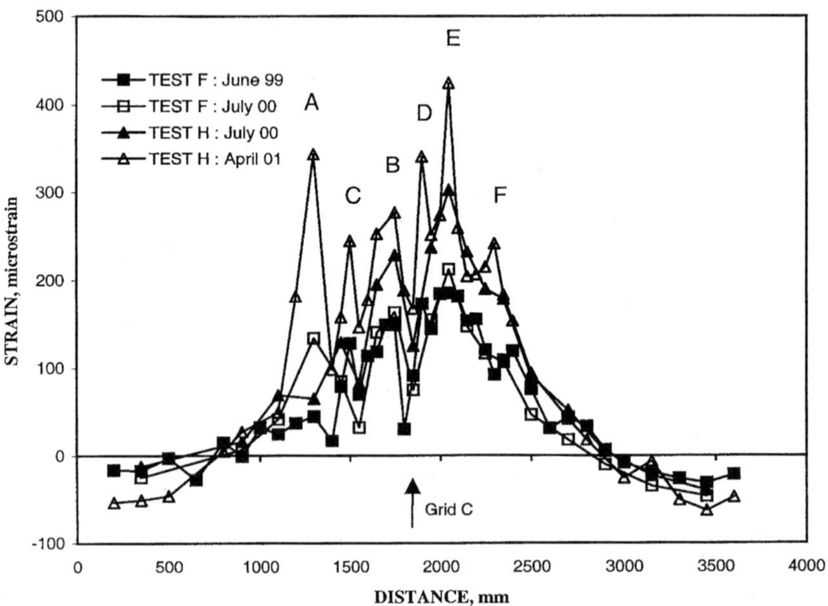

Figure 5 Strain distributions for Bar 7

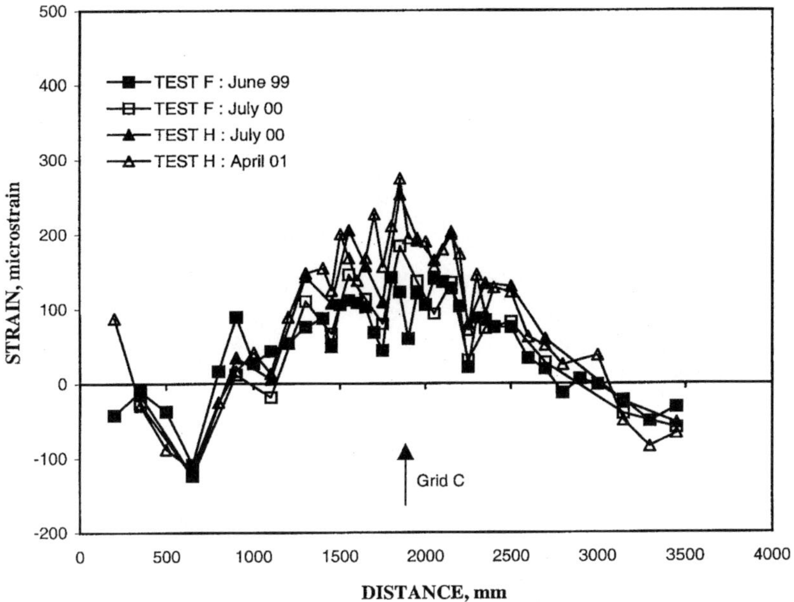

Figure 6 Strain distributions for Bar 8

The July 2000 load increase to form Test H raised strain levels in all these three bars, but the last site visit, in April 2001, indicated that localised changes had occurred during the intervening period. Two new peaks had appeared in the strains in Bar 6, points A and B in Figure 4, which were also present, and more prominent, in Bar 7 (Figure 5).

Bar 7 also had four other new peaks, at points C, D, E and F in Figure 5. Thus, Bars 6 and 7 had recorded the development of new cracks at positions A and B and the propagation of existing cracks at positions C-F. By contrast, there was little change in the strain distributions in Bar 8 over the same period, indicating that Bar 8 was beyond the zone of this new crack development.

Overall, the gauged bars reliably recorded slab strains over an extended time period. They were also sensitive to localised changes in slab behaviour, such as crack propagation. The next phase of the work is to correlate these measurements with predictions from a finite element analysis. This activity is currently being undertaken in association with Arup Research and Development.

CONCLUSIONS

1. Seven internally strain gauged reinforcing bars were successfully installed in April 1998 in the slab at sixth floor level of the full-scale reinforced concrete building at BRE Cardington.

2. A programme of long term service load testing was successfully conducted for two load intensities; 3.0kN/m^2 (Test F) and 4.7kN/m^2 (Test H). Up to 150 sand bags were used, each weighing 1 tonne.

3. The gauges bars gave detailed and reliable strain data over the two years of the test period.

Creep effects in the slab were found to be minimal due to its heavy reinforcement layout and age when loaded. However, the gauges clearly recorded the effects of crack propagation.

ACKNOWLEDGEMENTS

The authors thank the technical staff at the University of Durham and BRE Ltd for assistance with all aspects of the test programme and they are also grateful for the continuing support of Arup Research and Development. The Engineering and Physical Sciences Research Council (EPSRC) is thanked for funding both the manufacture of the strain gauged bars and the programme of site testing.

REFERENCES

1. ALLEN J.D., Reengineering the design and construction process, The Structural Engineer, 76, No. 9, May 1998, pp 175-179.

2. SCOTT R.H., Intrinsic mechanisms in reinforced concrete beam/column connection behaviour, Proceedings of the American Concrete Institute (Structural Journal), 93, No. 3, May-June 1996, pp 336-346.

3. BS8110, STRUCTURAL USE OF CONCRETE, Parts 1-3, British Standards Institution, 1985 (Part 1 republished 1997).

4. BEEBY A.W., SCOTT R.H. & ANTONOPOULOS A., Long term tension stiffening effects in reinforced concrete members, Challenges of Concrete Construction, Dundee, September 2002.

A STUDY ON MEASUREMENT SYSTEM FOR AUTOGENOUS SHRINKAGE OF CEMENT MIXES

T Nawa

Hokkaido University

T Horita

Hokkaido Electric Power Co Inc

H Ohnuma

Hokkaido University

Japan

ABSTRACT. We proposed a measurement method that can be measured immediately after casting in mold using the thin strain gauge having remarkably lower elastic modulus. The pretreatment of strain gauge and the temperature correction of strain were discussed. The effects of cement type, water cement ratio and the addition of superplasticizer on autogenous shrinkage of cement paste were investigated by using the proposed measuring method. The measurement results indicate that the autogenous shrinkage can be divided into four stages: initial period, induction period, the acceleratory period and a deceleratory slow period. The cement type affects strongly the autogenous shrinkage over all stages. The water cement ratio influences markedly the autogenous shrinkage within the age of 1day, particularly the acceleration-period shrinkage, but have less influence after the age of 1day. On the other hand, the superplasticizer influenced mainly the length of induction period of autogenous shrinkage. In comparison with temperature change due to cement hydration, it could be concluded that the acceleratory period of autogenous shrinkage and expansion of cement paste may be closely linked with the hydration of tricalcium silicate in cement.

Keywords: Autogenous shrinkage, Measurement method, Water cement ratio, Cement type, Superplasticizer, Hydration.

T Nawa is an associate professor in department of structural and geotechnical engineering, Hokkaido University, Sapporo, Japan. His research focuses mainly on development of cementitious material and superplasticizer to improve fluidity and autogenous shrinkage for high performance concrete.

T Horita is currently undertaking research into the utilization of fly ash to high performance concrete in the civil engineering section, department of research and development at Hokkaido Electric Power Co Inc.

H Ohnuma is a professor in department of structural and geotechnical engineering, Hokkaido University, Sapporo, Japan. His current research interest is the strength and the deformation of concrete under triaxial compression.

INTRODUCTION

High performance concrete such as high strength and self-compacting concrete has been rapidly moved from the laboratory to field use. For the high performance concrete with extremely low water binder ratio, however, the chemical shrinkage that occurs as the hydration of cement created the empty pores within cement paste, leading extensive reduction in its internal relative humidity. It is referred to as a self-desiccation. This causes remarkably large autogenous shrinkage of cement paste, thus providing the premature concrete with the intrinsic potential of cracks [1]. Indeed, several authors state that high performance concrete may crack as a consequence of restrained autogenous deformation [2-4].

This volume change involved by the hydration reaction of cement starts from the beginning of the making concrete. Aïtcin [5] pointed out the rate of autogenous shrinkage is highest up to an age of 1 day. Hence it is necessary to measure the autogenous shrinkage developed during first 24 hours from immediately after mixing in order to elucidate the mechanism of autogenous shrinkage. In spite of the obvious need for a comprehensive study of their behavior at early age, only limited information is available because early-age strength and elastic modulus of cementitious materials are so low that it is difficult to remove the deformations of the freshly mixed concrete from the volume change completely.

The deformation of fresh concrete hardly induced residual stress due to its plastic properties. On the other hand, the autogenous shrinkage occurs as cement hydrates, and thus the residual stress is surely generated in cement mixes although its value is smaller because of low elastic modulus of cement mixes. From these considerations, we propose a new measurement method that can measure the autogenous shrinkage from about immediately after mixing, by using the strain gauges covered with silicone with a low elastic modulus.

PROPOSED MEASURING SYSTEM

Strain gauge having low elastic modulus

For measuring the strain of the cement mixes, we used the thin strain gauge with 60 mm long covered with waterproof silicone sealant. Its elastic modulus is about 1.2 N/mm^2. The strain gauge could detect the strain when the elastic modulus of cement mixes exceeds the elastic modulus of gauges. According to Matsufuji et al [6], the dynamic modulus of cement paste made with OPC is around 1 N/mm^2 at 30 minutes after mixing at 0.45 of water cement ratio (W/C). When the W/C is around 0.27, the dynamic modulus is as high as 30 N/mm^2 at 10 minutes after mixing. Judging from the results by Matsufuji et al, it seems that the elastic modulus of neat cement paste reaches 1.2 N/mm^2, which is equal to that of the gauges, at about 30 minutes after mixing even for paste with a higher W/C. By this measurement method, therefore, it is possible to measure the autogenous shrinkage at earlier age than the initial setting time.

Outline of device

Figure 1 shows an outline of strain gauge and thermocouples equipped with a digital computer system. Strain gauge and thermocouples were vertically set in the center of cylindrical mold 50 mm in diameter placed and 100 mm in height as illustrated in Figure 1. A polytetrafluoroethylene (Teflon) sheet was placed between the cement mixes and mold to reduce friction between them. Cement mixes just after mixing was poured into the mold. The top surface of the sample was sealed after placing in order to prevent moisture dissipation. Then both shrinkage strain and temperature changes were continuously measured immediately after fabrication. Furthermore, whether a gauge was adequately embedded was confirmed by cleaving the specimen after demolding.

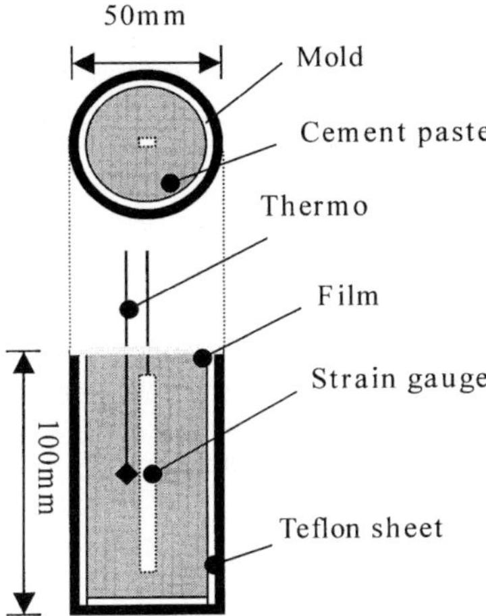

Figure 1 Measuring method of strain and temperature of cement paste

EXPERIMENTAL

Materials and mixture proportions

Ordinary Portland cement (OPC), high early strength Portland cement (HPC) and high belite Portland cement (LPC) was used as cement. The mineral cömposition calculated from Bogue's equations and physical properties of cements are shown in Table 1. A polycarboxylic-based superplasticizer was used as the chemical admixture. Neat cement pastes were proportioned with five levels of W/C: 0.25, 0.30, 0.35, 0.40, and 0.50. Neat cement paste specimens with a W/C of 0.25 were added the polycarboxylic-based superplasticizer at dosage of 0.1% and 0.3% by weight of cement.

Table 1 Mineral composition and physical properties of cements

TYPE OF CEMENT	MINERAL COMPOSITION (%)				DENSITY (g/cm^3)	BLAINE SURFACE AREA (cm^2/g)
	C_3S	C_2S	C_3A	C_4AF		
OPC	50	25	8.3	8.8	3.16	3390
HPC	65	58	8.5	7.6	3.14	4520
LPC	24	11	2.4	7.6	3.22	3490

Calculation of autogenous shrinkage strain

Cement actively hydrates during measurement of autogenous shrinkage, and the temperature of specimens is increased by hydration heat. Furthermore we should correct the strain for thermal expansion taking into account the difference in the thermal expansion coefficients between the cement mixes and gauges.

However, due to the striking differences between the elastic modulus and cross-sectional area ratios of gauges and samples, the effect of difference of thermal expansion coefficient between the cement mixes and gauge was able to be ignored, and the correction for thermal expansion of the autogenous shrinkage strain was carried on as shown in the following:

$$\varepsilon_a = \varepsilon - (\alpha_x - k)\Delta T \tag{1}$$

where ε_a is autogenous shrinkage strain after correction, ε is the measuring strain, ΔT is the increment of temperature, α_x is thermal expansion coefficient of cement paste to be 20×10^{-6}/°C, and k is the self-correction factor for temperature of strain gauge.

RESULTS AND DISCUSSION

Effect of pretreatment of strain gauges

Since the gauges produced in this study on a trial basis are covered with silicone gel, their pretreatment can affect the equilibrium moisture conditions between the sample and gauge, thereby affecting the strain to be measured. For this reason, the effect of pretreatment was confirmed by embedding gauges with two types of pretreatment: air-drying and 24-hour immersion in deionized water. Specimens were fabricated using neat cement paste made of ordinary Portland cement (density: 3.16 g/cm^3, specific surface area: 3350 cm^2/g) and deionized water as mixing water with a W/C of 0.30. The measurements are shown in Figure 2.

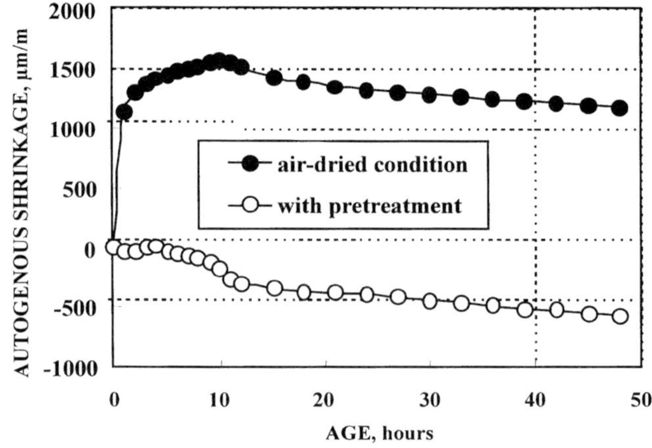

Figure 2 Influence of pretreatment of strain gauge on measuring strain

Gauges embedded in an air-dried condition significantly expanded from immediately after placing. This implies that silicone in air-dried gauges absorbs moisture from cement paste and causes the gauge to expand, leading to apparent expansion of paste. On the other hand, gauges immersed in deionized water for 24 hours led to a state of steady strain immediately after placing at about zero, suggesting no appreciable effect of moisture absorption of the gauges from paste. The data for gauges immersed in deionized water for more than 24 hours were similar to those immersed for 24 hours. Accordingly, 24-hour immersion in deionized water was found appropriate as pretreatment for gauges.

Effect of water cement ratio

Figure 3 shows the effect of water cement ratio on the autogenous shrinkage of neat cement paste. From the results, it could be found that the autogenous shrinkage can be divided into four stages: (1) initial shrinkage, (2) induction period, (3) the acceleratory period, in which the main shrinkage first begins to occur rapidly, and (4) a period of slow, continued shrinkage. Thus several different mechanisms must be involved in the autogenous shrinkage in order for the process to pass through these different stages.

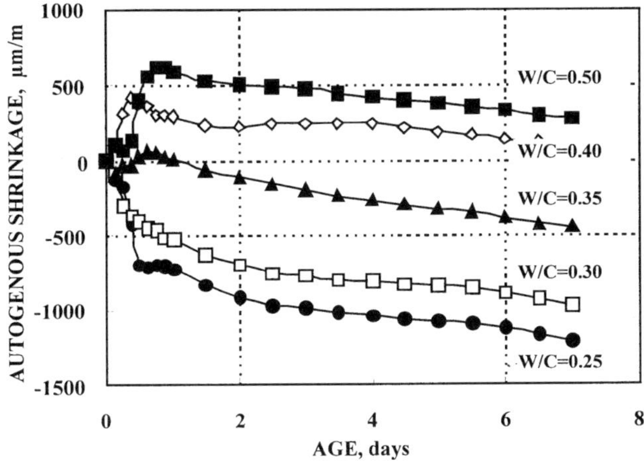

Figure 3 Influence of water cement ratio on autogenous shrinkage of cement paste

The autogenous shrinkage of neat cement paste is strongly affected by W/C, being higher as the W/C becomes lower. In particular, both the initial shrinkage and the acceleration-period shrinkage up to an age of 1 day were strongly influenced by W/C. After 1 day, the slowly continuing shrinkage of paste was similar for all paste regardless of W/C, with the increment of shrinkage strain being at about 500×10^{-6} between the age of 1 and 7 days.

The shrinkage is dependent on both the contracting force and rigidity of paste. Thus we could suppose that when rigidity of paste might overcome the force of contraction, the shrinkage of paste terminated despite of progressive cement hydration. Indeed Takahashi et al [7] reported that the starting point of induction period is related to the penetration time of a 3000g-loaded needle into paste. Therefore, it is deduced that initial shrinkage ends when a rigidity of paste overcome the contract force.

Meanwhile, a comparison the shrinkage behavior with the hydration of cement is useful to elucidate the mechanisms of autogenous shrinkage. Figures 4 and 5 show the strain and temperature of the neat cement paste as a function of time after mixing. Cement paste with a W/C of 0.25 exhibited temperature rise from immediately after mixing.

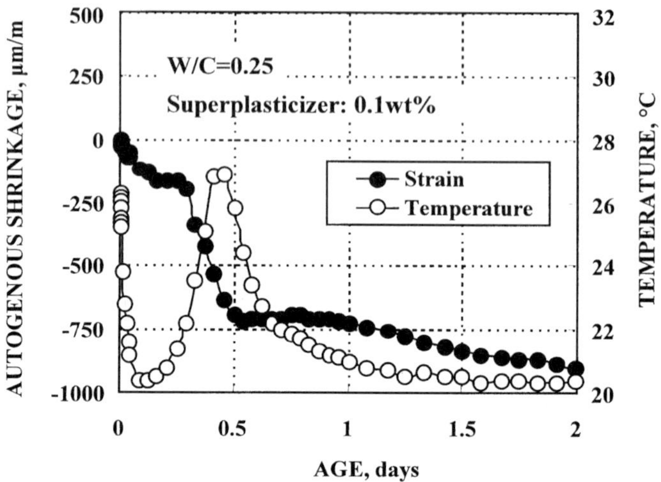

Figure 4 Changes in autogenous shrinkage and temperature of cement paste

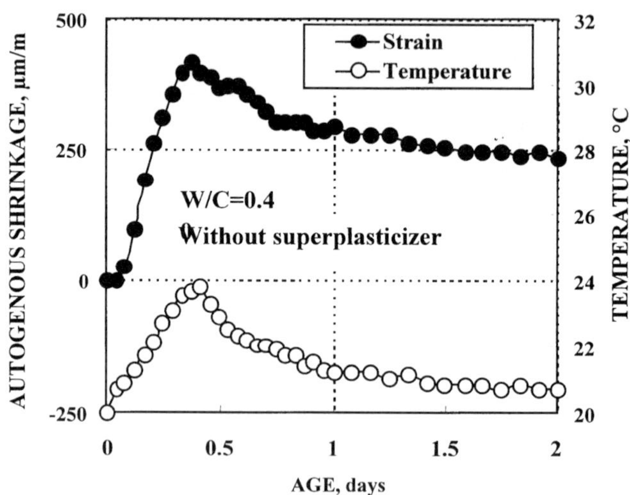

Figure 5 Changes in autogenous shrinkage and temperature of cement paste

This indicates that when W/C was 0.25, a large shrinkage occurred simultaneously with the temperature rise due to the active hydration of C_3S, whereas the specimens at W/C of 0.40 conversely expanded about the same time the rapid hydration of C_3S begins.

Ding et al. [8] has also been reported that the neat cement paste with a high W/C was expanded at most early. Hanehara et al [9] reported that the humidity in paste with a higher W/C of 0.50 was maintained around 100% while paste with a low W/C of 0.30 showed the remarkable reduction of internal humidity.

From these results, it seems that the acceleratory period of autogenous shrinkage and expansion of cement paste may be closely linked with the hydration of tricalcium silicate in cement: when W/C of paste is low, calcium silicate hydrates produced at acceleratory period in hydration of alite in cement promote to decrease humidity in paste, then inducing a shrinkage.

For paste with high W/C, however, calcium silicate hydrates exhibits swelling because the paste is wetting.

Effect of the cement type

The changes of length of pastes for OPC, LPC and HPC are illustrated in Figure 6. It is evident that autogenous shrinkage over all stages is different depending on the cement type. Compared with OPC paste, large autogenous shrinkage was observed for HPC paste at early age, whereas the LPC paste shows significant low autogenous shrinkage.

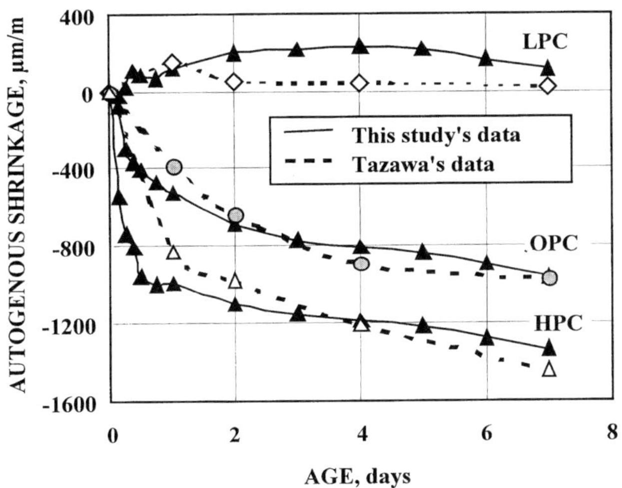

Figure 6 Influence of cement type on autogenous shrinkage of cement paste at w/c = 0.3

In order to verify the validity of the measurements by the method proposed in this study, we compared the data obtained in this study with prior data measured with contact gauge by Tazawa and Miyazawa [10] as shown in Figure 6. The broken lines indicate the data by Tarawa's method.

As shown in this figure, measurements by the method proposed in this study are well in agreement with that by Tarawa's method for all cement types, suggesting the validity of data measured by the method proposed in this study.

Effect of superplasticizer

Figure 7 shows the autogenous shrinkage and temperature of cement pastes at W/C of 0.25 with superplasticizer of 0.3 % to cement by weight. Compared with the results shown in Figure 4, a large dose of superplasticizer delayed significantly the rise of temperature for cement paste, suggesting that the hydration of C_3S in cement retarded by the addition of superplasticizer. It can also seen that the large addition of superplasticizer leads to a delay in the starting of acceleration period of shrinkage and the beginning of hydration of C_3S is well corresponded with that of autogenous shrinkage. In other words, the superplasticizer influenced mainly the length of induction period of autogenous shrinkage. Therefore we are sure that the autogenous shrinkage at acceleratory period of cement paste may be closely linked with the hydration of tricalcium silicate in cement.

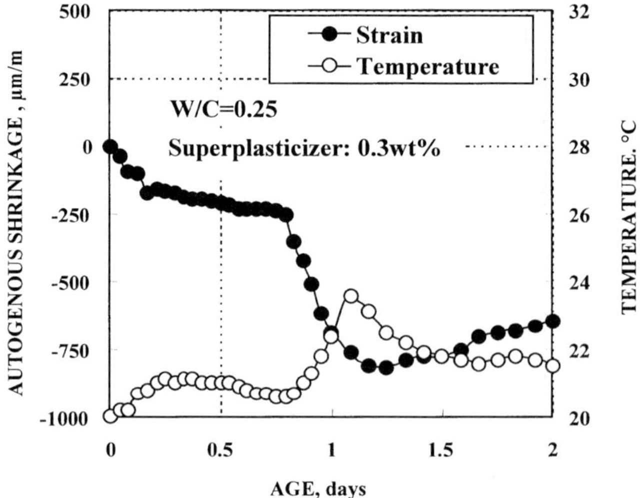

Figure 7 Influence of added superplasicizer on autogenous shrinkage of cement paste

CONCLUSIONS

In this study, the authors proposed a method of measuring autogenous shrinkage using low-elasticity strain gauges. After verifying the measuring conditions and measured data, this method was applied to the measurement of autogenous shrinkage of neat cement paste.

It was found that 24-hour immersion of strain gauges in deionized water is appropriate as pretreatment within the range of this study. It was also found that the temperature strain of the gauges is ignorable due to the large difference between the elastic moduli of the samples and gauges.

Compared with a prior study, the present method was found to lead to similar measurements, and thus it is considered sufficiently early for capturing autogenous shrinkage strain at very early ages.

Shrinkage study indicates that the autogenous shrinkage can be divided into four stages: initial period, induction period, the acceleratory period and a period of slow. The reason why the induction period appears in the shrinkage process is inferred that an increase in the rigidity of paste exceeds the increment of the shrinkage force.

The cement type affects strongly the autogenous shrinkage over all stages. The water cement ratio influences markedly the autogenous shrinkage within the age of 1day, particularly autogenous shrinkage at the acceleratory period, but have less influence at deceleration period after the age of 1day. On the other hand, the superplasticizer influenced mainly the length of induction period of autogenous shrinkage. In comparison with temperature change mainly due to C_3S hydration in cement, it could be deduced that the acceleratory period of autogenous shrinkage and expansion of cement paste may be closely linked with the hydration of tricalcium silicate in cement.

ACKNOWLEDGMENTS

We would like to thank Prof. Moriyoshi, A of Hokkaido University for invaluable advice and useful suggestions. We also acknowledge the support for the project by Kyowa Co Ltd.

REFERENCES

1. TAZAWA, E and MIYAZAWA, S and SHIGEKAWA, K, 'Macroscopic Shrinkage of hardening cement paste due to hydration, CAJ proceedings of Cement and Concrete', No 47,1991, p 528.

2. PAILLÈRE, A M et al, 'Duirabilité du béton à très hautes performances: incidence du reteait d'hydratation sur la fissuration au Jeune âge', Proceedings of the First International RILEM Congress on From materials Science to Construction Materials Engineering, Vol 3, 1987, pp 90-997.

3. TAZAWA, E and MIYAZAWA, S, 'Autogenous shrinkage of concrete and its importance in concrete technology', Proceedings of the Fifth International RILEM Symposium on Creep and Shrinkage of Concrete, Barcelona, Spain, 1993, pp 159-168.

4. SELLEVILD, E J, 'The function of condensed silica fume in high strength concrete', FIP Notes, Vol 4, 1987, pp 11-14.

5. AÏCTIN, P C, 'Autogenous shrinkage measurement', Autogenous Shrinkage of Concrete, edited by Tazawa, E, E & F N Spon, 1999, pp 257-268.

6. MATSUFUJI, Y, and KAWAKAMI,Y, 'Study on setting Process of cement paste by urtrasonic transverse wave velocity method', CAJ proceedings of Cement and Concrete', No 34 ,1980, pp 63-66

7. TAKAHASI, T, NAKATA, H and GOTO, S, 'An apparatus for measurement of autogenous srinkage of cement paste', Proceedings of the Japan Concrete Institute, Vol 18, No 1, 1996, pp 621-626.

8. CHEONG H M et al, 'Effect of high temperature history from hydration heat evolution on autogenous shrinkage in high strength concrete', Proceedings of thc Japan Concrete Institute, Vol 21, No 2, 1999, pp 1117-1122.

9. HANEHARA, S HIRAO, H and UCHKAWA, H, 'Relationship between autogenous shrinkage, and the microstructure and humidity changes at inner part of hardened cement paste at early age', Autogenous Shrinkage of Concrete, edited by Tazawa, E, E & F N Spon, 1999, pp 93-104.

10. TAZAWA, E and MIYAZAWA, S, 'Influence of constitutes and composition on autogenous shrinkage of cementitious materials' magazine of Concrete Research, Vol 49, No 178, 1997, pp 15-22.

EMISSIONS FROM CONCRETE FLOOR STRUCTURES AND COVERINGS DURING AND AFTER MOISTURE DAMAGE

V Penttala

L Wirtanen

Helsinki University of Technology

Finland

ABSTRACT. The emission of volatile organic compounds (VOC), ammonia, and formaldehyde and the moisture behaviour of different floor structures subjected to a moisture load were studied under laboratory conditions with the aim of designating the least emitting structures. The emitting compounds were collected primarily using a Field and Laboratory Emission Cell (FLEC). Additional samples were taken from the indoor air and from inside the structures. The emission rate of the total amount of VOC (TVOC) was in general 80-100 $\mu g/m^2 \cdot h$. It exceeded 200 $\mu g/m^2 \cdot h$ principally at the first measurement, i.e. one month after the installation, of vinyl carpet or mosaic parquet coatings. The external moisture load did not prominently influence the TVOC-values, but excess moisture caused changes in the emission rates of single compounds. The results indicated that it is important to choose low-emitting materials in order to guarantee low emission rates also during and after possible moisture damage.

Keywords: Emission, VOC, Ammonia, Moisture, Floor structures.

Mr L Wirtanen, is currently working on his doctorate thesis on the influence of moisture on the chemical emission from wall structures.

Professor V Penttala, has been working as the Professor of Building Materials Technology at Helsinki University of Technology in Finland since 1982. His main research interests have been durability aspects of concrete and microstructure of building materials. During the last years he has done research on the freezing and thawing durability of concretes, the emissions from building materials during and after moisture damage, very high-strength concretes, and smart, multifunctional building structures.

INTRODUCTION

Statistically more than 60 % of the residential buildings in Finland are at least once damaged by excess moisture during their service life. A moisture load is caused e.g. by improperly performing structural components or by a leaking plumbing. The severest consequence of a moisture load is hydrolytic degradation reactions in the material, which leads to the emission of different material specific chemical compounds.

The Concrete Association of Finland has set guideline values below which the coating of a concrete floor is considered safe [1]. The guideline values, which are given in Table 1, are a compromise between the coating application work and the functionality of the coating. When the relative humidity of the concrete floor is below the given maximum humidity value, the prerequisites for a successful coating are given.

Table 1 Guideline values for the successful coating of a concrete floor [1]

MAX. RELATIVE HUMIDITY OF CONCRETE (%)	COATING	COMMENTS
80	Mosaic parquet	The wooden coating loosens because of movements caused by its hygroscopicity
85	Plastic carpets with a felt or cellular plastic based substructure Rubber carpets Water proof cork plates Textile carpets with a plastic sub-structure Carpets made of natural materials without a plastic substructure	Micro-organisms, moisture durability of the adhesives
90	Plastic plates Plastic carpets without a felt or cellular plastic based substructure Linoleum carpets Wooden floors that are not attached to the substructure, a moisture barrier between the concrete and the wood Polyurethane based plastics Synthetic textile carpets without a substructure	Most adhesives deteriorate in high relative humidities and there might occur changes in the coating The moisture barrier under a wooden floor is e.g. a 0,2 mm thick plastic sheet with taped overlapping joints.
97	Epoxy, acryle, and polyester based plastic moulding compounds	The surface of the concrete floor has to be dry and warm enough

The emitted compounds decrease the quality of indoor air and might even cause adverse health effects. Saarela presented in her report [2] different material specific humidity levels for moisture and alkaline moisture induced degradation reactions and chemical emissions caused by these. The results allocated for this work are shown in Table 2.

Table 2 Examples of humidity levels that cause moisture induced degradation reactions in materials and the emissions resulting from them [2]

MATERIAL	RH (%)	DAMAGE AND EMISSION
PVC-carpets	75-85	Dyeing, degradation reactions, 2-ethyl-1-hexanol
Waterborne adhesives	>95	Saponification, material specific emissions
Urea-formaldehyde-based resins in lacquers and binders in insulating	>60-70	Formaldehyde
Micro-organisms	>85	MVOC

The emissions from a floor diffuse into the indoor air from the surface of the structure and also from cracks and joints between the floor and the wall, as shown in Figure 1.

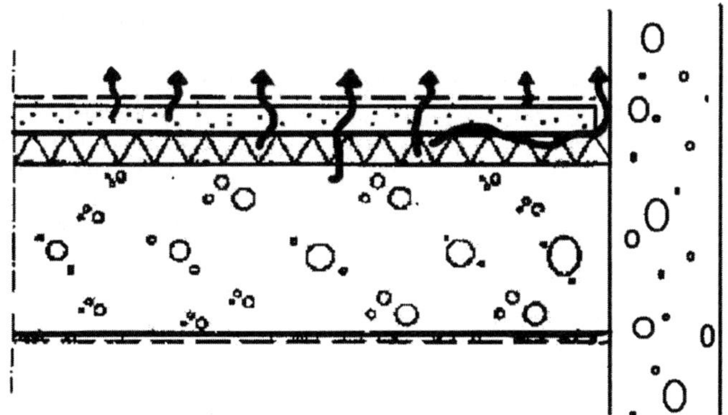

Figure 1 Diffusion of emitted compounds into the indoor air

The objectives of this study were to survey the mechanisms of moisture damage and the subsequent chemical emissions of common floor structures, and to find out which structures have the least emissions. This was accomplished by exposing the structures to an external moisture load, measuring the moisture content of the structures, and collecting the emitting compounds (VOC, ammonia, and formaldehyde) periodically. The FLEC and an overview of the test samples are shown in Figure 2.

METHODS

Five different ground floor structures and ten different intermediate floor structures were studied in this research project. The structures were built according to the Finnish building code taking the material manufacturers' recommendations into consideration. One test sample of each structure was exposed to the primal external moisture load, while the other was used as a reference sample.

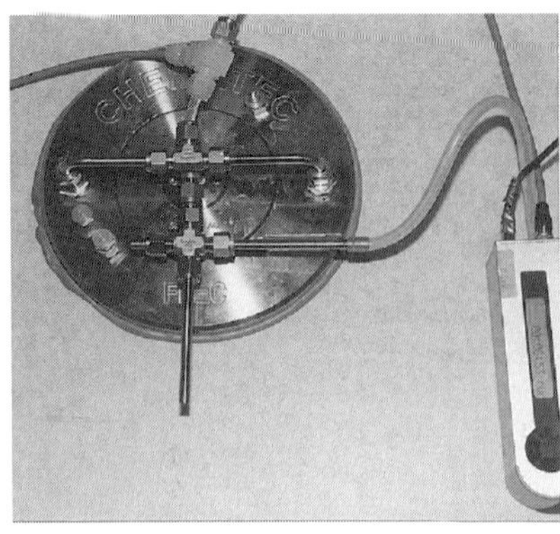

Figure 2 The FLEC (to the left) and an overview of the test samples (right)
The floor structures are on the left-hand side in the picture

The structures used in this study had authentic cross-sectional dimensions. Therefore, both the material specific emissions and the effects of the external moisture loads could be investigated. The material specific moisture load constitutes mainly water, which is added to the concrete and the levelling compound. This water diffuses later into the other construction components and is then usually very alkaline. The levelling sand underneath the ground floor structure was kept wet throughout the whole testing period and this formed thus an additional external moisture source. The primal external moisture load constituted tap water, 10 litres per m², which was poured on the structures after 6 months of storage. The coatings were slit before the pouring of the tap water. The cutting was 5 mm wide and 300 mm long. This enabled the water to intrude under the coating. Eight days later the coatings were removed. All structures, except the one that was coated with hardwood mosaic parquet, were repaired with the same covering material as was originally used. A vinyl carpet replaced the hardwood mosaic parquet. New coatings were installed after a six-week drying period. The installation of the vinyl carpet coating was preceded by the application of a levelling compound layer two weeks earlier. The vinyl carpet was glued with a water-soluble dispersion-type acrylic-based adhesive. The hardwood mosaic parquet was glued with a two-component solvent free polyurethane-based adhesive. The lacquer used was a water-soluble dispersion-type acrylate-polyurethane-based lacquer. The material combinations of the different floor structures used are shown in Table 3 and Figures 3-5. EPS denotes expanded polystyrene, VC is vinyl carpet, and LPC stands for laminated parquet composite.

The moisture content of the structures was recorded using temperature and relative humidity measuring probes. The measuring probes were installed into PVC-tubes that were situated at appropriate locations in the structure as shown in Figure 6.

Table 3 Material combinations of the different floor structures
Structures A-E were ground floor structures (a concrete slab on top of a thermal insulation)

	THERMAL / SOUND INSULATION	FLOATING STRUCTURE	COATING
A-D	rock wool / EPS	-	VC / LPC
E	rock wool	-	hardwood mosaic parquet
F-G	concrete slab	-	VC / LPC
H-K	rock wool / glass wool	concrete slab	VC / LPC
L-M	rock wool	plasterboard	VC / LPC
N-O	rock wool	mortar slab	VC / LPC

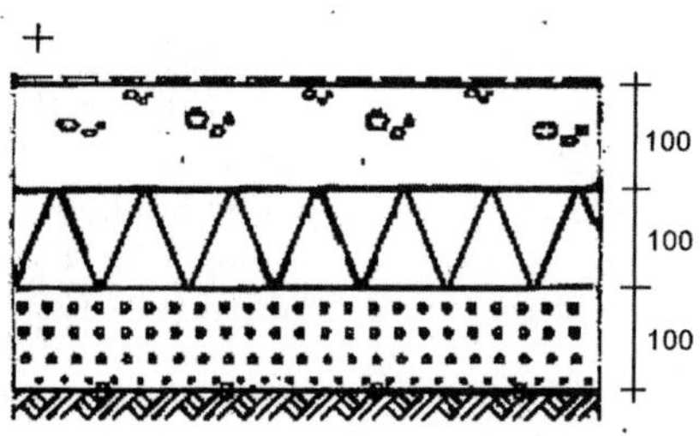

Figure 3 A ground floor structure (structure A-E).

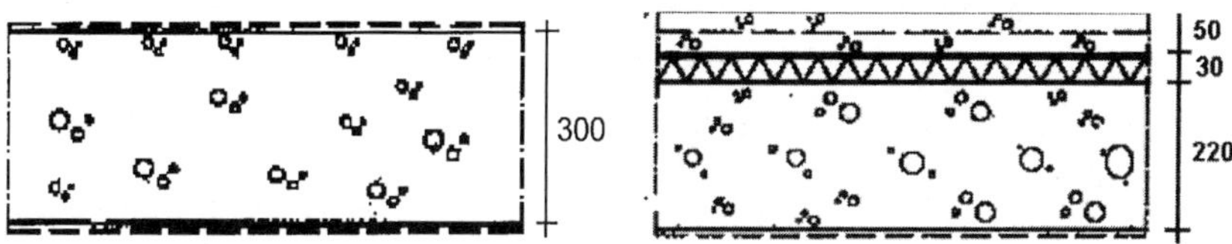

Figure 4 Intermediate floor structures
Structure F-G on the left and structure H-K on the right

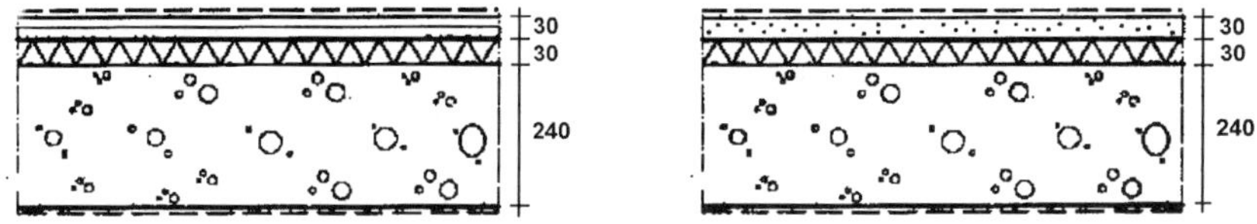

Figure 5 Intermediate floor structures
Structure L-M on the left and structure N-O on the right

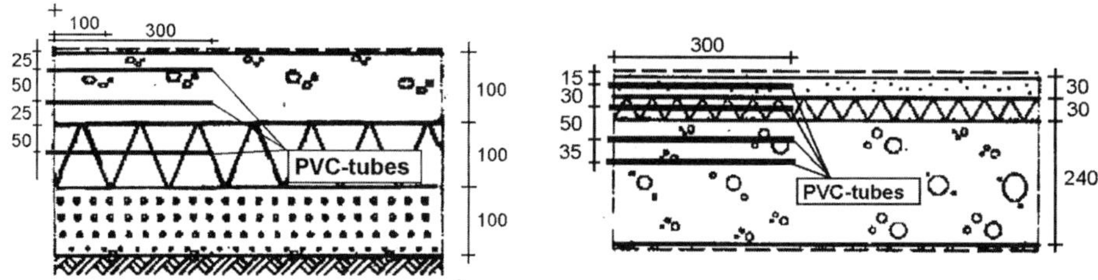

Figure 6 PVC-tubes installed into a ground floor structure (left) and an intermediate floor structure (right) with a floating mortar slab

The emissions of VOC, ammonia and formaldehyde from the surface of the structures were collected using a FLEC and an area specific emission rate was calculated according to a European standard [3]. VOC and ammonia in the indoor air and from inside the different structural components were collected by drawing air directly through test tubes. The tubes were placed near the test specimens for the sampling of the indoor air and into stainless steel tubes that were put inside PVC-tubes used for the humidity measurements for the sampling of the different materials. The orifice of the test tube was thus in the vicinity of the material in question. The air flow rates and the sample sizes used are shown in Table 4. VOC-samples were collected into Tenax TA-tubes and analysed using the GC/MS method. Ammonia samples were collected into a dilute (0.1 M) sulphuric acid solution using an impinger tube and analysed with an ion-specific electrode. Formaldehyde samples were collected into a sodium-disulphite solution using and impinger tube and analysed according to a Finnish standard [4]. The analysis of the samples was carried out at the Institute of Occupational Health in Helsinki.

Table 4 Airflow (ml/min) and sample volume (l) used for samples collected from the indoor air and from inside the structures

	AIR SAMPLES	SAMPLES COLLECTED FROM INSIDE THE STRUCTURES
VOC	250 ml/min, 10 l	100 ml/min, 3 l
Ammonia	500 ml/min, 50 l	500 ml/min, 50 l
Formaldehyde	500 ml/min, 50 l	500 ml/min, 50 l

RESULTS

The increased moisture content of the insulating material due to an external moisture load increased also the humidity level of the overlying structures irrespective of the structure or material in question. Changes in the humidity distribution of the different materials correlated with their porosity characteristics. The porosity of concrete, mortar and gypsum measured by mercury intrusion porosimetry were 10 %, 30 %, and 60 %, respectively. The humidity distribution of the three floating structures is shown in Figure 7. In Figures 8 and 9 is shown the TVOC emitted from the surface of different structures.

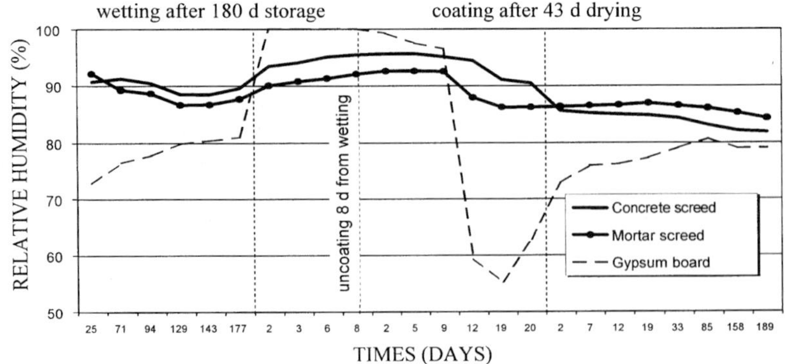

Figure 7 The relative humidity history of floating floor structures made of concrete, gypsum, and mortar at different stages of the study. Observe that the time axis is not pitched

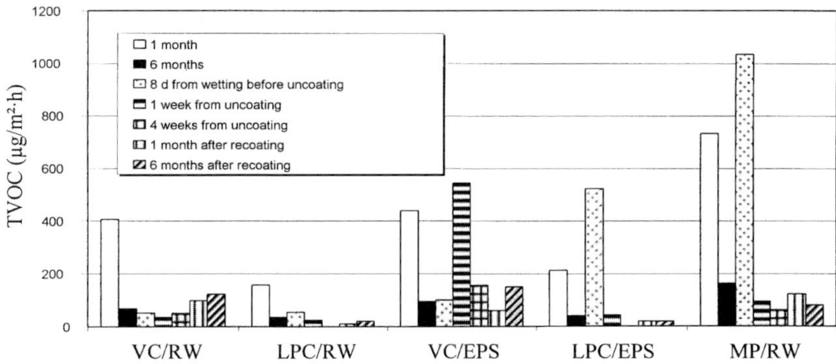

Figure 8 The emission rate of TVOC emitted from the surface of ground floor structures at different stages of the study. VC denotes vinyl carpet, LPC is laminated parquet composite, RW is rock wool, EPS stands for expanded polystyrene, and MP is hardwood mosaic parquet

The primal external moisture load did not influence the TVOC emitted from the surface of the structures the mosaic parquet coating excluded. Its influence could, on the other hand, be observed when examining single chemical compounds. The emission of ammonia was highest when the coatings had been stripped, and the rate of drying of the underlying structures was most rapid.

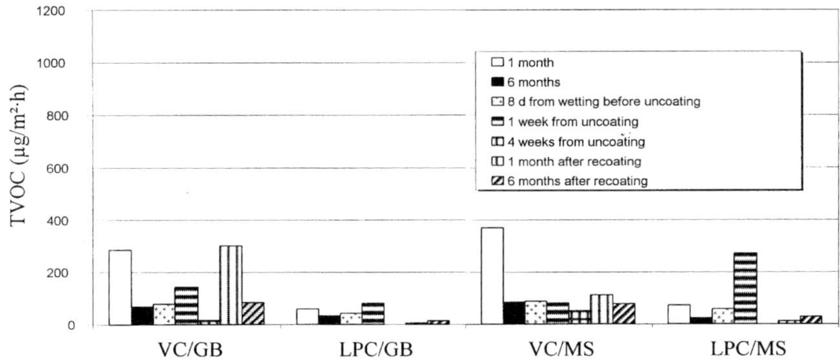

Figure 9 The emission rate of TVOC emitted from the surface of intermediate floating floor structures made of gypsum board or mortar slab at different stages of the study. VC is vinyl carpet, LPC denotes laminated parquet composite, GB is gypsum board, and MS stands for mortar slab

Results including TVOC, some single organic compounds, and ammonia emitted from the surface of the structures covered with vinyl carpet or laminated mosaic parquet are shown in Figures 10-13. The results have been calculated as an average of all the different substructure types.

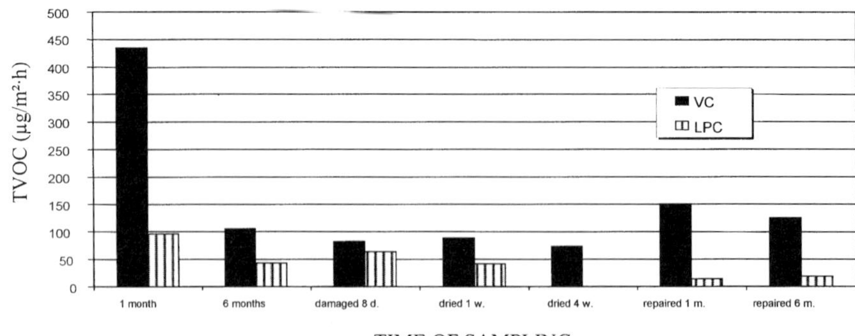

Figure 10 The emission rate of TVOC emitted from the surface of the structures covered with a vinyl carpet (VC) or a laminated parquet composite (LPC) at different stages of the study. The emitted compounds were collected using a FLEC

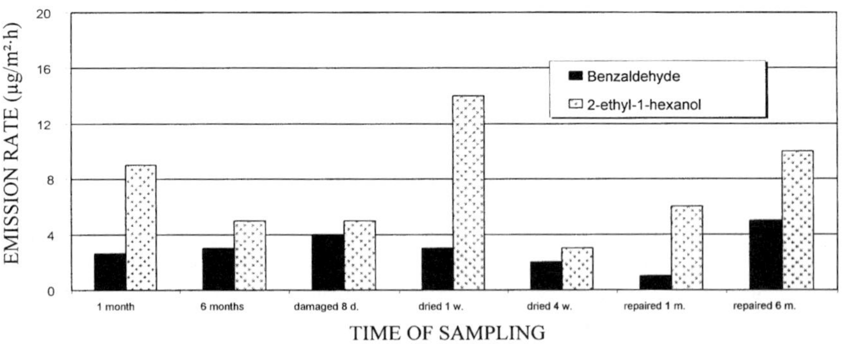

Figure 11 The emission rate of benzaldehyde and 2-ethyl-1-hexanol emitted from the surface of structures covered with a vinyl carpet at different stages of the study. The emitted compounds were collected using a FLEC

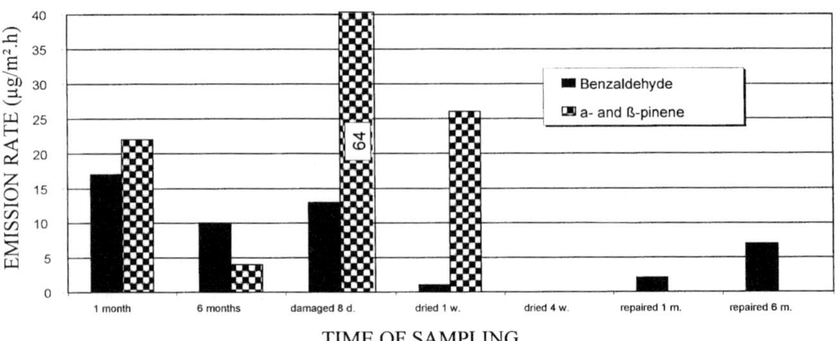

Figure 12 The emission rate of benzaldehyde and α- and β-pinenes emitted from the surface of structures covered with a laminated parquet composite at different stages of the study. The emitted compounds were collected using a FLEC

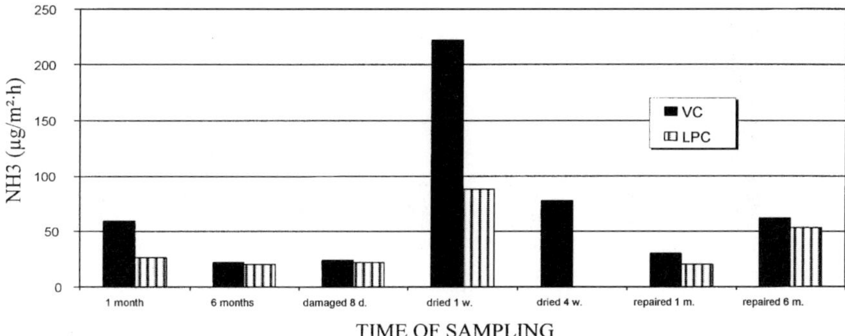

Figure 13 The emission rate of ammonia emitted from the surface of the structures covered with a vinyl carpet (VC) or a laminated parquet composite (LPC) at different stages of the study. The samples were collected using a FLEC

DISCUSSION

The chemical emissions from the floor structures covered with vinyl carpet or hardwood mosaic parquet were higher than from those covered with a prefabricated laminated parquet composite. The differences were mostly due to the adhesive and lacquer that was used for the installation of the vinyl carpet and the mosaic parquet.

The external moisture load did not affect the TVOC-values prominently, the mosaic parquet coating excluded. Some material specific emissions, however, could be noticed. The moisture load caused an increase in the emission of terpenes originating from the laminated parquet composite. Uncovering the structures after the induced water damage caused, on the other hand, an increase in the emission of e.g. 1-butanol, and 2-ethyl-1-hexanol, originating from the adhesive used for the installation of the vinyl carpet. To which extent this was caused by the external moisture load is difficult to estimate because the vinyl carpet creates a rather impermeable coating. The water that was poured on the surface of the structures ran, in the case of structures covered with a vinyl carpet or mosaic parquet, usually into the insulating layer and caused an increase in the humidity level of the substrate, thus affecting the coating from underneath. In the case of the laminated parquet composite the water intruded also into the space between the substrate and the coating. The TVOC-value of the mosaic parquet cover increased sevenfold because of the moisture load, being 150 $\mu g/m^2 \cdot h$ after 6 months of storage and 1100 $\mu g/m^2 \cdot h$ 8 days after the induced moisture load before uncovering the structure. The emissions originated mainly from the paint and the lacquer. Its TVOC-level did not, however, differ from that of the other structures TVOC-level one week after the uncovering.

The amount of TVOC and ammonia of the air samples correlated well with the emissions collected from the surface of the structures. The TVOC values showed, thus, a peak after one month of storage and the ammonia values after uncovering the structures, when the emission rates of the samples collected from the surface of the structures were also largest. The amount of formaldehyde did not, with a few arbitrary exceptions, exceed the limit of detection.

The TVOC values of the samples collected from inside the structures varied a lot. The expanded polystyrene used in the ground floor structures had an average TVOC value of 2800 $\mu g/m^3$ while the rock wool used in the corresponding structure had an average TVOC value of 70 $\mu g/m^3$. The rock wool used as sound insulation in the intermediate floor structures with a floating gypsum board or mortar slab, as well as the gypsum board itself, had TVOC values ranging between 1000 and 4000 $\mu g/m^3$. The composition of the different rock wool products was the same, but the relative humidity of the rock wool in the ground floor structure was close to 100 % while the relative humidity in the intermediate floor structure was 90 %.

Concrete and the mortar slabs had TVOC values ranging between 20 and 250 $\mu g/m^3$. The main components of the expanded polystyrene were styrene and pentane, while those of the rock wool sound insulation and the gypsum board were acetone, 1-butanol, and 2-ethyl-1-hexanol. The expanded polystyrene and the gypsum board had been classified as low emitting materials, while the rock wool was non-classified, according to the Finnish guideline values [5]. The rock wool is low emitting according to the material manufacturer and it has also been classified as such after this project was finished.

The emitting compounds that are generated within a structure diffuse into indoor air through the different material layers and also through cracks and joints. If all the compounds that are generated inside the structural components diffuse into indoor air via the joint between the floor and wall the emission rate in a model room [6] would be according to the results shown in Table 5. The values in Table 5 have been calculated with a specific airflow rate of 1.3 $m^3/m^2 \cdot h$. The characteristics of the model room is shown in Table 6.

Table 5 The emission rate of TVOC for the different materials inside the test samples

MATERIAL	EMISSION RATE ($\mu g/m^2 \cdot h$)
Expanded polystyrene in ground floor slab	3600
Rock wool in ground floor slab	91
Rock wool in intermediate floor slab	1300-12400
Gypsum board	1300-12400
Concrete	30-350
Mortar slab	30-350

Table 6 Characteristics of the model room [6]

MODEL ROOM	AREA SPECIFIC AIRFLOW RATE ($m^3/m^2 \cdot h$)
Room volume 17.4 m^3, $n = 0.5h^{-1}$	
Floor area = 7 m^2	1.3
Wall area = 24 m^2	0.4
Sealant area = 0,2 m^2	44

The emission rate measured at the surface of concrete, gypsum board, and mortar screed one week after the removal of the laminated parquet composite after the induced external moisture load was 40 μg/m²·h, 80 μg/m²·h, and 270 μg/m²·h, respectively. On the basis of these results it can be observed that the method of collecting samples from inside the structure gives results of the same order of magnitude compared with the results that were collected using the FLEC as far as concrete and the mortar slab is concerned. It could also be observed that the compounds that were emitted from the rock wool were at least partly absorbed by the gypsum board. The results shown in Table 5 also indicate that the emission rate of the expanded polystyrene and the rock wool used in the intermediate floor structure were strongly affected by the moisture conditions considering that the materials in question have been classified as low emitting materials according to [5]. The emission rate limit in [5] for low emitting materials is 200 μg/m²·h. The method of sampling from inside the structures is not standardised. One factor that has impact on the results is the airflow rate. Its influence was not, however, tested in this research project.

CONCLUSIONS AND IMPLICATIONS

The aim of the study was to collect information on the moisture behaviour and emissions from real structures during and after a moisture damage. A moisture damage affects the whole structure and may result in additional chemical emissions. Its influence can be observed especially when studying single chemical components. This urges on including the operational humidity levels in the labelling tests of the different building materials.

Taking the moisture behaviour and chemical emissions into consideration, a low emitting ground floor structure according to the results from this study has rock wool thermal insulation and is covered with laminated parquet composite. Using the same criteria, a low emitting intermediate floor structure has a floating concrete slab and is covered with laminated parquet composite. The materials used should be low emitting materials in order to ensure low emissions also in connection with a water damage. Detailed description of this research project has been presented in Leif Wirtanen's licentiate's thesis [7]. Part of the results of the research project has also been presented in Eronen et al [8].

ACKNOWLEDGEMENTS

The authors wish to thank the National Technology Agency in Finland (Tekes), Gyproc Oy, Lohja Rudus Oy, Optiroc Oy, Paroc Oy, Saint-Gobain Isover Oy, Skanska Oy, Upofloor Oy, and VVO-rakennuttaja Oy for their financial support.

REFERENCES

1. BLY 4/BY 31, Concrete floors: Classification, coating, design and construction code of practise. The Finnish Concrete Association, The Finnish Concrete Floor Association, Helsinki, Finland, 1989, p 127.

2. SAARELA, K, Organic emissions from building materials and their reduction (in Finnish), The Finnish Ministry of the Environment, Report 3, 1992, p 68.

3. ENV 13419-2, 1999, Building products – Determination of the emission of volatile organic compounds – Part 2, Emission test cell method, Brussels, Belgium, p 19.

4. SFS 3862, 1981, Air quality: Workplace atmosphere, determination of formaldehyde with chromotropic acid method (in Finnish), The Finnish Standards Association, Helsinki, Finland, 1981, p 9.

5. CLASSIFICATION OF INDOOR CLIMATE, 2000, Requirements for building products: The emission classification of building materials, The Finnish Society of Indoor Air Quality and Climate: Publication 5, Espoo, Finland, p 40.

6. DANISH STANDARD / INF 90, 1994, Directions for the determination and evaluation of the emission from building products (in Danish), Danish Standard, Copenhagen.

7. WIRTANEN, L, The moisture content and emissions of ground slabs and floors between storeys subjected to a moisture load (in Finnish), Licentiate's thesis, Helsinki University of Technology, Espoo, Finland, 2001, p 161.

8. ERONEN, J, WIRTANEN, L, RÄSÄNEN, V, PENTTALA, V, The influence of coatings on the drying of concrete (in Finnish), Helsinki University of Technology, Building Materials Technology, Report 15, Espoo, Finland, 2001, p 71

PRECAST FLOOR DIAPHRAGMS ACTION TO ENSURE THE STRUCTURAL INTEGRITY OF MULTISTOREY BUILDINGS

A Cholewicki

Building Research Institute

Poland

ABSTRACT. Hollow core units, widely manufactured in the world, offer possibilities for construction of very attractive modern structures composed of floor diaphragms, framed elements and shear walls. The tying of the individual units into one integrated superstructure is the main objective of this contribution. Ties of the hollow core units often are dimensioned according to certain general principles (e.g. recommendations based on Eurocode 2). The author's concept is to show the calculation procedures for dimensioning of the tying steel bars in order to improve the interaction of those hollow core slabs in the over-all behaviour (diaphragm action, alternative model action) and as the members of the composite beams. Some of these procedures has been already published (e.g. Chapter 4 of the newest document [1] of FIB Commission "Prefabrication").

Keywords: Hollow core floors, Diaphragm and composite action, Shear connections, Alternative model action.

Professor A Cholewicki, is Professor at the Building Research Institute in Warsaw. He is specialising in research on three dimensional analysis of multistoried buildings, behaviour of connections of precast structures, behaviour of buildings under accidental loads (soil tremors or gas explosions). Professor Cholewicki is representative of Poland to the FIB Commission on Prefabrication.

INTRODUCTION

In a modern precast framed building, like the one shown in Figure 1 with floors made of hollow core slabs, very typical is the connection at the slabs support on the internal beam.

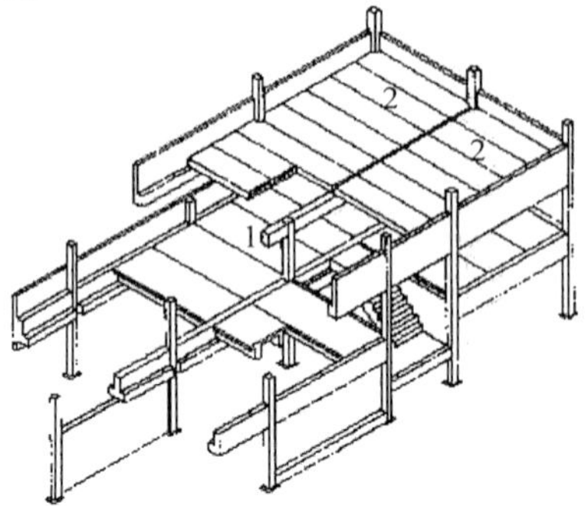

Figure 1 Multistorey skeletal structure [2]; 1-internal supporting beam, 2-hollow core floor slabs

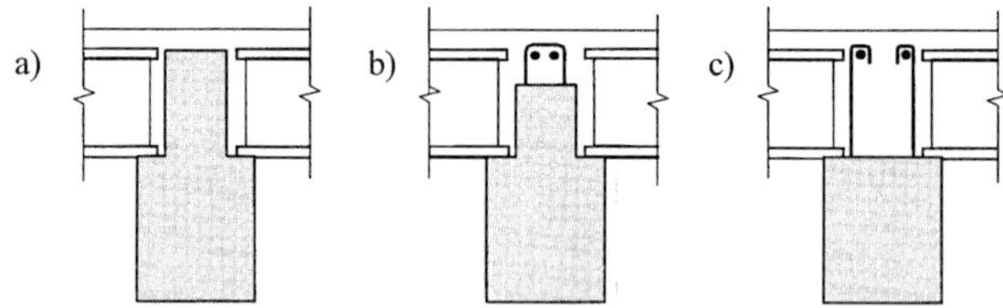

Figure 2 Composite concrete sections with different shapes of the internal supporting beam [3]

The shape of upper part of the beam cross section (Figure 2) determines the location of the main tying reinforcement which is necessary to achieve the diaphragm action of the precast floors and of the position of the coupling ties (Figure 3).

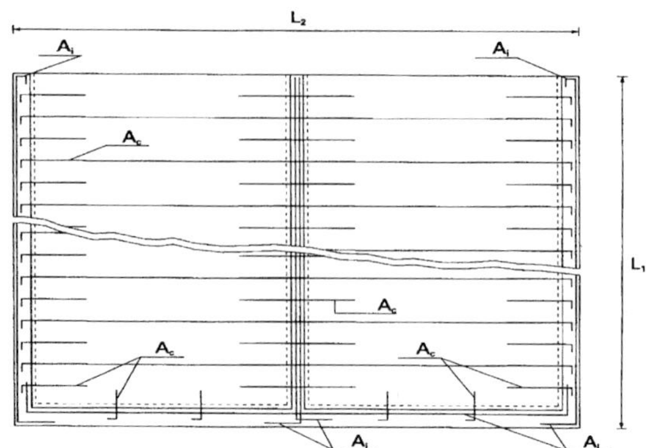

Figure 3 Tying in a floor diaphragm

In the present European practice there is a strong tendency to reduce the height of the total composite internal beam (see examples in Figure 4).

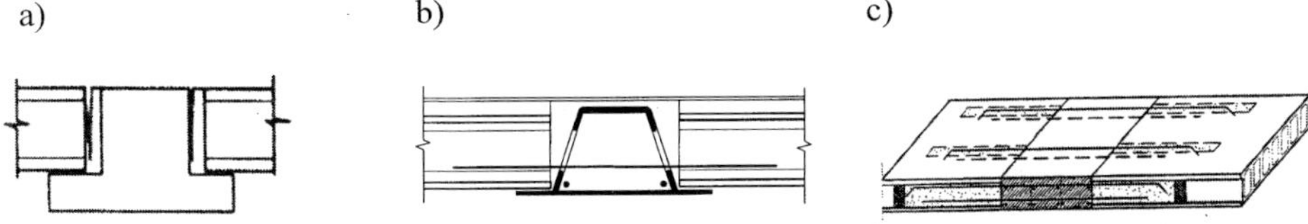

Figure 4 Examples of the composite internal beams with a reduced height of the section [4] ÷ [6]

Tying and coupling ties (Figure 3) placed into joint gapes of the floor diaphragm can be either:

- dimensioned according to general principles settled in several international (like EC2 [7] and [12]) or national documents

or

- calculated in order to achieve the desired effects of the structural interaction.

This second approach has been the objective of author studies in recent years, particularly concentrated on:

- diaphragm action,
- composite action,
- alternative model action (aimed at prevention against progressive collapse).

Some of author's concepts have been briefly presented in this contribution. It should be marked however that even within mentioned here three groups of the structural functions the particular demands can be contradictory (as an example it can be mentioned the position of the coupling ties which is determined for the transmission of supporting moments and for an accidental situation to ensure the interaction of the floor in secondary bearing system).

DIAPHRAGM ACTION

Hollow core slabs connected together into one large size diaphragm can be subjected to various in plane loads appearing due wind and also due to accidental situations like fire or an external explosion. All these loads and situations generate the occurrence of the shear force V along the longitudinal joints of the floor slabs (Figure 5). Following Formula for the tie force F_{tv} in the main tying reinforcement crossing a longitudinal joint was proposed by Cholewicki and Elliott [1]:

$$F_{tv} = \frac{V - V_{sk}}{n_t \mu} \tag{1}$$

where: V_{sk} part of the shear force transferred directly by the concrete in the connection (acting as a shear key)

n_t - total number of the tying bars in all perimeter and internal tie beams, which can be reckoned as transferring the shear force V

μ - friction factor (observe, not friction coefficient)

The model experiments carried out by Chalmers University of Technology [8] have showed that only the reinforcement bars laying in the diaphragm plane can be reckoned as the effective ones for the tying function. The experimental works conducted by Elliott [9] gave the background for the specification of the values of friction factor μ. The information about μ and the general shear wedging mechanism were presented in [1].

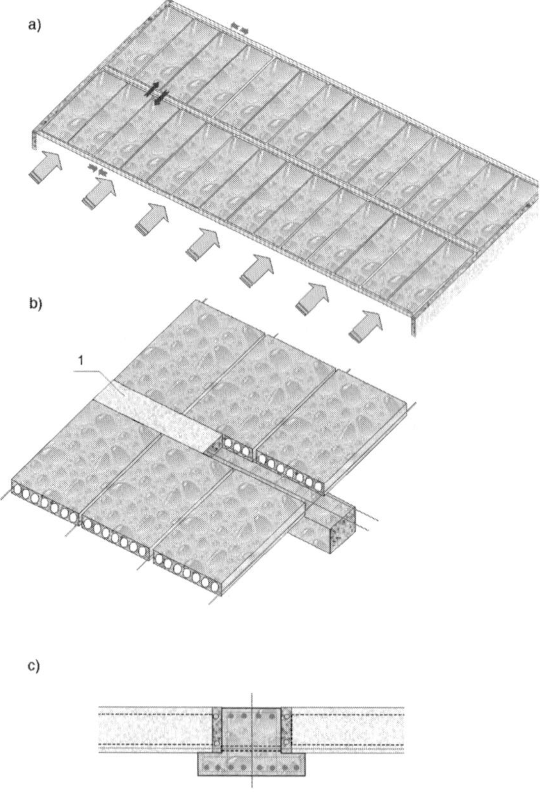

Figure 5 In plane action of the floor diaphragm
a) schematic structural system
b) internal beam – local shear key (sign 1)
c) inverted tee beam (also local shear key)

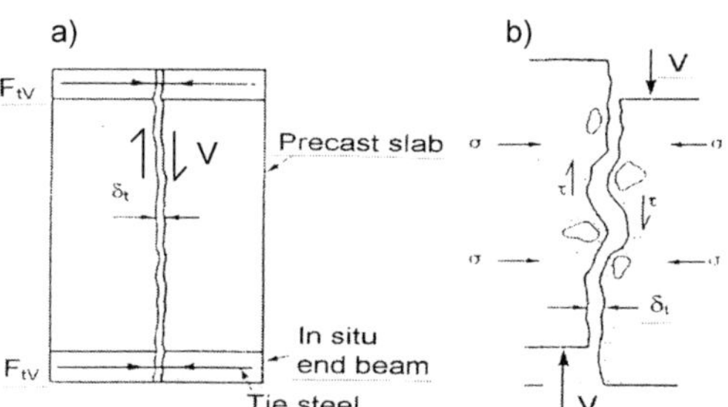

Figure 6 Definitions of shear transfer mechanism
a) initial precracked condition
b) aggregate interlock mechanism

According to [1] the tying reinforcement should be dimensioned to satisfy following shear displacement limitation:

$$\delta_{tv} = \frac{F_{tv}}{K_t} + \delta_{ti} \leq \delta_{t\max} = 0{,}5\ \text{mm} \tag{2}$$

where: K_t – axial stiffness of the tie beam (see [1])
δ_{ti} – initial crack width
$\delta_{t\,max}$ – limiting crack width for aggregate interlock [1]

The condition (2) leads to application of relatively high values of μ, even higher than $\mu = 5$ (see calculation example in [1]). The limitation of the crack width influences the global shear stiffness of the floor diaphragm, because the shear displacements of the individual longitudinal joints are negligible small (see Figure 7). The very effective shear-friction mechanism was also observed in tests reported by Moustafa [10]. On their background in PCI manual [11] so called effective coefficient of friction $\mu_e \leq 2{,}2$ was recommended.

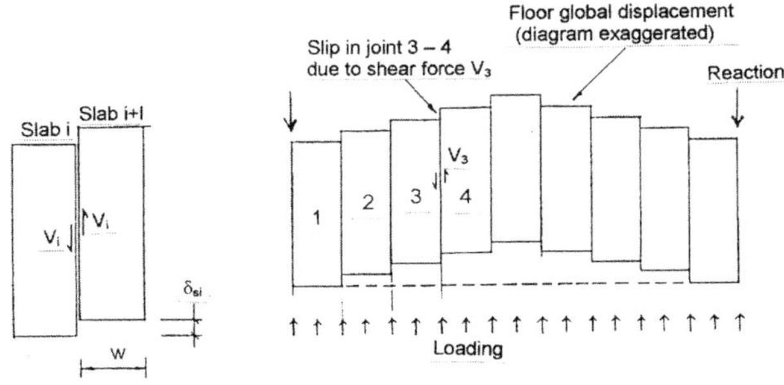

Figure 7 Shear displacements in the floor diaphragm

COMPOSITE ACTION

In the newest draft of ENV document [12] following motto is expressed "In addition to satisfying the requirements of cross-section design, the analysis of precast concrete structures shall take into account the behaviour of the connections between elements". This fits well just to the point below discussed, whereas the role of the stiffness C_c of the shear subjected connection has been here particularly emphasized. Cholewicki [13] presented a solution according to two-beams model, which describes the effects of the shear deformability of the interface connection on the distribution of the unit shear forces $V_c^{'}$ and the sum of those force along the half a length of the beam (Figure 8a). The diagram of the unit shear forces $V^{'}(\xi)$ is governed by the $\eta^{'}(\xi)$ coefficients (Figure 8b) and given by

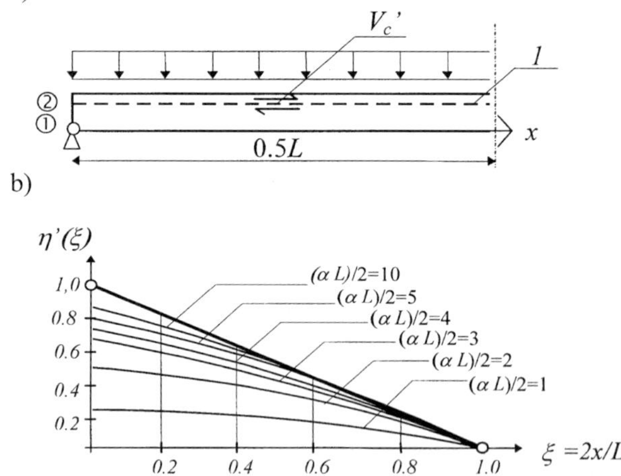

Figure 8 Shear stresses distribution upon the $\frac{\alpha L}{2}$ value [13]
a) two-beam model of the composite structure,
b) diagram of unit shear forces V_c',
1- connection with stiffness C_c

following Formula:

$$V_c^{'} = V_{c\max}^{'} \eta^{'}(\xi) \tag{3}$$

where: $V_{c\max}^{'}$ is the maximum unit shear force calculated under assumption of non-deformable shear connection

In the author solution in all relationships the characteristic value $\frac{\alpha L}{2}$ appears, which expresses effects of the structural parameters and thus

$$\frac{\alpha L}{2} = \frac{L}{2}\sqrt{\left(\frac{1}{E_{c1}A_1} + \frac{1}{E_{c2}A_2} + \frac{a^2}{E_{c1}I_1 + E_{c2}I_2}\right)C_c} \tag{4}$$

where: E_{c1}, E_{c2} - respectively, concrete elasticity modulae in parts 1 and 2,
A_1, A_2 - respectively, cross section areas of parts 1 and 2,
I_1,, I_2 - respectively, moments of inertia of parts 1 and 2,
a - distance between the central points of parts 1 and 2.

Cholewicki [13] proved that when

$$\frac{\alpha L}{2} \geq 5 \tag{5}$$

the composite structure can be considered as a homogenous one with the shear connection as an undeformable (under shear) medium.

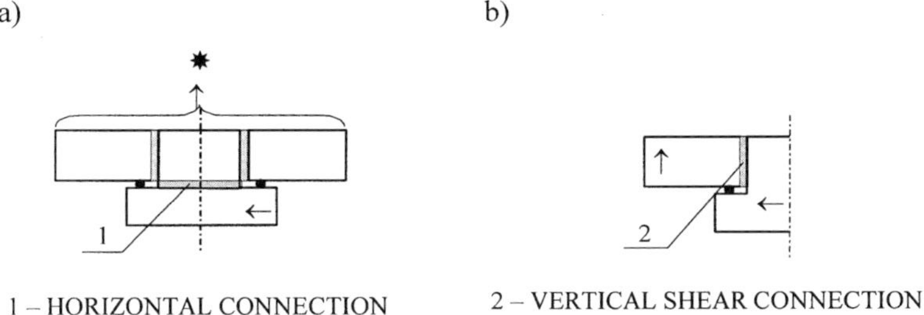

Figure 9 Distinguished parts of 1 and 2 in the composite floor
a) case of horizontal connection, b) case of vertical connection

The presented two-beams model and derived according to it Formulae can be applied in order to answer several important questions in considerations of the effectiveness of tying reinforcement projected to the connection. These questions are e.g.:

1. minimum shear stiffness C_c of the ties projected to joint gape in order to guarantee the fulfilment of condition (5),in consideration of the shear deformability along the horizontal (Figure 9a) and vertical (Figure 9b) shear interfaces.

2. detailing provisions necessary to be fulfilled in case of the structural topping (not discussed here) interaction in the composite floor.

ALTERNATIVE MODEL ACTION

In case of an accidental situation which leads to complete failure of one structural member (column) the appearance of the alternative bearing system is expected. The framed system acting as a whole must be able to transfer, through the bridging over the damaged column, the extra loads on the surrounded columns.

So called global model for the numerical analysis should be applied in order to distribute the vertical force from that one column which is supposed to be totally damaged. The distribution of that force on the adjacent undamaged columns follows in two planes i.e in the plane of main internal beam and in the one being perpendicular to the main beam (an example is shown in Figure 10). The symbolic presentation of those two distribution directions is shown in Figure 11.

The purpose of global analysis is the checking of the increase of forces in vertical bearing elements. The next step of analysis is concentrated on the horizontal structural members bridging the lost support it is a composite structure created by beam (or beams) with floor slabs (Figure 12).

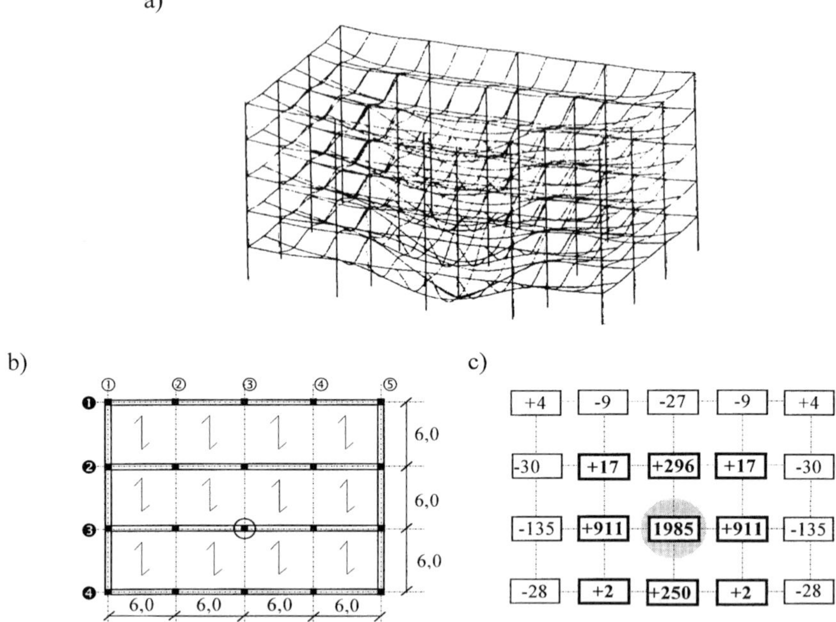

Figure 10 Global model of locally damaged skeleton building
a) Displacements after removal of one support
b) Calculation example (author's analysis) of building with longitudinal beams
c) Distribution of the vertical force from the damaged column into the sourrended ones (values given in kN)

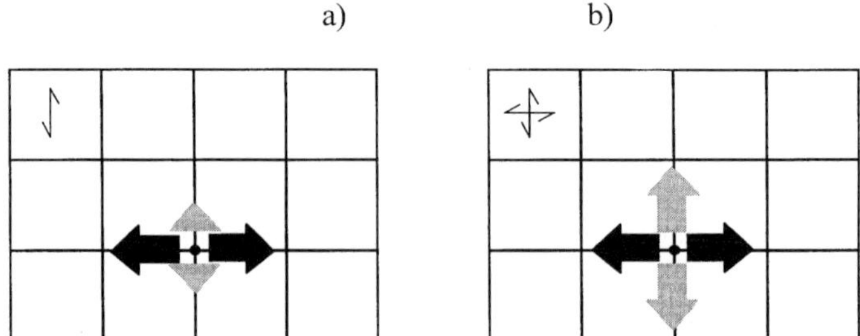

Figure 11 Distribution directions of the axial force from the damaged column
a) In the plane at the main beam
b) Activated distribution in the plane perpendicular to the main beam

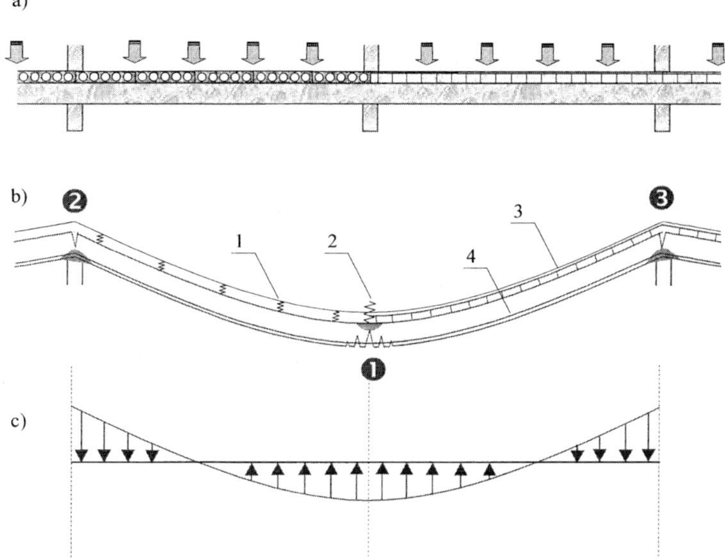

Figure 12 Out of plane action of the floor diaphragm
a) b) alternative model of the beam
b) diagram of correction reactions along the slabs supports on the beam

❶- *plastic hinge*
❷❸- *possible locations of plastic hinges*
1 - ties giving the extra support along the plane perpendicular to main beam
2 - suspension effect (from the storeys above the damaged column) and extra support given by a second order beam
3 - longitudinal reinforcement in the tie beam
4 - main flexural reinforcement

The size of the numerical task can be in this way reduced and the individual tying members can be modelled in more detail. Two alternative bearing systems can be assumed:

1. catenary model in which the connections in potential plastic hinges , and (Figure 12b) must fulfil very high demands as concerns the ductility and continuity

   or

2. elastic – plastic model characterized by small and controlled values of displacements and rotations in all hinges and ties (with except that only one which must remain in elastic stage).

The author has carried out a theoretical study which showed that:

1. through modification of design principles of flexural reinforcement in the main beam (sign 4 on Figure 12b); and

2. through modification of design of the reinforcement placed (sign 3 on Figure 12b) in the cast in situ tie beam

the increase of the bearing capacity of the alternative bearing model can be achieved. The modification 1° lies on the principle of the constant amount the reinforcement along the full span of the beam and with equivalent cross sections of ties in the zone "through the column". Interaction of the reinforcement in the tie beam (modification 2°) depends on the effectiveness of vertical ties ensuring the composite action along the beam interfaces. These vertical ties are subjected to splitting shear forces and under, certain conditions, even to real vertical tensile forces; these may appear is because the beam is a flexible supporting element (redistribution principle of the vertical reactions along beam and tie beam interfaces explains Figure 12c).

Study has showed that the capacity of main beam after its span would be doubled, in spite of the described above modifications, is still insufficient. The lack of the capacity requires more activated interaction of the second order reinforcement being placed perpendicular to the main beam (Figures 11b and 13).

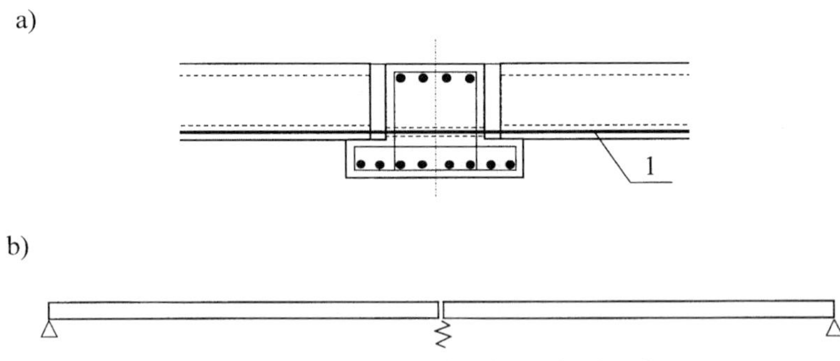

Figure 13 Floors interaction in alternative bearing system
a) ties as continuity reinforcement (1)
b) ficticious beams appearing due to composite action of slabs and ties

The safety condition for the q_{acc} loading in the accidental situation can be following:

$$q_{acc}\,(R_M, R_{fl}, R_{oth}) < q_k\,\theta_{red}\,\theta_{dyn} \tag{6}$$

where: R_M - carrying capacity ensured by the beam and the tie beam flexural reinforcement (after its modification).

R_{fl} – increase of the capacity due the suspension effects imposed by the floor slabs.

R_{oth} – increase of the capacity due the other, mentioned above, parameters.

q_k – characteristic value of the loading applied for dimensioning of the main beam for the exploitation stage.

θ_{red} – reduction coefficient (less than 1) due to redistribution of the vertical reactions between the slabs and the main bean (supposed diagram of the correction forces reactions – in Figure 12c which are added to the steadily distributed reactions).

θ_{dyn} – dynamical coefficient (larger than 1).

In case when damaged column is situated at the perimeter and the spandrel beams have relatively high cross sections the diagonal model of the secondary bearing structure can be adopted (Figure 14).

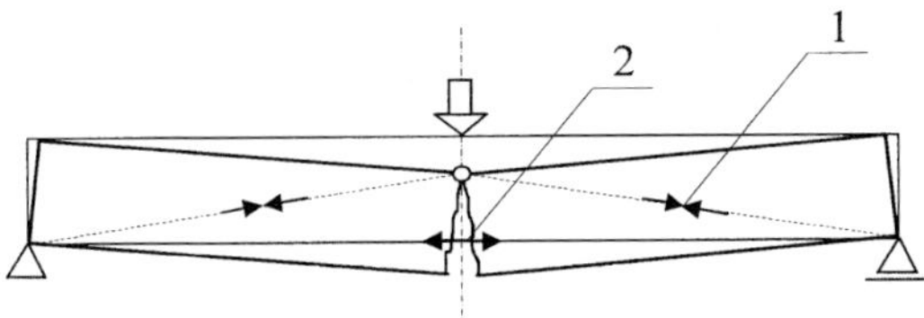

Figure 14 Alternative bearing system along the perimeter in case of higher sections of the spandrel beams
1 – ficticious compressive strut
2 – reinforcement and diaphragm subjected to tension

CONCLUSIONS

1. In case of modern framed systems built with hollow core floors there is a necessity of a very precise dimensioning and detailing of ties, because the floor elements have relatively large spans and the gapes for the in situ grouting concrete are small.
2. The same ties fulfil very differentiated structural functions, author has discussed them on examples of three models and particularly emphasized the effects of the stiffness of connections (diaphragm and composite action models).
3. The recommendations for tying formulated in national documents and authorized also by Eurocode 2 should be treated as the general guidance and they need development towards a system of design requirements.

ACKNOWLEDGMENTS

The author would like to acknowledge valuable discussions and cooperation with Doctor K S Elliott, Senior Lecturer, Department of Civil Engineering, Nottingham University. The result of these discussions and co-operation was Chapter 4 of the FIB Document [1].

REFERENCES

1. FIP GUIDE TO GOOD PRACTICE, Special design considerations for precast prestressed hollow core floors, Chapter 4, Diaphragm action, 2000.

2. ELLIOTT, K. S., Multistorey precast concrete framed structures, Blackwell Scientific Oxford, May 1996.

3. STRUCTURAL PRECAST CONCRETE HANDBOOK, Construction Industry Development Board, Singapore, March 1999.

4. FIP COMMISSION, Planning and design handbook on precast building structures, Prefabrication, 1994.

5. DELTATEK, O. Y., Delta – composite beam, Information paper, Lahti Finland.

6. DELLA, B. B., Precast prestressed hollow core slabs in continuous floor: design and detailing of composite and restrained support, 4th International Symposium Developments in Prestressing & Precasting, Singapore, July 1997.

7. ENV 1992-1-3, Eurocode 2 – Design of concrete structures, Part 1-3, Supplementary rules for precast concrete elements.

8. CEDERWALL, K., SVENSSON, S., ENGSTROM, B., Diaphragm action of precast floors with grouted joints, Nordisk Betong, February 1986, p 123-128.

9. ELLIOTT, K. S., DAVIES, G., OMAR, W., Experimental and theoretical investigation of precast hollow cored slabs used as horizontal diaphragms, The Structural Engineer, 1992, Vol 70, No 10.

10. MOUSTAFA, S .E., Effectiveness of shear-friction reinforcement in shear diaphragm capacity of hollow-core slabs, Journal of the Prestressed Concrete Institute, Vol 26, No 1, January-February 1981, p 118-132.

11. MANUAL FOR DESIGN OF HOLLOW CORE SLABS, Prestressed Concrete Institute, Chicago, 1985.

12. ENV 1992-1-3, (1st Draft 2001) Chapter 12.

13. CHOLEWICKI, A., Structures composed of precast elements (in Polish), Building Research Institute, Warszawa, 2001.

HYBRID STRUCTURES HOLLOW CORE SLABS AND CHALLENGING ARCHITECTURE

J N J A Vambersky

Delft University of Technology

Netherlands

ABSTRACT. New building materials and new construction technologies make the building industry today look different from the one of some years ago. Structural steel, reinforced and pre-cast concrete are now common and compete with each other on the construction market. Joining forces, seeking synergy rather than confrontation, is the emergent trend. The result of this trend is the growing number of hybrid structures in the today's daily building practice. Hybrid or mixed construction with pre-cast concrete means combined use of pre-cast concrete with other materials, such as steelwork, timber, cast in situ concrete and glass, for the benefit of the whole. In The Netherlands two recent examples of the above development are the office tower "Malietoren", constructed over the motorway entering the city of The Hague and the 100 m high sloping office tower "Belvedere", in Rotterdam, where the ambitions of the architect Renzo Piano to make appealing but still market price competitive building were made possible because of hybrid structure. The considered concepts, alternatives, final solutions and detailing of these two architecturally appealing buildings are dealt with in this paper, to illustrate this new development. Development, in which environmental considerations and pre-cast concrete hollow core slabs are playing an important role.

Keywords: Hybrid structures, Pre-cast concrete, Hollow core slabs, Structural design of buildings and appealing architecture.

J N J A Vambersky, is a Professor at Delft University of Technology and Consulting Engineer and partner at Corsmit Consulting Engineers, Rijswijk, the Netherlands.

INTRODUCTION

Benefits of Hybrid Construction

Hybrid construction maximises the structural and architectural advantages in combining components made of different materials. To achieve this, good co-operation of the architect with structural engineer, services engineer, manufacturer, supplier and the contractor is required. Hybrid or mixed construction must be distinguished from "composite" construction where different materials are constructed to act homogeneously.

In hybrid construction the different materials may work together or independently, but will always provide advantages over the use of a single material. Today the engineering practice, the builders and the users are discovering that the hybrid construction is essential to meeting architectural requirements, providing high surface finishes, minimising structural floor depths, achieving better sustainability and ensuring rapid construction, all of which translate into substantial savings and better quality of the end product. Hybrid construction methods vary considerably with the type of construction and building function. These reflect local trends, environmental and physical conditions, relative material and labour costs and local expertise. The fib-commission 6 -Prefabrication, Working group on Mixed Construction[1], reports that the permutations for hybrid construction are numerous.

The same working group reports that:

- Hybrid construction is today being used in more than 50 per cent of new multi-storey buildings.
- Pre-cast concrete is ideally suited to hybrid construction as it may readily be combined with other materials and technologies.
- Using pre-cast concrete and structural steel as the dominant material in hybrid construction, on-site operations are considerably reduced.
- Various case studies claim that hybrid construction can save between 10% and 20% construction time.
- Hybrid construction is, by definition, cost effective because it maximises the beneficial structural and architectural advantages in using components made of different materials.
- In some cases client and architectural demands can only be satisfied using mixed construction.

Different Floor Types and their Environmental Costs

When analyzing the environmental load, it turns out that the bearing structure is responsible for the second largest part of the environmental load of the building (the energy consumption is responsible for the greater part, assuming the life span of a building is 75 years)[7].

In his research 'Environmental comparison of bearing structures at Delft University of Technology Arets [7] concludes that the horizontal elements of the bearing structure, consisting of the floor and roof structure and beams, are causing 75-95% of the environmental load of the total bearing structure. Especially the floor and roof structure play a very important role. Therefore, optimisation of the floor structure in this respect can lead to an important environmental improvement of the whole building.

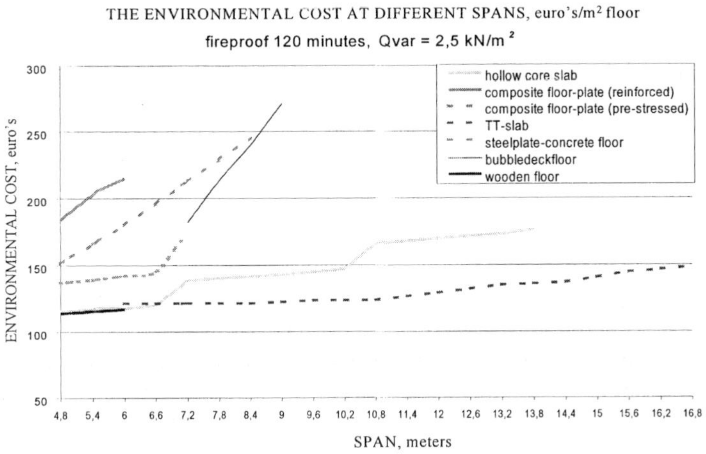

THE ENVIRONMENTAL COST AT DIFFERENT SPANS, euro's/m² floor
fireproof 120 minutes, Qvar = 2,5 kN/m 2
ENVIRONMENTAL COST, euro's
300
250
200
150
100
50
hollow core slab
composite floor-plate (reinforced)
composite floor-plate (pre-stressed)
TT-slab
steelplate-concrete floor
bubbledeckfloor
wooden floor
4,8 5,4 6 6,6 7,2 7,8 8,4 9 9,6 10,2 10,8 11,4 12 12,6 13,2 13,8 14,4 15 15,6 16,2 16,8
SPAN, meters

Figure 1 The environment costs at different spans

For the different floor types and different spans between 4,8 and 16,8 m, assuming there are no intermediate beams, Arets [7] calculates the environmental load, in terms of costs (euro's). The graphic shows that, based on the same requirements, the wooden floor has the lowest "environmental costs" for short spans and for larger spans the concrete TT-slab has the lowest "environmental costs". When the costs of the different floor types are taken into account and added to the "environmental costs" the "integral costs" arise. The graphic below shows that de hollow core slab and for larger spans the TT-slab have the lowest "integral costs", indicating that these floors from environmental and costs point of view together perform the best.

THE INTEGRAL COST AT DIFFERENT SPANS, euro's/m² floor
(environmental cost + cost = integral cost)
fireproof 120 minutes, Qvar = 2,5 kN/m 2
INTEGRAL COST, euro's
400
350
300
250
200
150
hollow core slab
composite floor-plate (reinforced)
composite floor-plate (pre-stressed)
TT-slab
steelplate-concrete floor
bubbledeckfloor
wooden floor
4,8 5,4 6 6,6 7,2 7,8 8,4 9 9,6 10,2 10,8 11,4 12 12,6 13,2 13,8 14,4 15 15,6 16,2 16,8
SPAN, meters

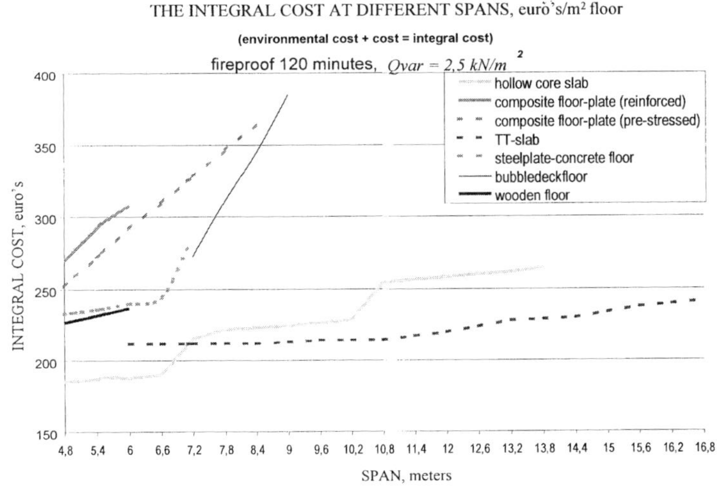

Figure 2 The integral costs at different spans

OFFICE TOWER "MALIETOREN", THE HAGUE

Growing human population and its demands for living and working space are often in contradiction with our striving for saving grasslands, forests and natural resources. Multiple use of space for our building activities is one of the answers to this dilemma.

Office tower "Malietoren" (Figure 3) is such a case. It is situated over the motorway "Utrechtse Baan" entering the city of The Hague. The building is almost square in plan, 40 m long and 32.2 m wide (Figure 6). The ground floor is designed as entry and reception area (Figure 5). Here above are five car park floors reached by a spiral ramp cantilevering half way over the motorway on the north face of the building. The sixth and seventh floors are conference facilities whilst the remaining 13 floors are designed as offices. The building services are concentrated at the top floor of the building, bringing the total height to 74 m.

The Structure

Building over an existing motorway is never easy. The motorway – effectively a watertight reinforced concrete trough sunk into the ground – cannot be closed without severe effects on the life of the city. As a result, it had to be bridged over to prevent any disturbance including additional loads or penetrations to the trough.

Figure 3 Malietoren office tower

Figure 4 Reception area

Figure 5 Ground floor construction

A solution adopting a composite concrete truss transfer structure (Figure 6) with a height of 8.2 m and a span of 32.2 m at the entrance level proved to be the best.

It also satisfied the architectural perception in terms of structural demands and economy. The 2m-deep pre-cast, pre-stressed and post-tensioned lower chord of the truss was designed to function in the erection stage as a simply supported beam to carry the weight of the ground floor acting as construction working area. Being prefabricated, the beams and working floor were placed very quickly (in a single night) (Figure 4). Diagonals and the upper chord were then added in in-situ concrete B65. The upper floors are pre-cast hollow-core slabs on pre-cast pre-stressed concrete beams - very economical and fast to erect and so as we learn now, also rather attractive from the sustainability point of view. (Figure 6).

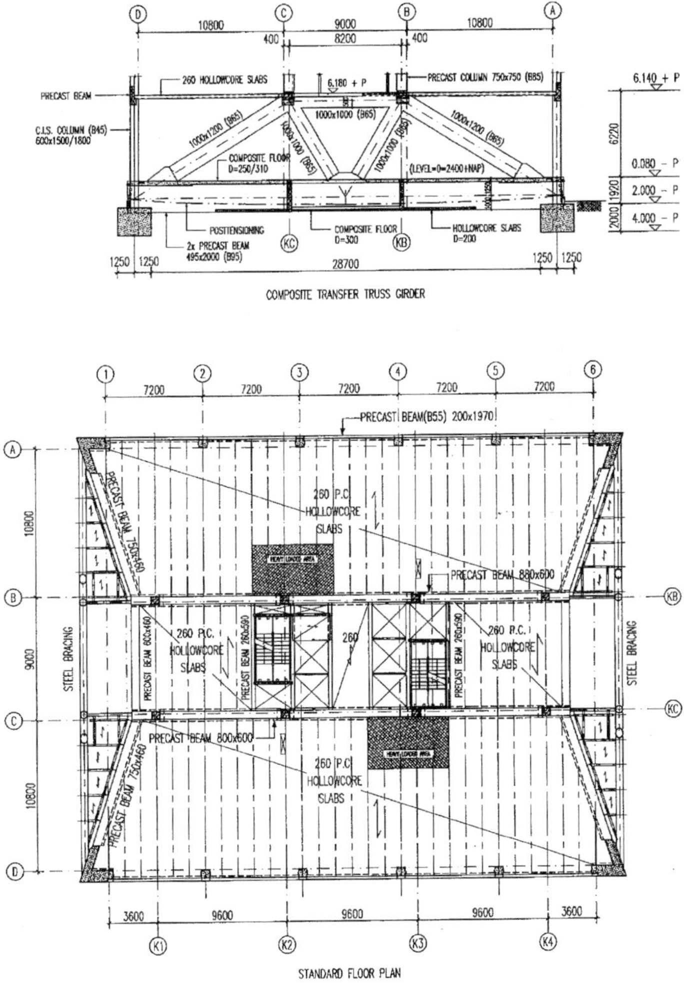

Figure 6 Floor plan and truss structure of Malietoren office tower

Innovation

For the high-strength (B85) two-storey pre-cast concrete columns, an innovative, but simple and cheap, connection (Figure 7) was developed using steel plates and epoxy resin injection resulting in an enhanced speed of erection and a minimum column cross-section area (8% reinforcement). A composite pre-cast concrete facade and structural steel bracing were integrated (Figure 8) in a simple way by in-situ concrete columns in a stabilising facade tube. (Figure 6). The result is a high quality building respecting the environment and ecology at a very competitive price.

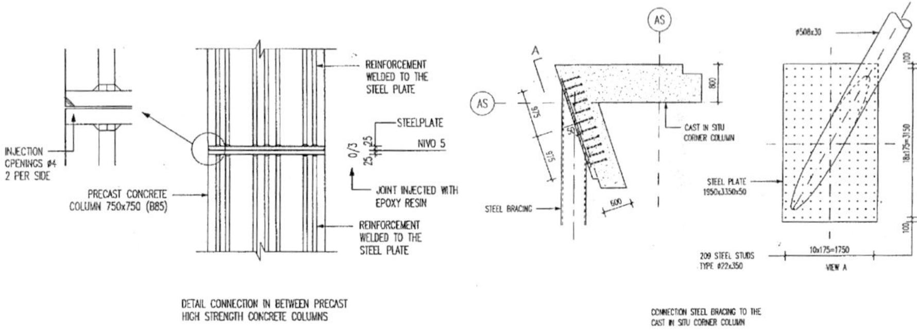

Figure 7 Connection details

Figure 8 Façade details

The Sloping Tower "Bélvèdere" Rotterdam

The old harbours in the centre of Rotterdam have lost their function already some time ago. Today these areas are very valuable locations in the city urban developments. Wilhelmina pier, from where the most of European emigrants in the past century boarded the ships on their way to America, is one of them. Located on the river Maas it is connected to the centre of the city with the already well known Erasmus bridge, "The Swan bridge". Development of high quality buildings on this location is endeavoured by as well the Municipality of Rotterdam as by the designers involved.

The Building

With the above city ambitions and the urban zoning plan as background, the ideas of Renzo Piano, who was commissioned by the Real Estate Developer William Properties BV to make the architectural design, were inspired by the Erasmus bridge adjacent to the very location of the building site. The 100m tall building, consisting of ca. 20000m^2 floor area office space, 5000m^2 retail and commercial space and two levels underground car park for ca 250 cars, is following the inclination of the cable stays of the Erasmus Bridge (Figures 9 and 10). Also in this building pre-cast pre-stressed hollow core slabs are used.

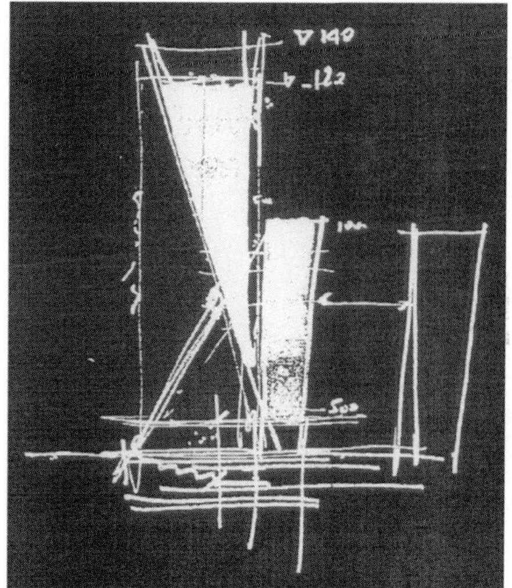

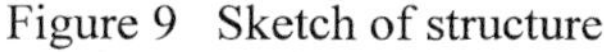

Figure 9 Sketch of structure

Figure 10 Impression of building and Erasmus Bridge

The Inclination

The 6 degrees inclination of the building, or better to say of its eastern facade close to the Erasmus bridge, was the first challenge to the structural engineer to solve. The position of the columns (vertical supports), in relation to the inclined side of the building, influences the horizontal force components of this inclination. Namely, the bigger the span, the bigger the vertical floor support reaction to be carried by the inclined side and the bigger the horizontal load caused by this inclination on the stabilising structure (Figure 15). However, even with the best possible position of the columns in the given architectural design, this horizontal force component was still too big to be taken by conventional structural measures like cores or shear walls only. As an excellent architect also Renzo Piano had the same feeling and has drawn already in his first sketches on the back side of his cigar box a reversely inclined compression strut to compensate and support the leaning side of the building (Figure 9). This synergy of form and structure and the chemistry of understanding in between the different disciplines had a big potential in it to come to a strong architectural and structural concept for this building.

The Inclined Buildings in General

Before the structural design of the Belvédère tower will be dealt with it might be interesting to focus first on the phenomena "inclined building" and what for consequences the inclination will or will not have on the price of the building depending on the structural scheme which has been chosen.

The following static schemes A, B and C (see Figures 12, 13 and 14) for an inclined building with the top of the building being 1/2 of the base width "a" out of plumb, are very illustrative.

Figure 11 Impression of Belvedere tower

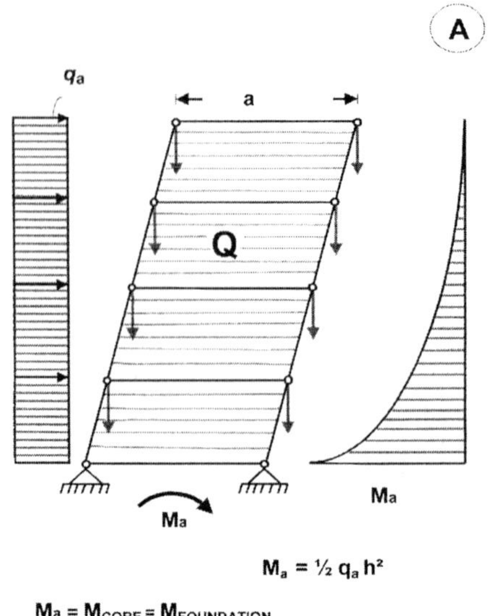

Figure 12 Static scheme A

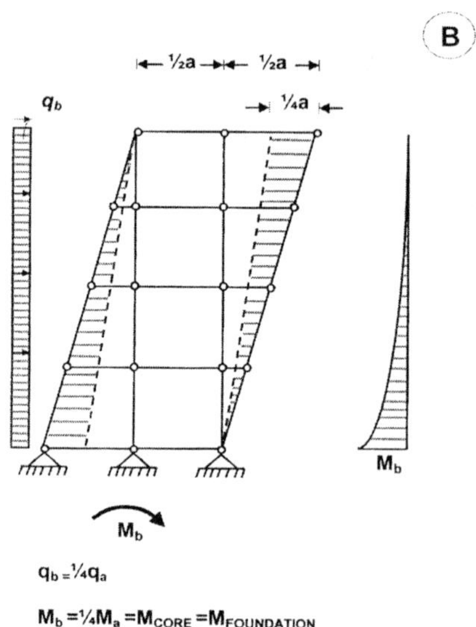

Figure 13 Static scheme B

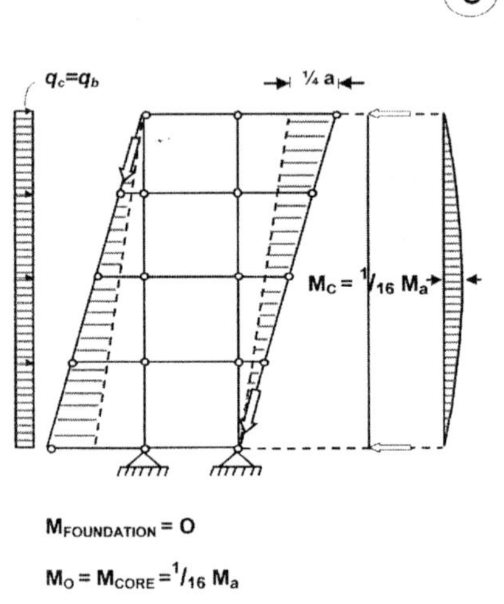

Figure 14 Static scheme C

The structural scheme A (see Figure 12) shows a situation where it has been chosen for floors spanning from one inclined facade to another. This has as a consequence that the total weight Q of the whole grid will generate horizontal force components as a function of the inclination of the columns supporting the floors. These horizontal forces on their turn will cause moments as well in the stabilising elements such as cores and shear walls as in their foundations. The total moment, which has to be resisted, is Ma.

The structural scheme B (see Figure 13) shows a situation where two vertical inner columns have been introduced as close to the facades as possible, in order to reduce the reactions of the floors supported by the inclined facades. The smaller the reactions on the inclined facades will be, the smaller will be also the horizontal components of these reactions which on their turn will have to be resisted by the stabilising elements. In the given example this results already in a considerably reduced moment Mb=1/4Ma, which again has to be resisted by as well the stabilising elements as by their foundations.

The scheme C (see Figure 14) finally shows a situation, which is exactly the same as scheme B, except for the foundation support of the left-hand side facade. This foundation support is here omitted and the facade is designed as a hanger which is transferring the gravity forces first to the top of the building, meeting there the vertical column which is on its turn taking these forces to the foundation. This static scheme has the following consequences:

- The moment Mc due to inclination which has to be taken by the stabilising elements, is only 1/16Ma.
- This moment has to be resisted only by the stabilising elements and not by their foundations as the moment in this scheme at the foundation level is zero.
- As the gravity forces at the left hand side facade are hung up, there is no need of supports what gives more freedom to the architect and more column free space at ground floor level.

It is clear that there is a great difference in price for a building with the scheme A and a building with the scheme C. This in spite of the fact, that the buildings and their inclination are the same.

The Static Scheme

The reversely inclined strut, depending on its angle of inclination, the magnitude of the vertical load assigned to it, the place where it is attached to the building and the way how it is designed and detailed to function in the total structural scheme can:

- Compensate (counterbalance) the horizontal forces caused by inclination of the building
- Form part of (and function as) an outrigger
- Perform both

An analysis of different structural schemes revealed that the most economical solution was the counter balancing only, in combination with structural core for overall stability (Figure 16).

For counter balancing the central column in the inclined facade does not continue to the foundation but stops at the level +10.50m above ground level. The total force in this column is transferred to the strut and together with the tuned angle of the strut it counterbalances from the level +46.55m the influence of the slope of the building.

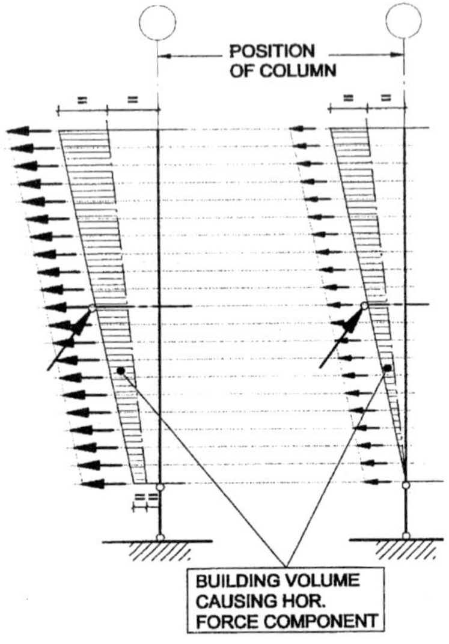

Figure 15 Inclination of structure

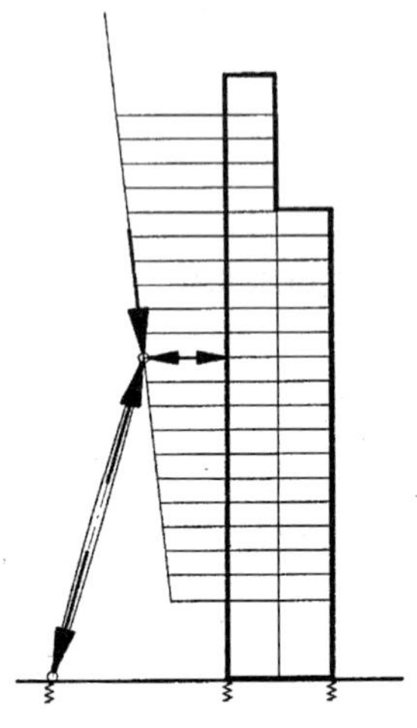

Figure 16 Horizontal and vertical forces

The Structure

Once established, this static concept appeared to be a very strong one. Different alternatives in different structural technologies could be realised. Out of these alternatives two economically equal ones were considered and maintained up to the tender stage. Both alternatives were alternatives with hybrid structures:

Alternative 1: Cast in situ core and load bearing facade in grid 8 with pre-cast pre-stressed hollow core slabs in between grids 6 and 8. The inclined part of the building in between grids II and 6 being made of structural steel with composite steel concrete floors.

Alternative 2: Cast in situ core and load bearing facade in grid 8 with pre-cast pre-stressed hollow core slabs in between grids 6 and 8, the inclined part of the building in between grids II and 6 being made with two storey high pre-cast concrete columns and composite pre-cast concrete beams bearing pre-cast pre-stressed hollow core slabs (Figures 17 and 18).

In both alternatives the compression strut has been designed as steel concrete composite member with structural steel tube diameter 800mm at the ends (2000mm in the middle). Because of the required fire resistance, the strut was at both ends partially filled with concrete over a length of 5 m. The floor at level +46.55m where the horizontal balancing force from the strut is transferred to the structure is designed as massive 260mm thick cast in situ concrete floor. Finally on grounds of contractors experience and preference, as there was no difference in price, the alternative 2 was chosen.

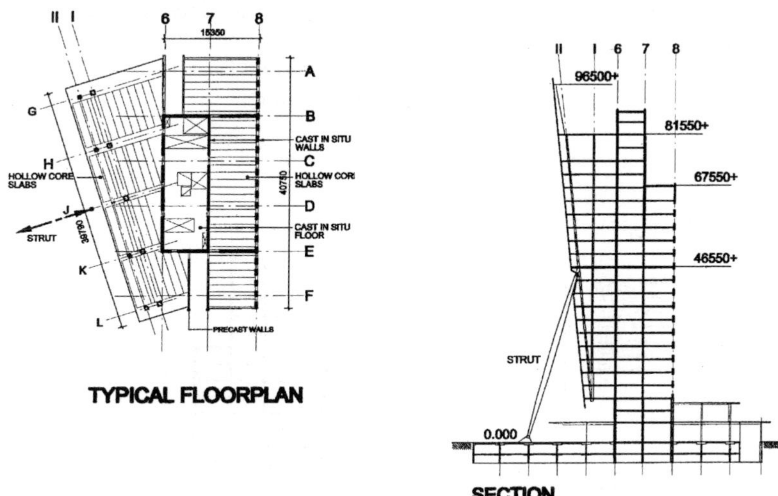

Figure 17 Floor plan of structure

Figure 18 Typical section of structure

As in the required architectural expression the emphasis was on the strut as an important element, it was for the architect very important that the remaining columns under the building should be visually almost nonexistent. This especially at the top of these columns where the high-rise part of the building is transferring its gravity forces to these columns. The solution was a hybrid construction in which the columns diameter 800mm were executed in high strength concrete grade B90 and provided with massive steel column caps reducing the diameter of the column to only 300mm there, where it meets the upper structure. To transfer the full column force to the 350mm thick concrete wall in concrete grade B65 situated above this strongly reduced column section, a massive steel transfer block was introduced in this wall to spread this concentrated force over a sufficient length of the wall (see Figure 19 and Figures 20 and 21).

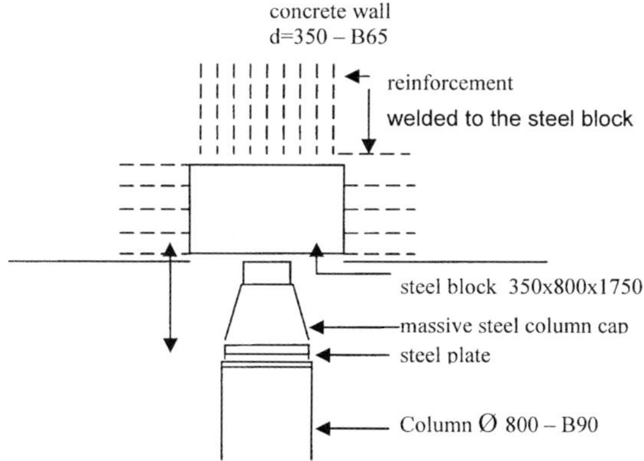

Figure 19 Details of massive steel transfer block

Figure 20

Figure 21

General information office tower “Malietoren”, The Hague

Client:	Multivastgoed (Real Estate Development), Gouda, The Netherlands
Contractor:	Wilma Bouw, The Hague, the Netherlands
Architect:	Benthem Crouwel Architecten, Amsterdam, The Netherlands
Structural Engineer:	Corsmit Consulting Engineers, Rijswijk, The Netherlands. In cooperation with Ove Arup & Partners, London

General information sloping tower "Belvedere", Rotterdam

Client:	William Properties BV, Rotterdam, The Netherlands
Contractor:	HBG Rijswijk (ZH), The Netherlands
Architect:	Renzo Piano, Genova Italy / Paris France
Structural Engineer:	Corsmit Consulting Engineers, Rijswijk, The Netherlands

REFERENCES

1. ELLIOT, Kim S., Pre-cast Concrete in Mixed Construction, The Second International Symposium on Prefabrication, 17-19 May 2000, Helsinki, Finland.

2. FONT FREIDE, J.J.M., PETERS, P., Een hoogwaardige constructie, het VNO-kantoor boven de Utrechtsebaan in Den Haag, Cement, 12/1995

3. VAMBERSKY, J.N.J.A., “Mortar joints loaded in compression”, proceedings of the International Seminar Prefabrication of Concrete Structures, Delft, the Netherlands, 25-26 October, 1990.

4. VAMBERSKY, J.N.J.A., "Hybrid Structures Solutions and Opportunities", fib Symposium "Structural Concrete - The Bridge Between People", Prague, Czech Republic, 12-15 October 1999.

5. VAMBERSKY, J.N.J.A., Hybrid structures, solutions and opportunities, www.corsmit.nl/vam.html

6. ARETS, M.J.P., Milieuvergelijkingen van draagconstructies ('Environmental comparison of bearing structures')- Master thesis, Delft University of Technology, Faculty of Civil Engineering and Geosciences, Department of Building Engineering, 16 October 2001.

CLOSING PAPER

SOME ASPECTS OF THE SERVICEABILITY OF SLAB SYSTEMS

A W Beeby

University of Leeds

United Kingdom

ABSTRACT. Serviceability issues, particularly the control of deflections, is becoming an increasingly important factor in the design of slabs. This issue is addressed by a number of the authors in this seminar. A number of aspects of this problem are discussed. There is some evidence that the deflections of actual structures exceed those calculated by current methods. Recent work carried out at Cardington has shown that in very many, if not most cases, the most critical loading to which a slab may be subjected is during construction; this is generally not considered in design. The effect of this high early loading on the future performance is considered. In slabs, the accurate assessment of deflections depends to a great degree on the assessment of the effects of tension stiffening and tensile strength. This has recently been studied in a major joint research programme reported in outline in another paper in this seminar. It is pointed out that current code methods give very different answers for the deflections of many practical slab situations.

Keywords: Slabs, Deflections, Construction loads, Tension stiffening, Code methods, Criteria.

A W Beeby, is Professor of Structural Design in the School of Civil Engineering at the University of Leeds. Prior to 1991 he had worked for the British Cement Association, formerly the Cement and Concrete Association, for 27 years. His main interests have always been the study of the behaviour of reinforced concrete and the introduction of research into design codes.

INTRODUCTION

Over the last 30 years or more there has been a steady trend towards longer spans and reduced construction depths. In commercial developments this has arisen from two main changes in the usage of structures. Firstly, a demand for more open-plan space giving greater flexibility in the use and, secondly, a desire to limit structural depth, giving more freedom for the arrangement of services and, in tall buildings, possibly the provision of a greater number of floors. A further trend towards the greater use of flat slab construction, due to its ease and potential speed of construction together with minimum construction depth, should also be noted.

Ever since the introduction of higher strength reinforcement with service stresses (permissible stresses in the terminology of our codes up to 1972) in excess of 200 N/mm^2, the dimensions of solid slabs has been dominantly defined by considerations of deflection control rather than strength. A possible exception to this is where, in the design of flat slabs for punching shear, it has been felt desirable to avoid the use of shear reinforcement and this has influenced the chosen depth of slab.

Early methods of designing to limit deflection were simple and highly approximate, based usually on allowable span/depth ratios. These are still extensively used, though their formulation is frequently more sophisticated than in the past. Increasingly, however, design is based on the explicit calculation of deflections. These calculations tend to be fairly complex but this is becoming less of a problem with the increasing use of computers.

One major problem with the explicit calculation of deflections is that the agreement between the calculated deflection and what actually occurs may be checked. It is commonly found that the calculated and measured values differ greatly. This is not an issue that arises for calculations for the ultimate limit state because the actual ultimate strength cannot be known except in the most unfortunate circumstances. Probably the assessment of the ultimate strength is no more accurate than the assessment of the serviceability performance but the comparison cannot be made.

One reason why the measured and calculated values should not be expected to agree is that concrete structures creep and the calculated design deflection would not be expected to be reached until, possibly, 30 years from construction. Furthermore, the design loads rarely occur. We would generally expect, therefore, that deflections measured soon after construction would be considerably below the specified limiting values. Usually, this is the case and this has provided ammunition for those arguing for less conservative methods of calculation. Unfortunately, there are an increasing number of cases reported where this is not so. Examples are available in papers in this seminar (Boyce [1]) and from elsewhere (e.g. Gilbert [2], Vollum et al [3]).

There is a need to understand all aspects of design for serviceability better as its importance in the design process increases. This includes external factors which may effect deflections, the accuracy of the methods of calculation and the criteria defining adequate performance.

CONDITIONS DURING CONSTRUCTION

The designer rarely, if ever, takes account of what happens to a structure during construction when designing for deflection control. It is probably most commonly assumed that the structure starts to be loaded at 28 days, that there is a permanent load defined by the dead weight of the structure, finishes, cladding, partitions and some proportion of the imposed load and that there is a short term, variable load which may occur at any time subsequent to the application of the dead load. Studies [4], [5] carried out during the construction of the seven storey flat slab structure as part of the European Concrete Building Project at Cardington have thrown more light on loading conditions at early ages.

The first part of the study, reported in [4], was concerned with the definition of criteria for the striking of formwork. The study resulted in a convenient factor for defining the loading conditions at different ages in relation to serviceability performance. Briefly, the factor was developed as follows. It is assumed that the deformation of a slab will be dominantly influenced by the degree to which the slab is cracked. The design assumes a given service load, w_{ser}, and a tensile strength which will be a function of the specified characteristic concrete strength, f_{cu}. In the study, it is assumed that the tensile strength is proportional to $f_{cu}^{0.6}$. It follows that the deflection implicit in the design will be a function of $w_{ser}/ f_{cu}^{0.6}$. The equivalent conditions at any particular time and under any particular loading at any particular time will be the same function of w/f_c where w is the load at the time considered and f_c is the actual compressive strength of the concrete in the slab. The state of cracking and deflection relative to that implicit in the design can now be expressed by the factor: $(w/w_{ser})(f_{cu}/f_c)^{0.6}$. This factor has been designated as the Cracking Factor, F_{cr}. If F_{cr} is less than 1.0 then the conditions in the slab are less severe than are implicit in the design; if $F_{cr}>1$ then the loading conditions are more severe. This expression is convenient since it allows the state of loading to be judged without any actual knowledge of what the deflection under service loads was actually assumed to be. It is thus a useful parameter for judging on site whether the slab is being overloaded to an extent where the service performance was likely to be impaired. The concept was that, provided at the time of striking, $F_{cr}<1$, then striking the formwork would not result in cracking or deflection of the slab greater than was implicit in the design and hence the service performance would not be impaired. The cracking factor, however, gives a means of judging the loading conditions at any time and not just at the time of striking the formwork.

The second part of the study, reported in [5], was concerned with monitoring the loadings on the slabs within the structure and the forces in props and backprops during construction. Since there are differences between the terminology used in different countries, it may help if the terms used here are defined.

Formwork describes the shuttering to a slab.

Falsework is the temporary structure necessary to support the formwork from the previously cast slab below.

Backprops are installed below the slab supporting the formwork and falsework to distribute some of the load applied to the lower slab by the falsework and the slab being cast. Backprops may be required at more than one level.

Reprops are installed to replace the formwork and falsework when this is stripped out and it is considered that the freshly stripped slab is incapable of supporting its self-weight and any construction loading.

Props are adjustable vertical supports which may be used as part of the falsework or as backprops or reprops.

At Cardington, reprops were never used as application of the Cracking Factor showed that the slabs could safely support their self-weight at times as short as 19 hours. This is believed to be generally the case. Backprops were installed prior to the erection of falsework on the fresh slab but were only tightened to finger tight so that the backpropping system was not subjected to significant loading from the dead weight of the fresh slab but only from additional loads applied to the fresh slab as the formwork was erected and the next slab cast.

The forces in the props and backprops were measured by means of load cells placed under the props supporting the formwork and the props forming the backpropping system. The load cells were installed and the readings taken by Building Research Establishment staff. The results were placed on a web page where they could be accessed by any of the research groups involved in the project.

The standard way of designing backpropping systems is to assume that:

(a) the props are rigid.
(b) that all slabs have the same stiffness.
(c) that the props supporting the formwork are directly over the backprops.

Application of these assumptions suggest that any loads applied by the falsework to the uppermost slab are shared equally by the uppermost slab and any slab supporting backprops. Thus, if one level of backprops is provided then the uppermost slab and the slab below supporting the backprops both carry a load equal to their dead weight, any construction loads applied directly to the slabs and half the loads arising from the concreting of the next slab above.

The study showed that the assumptions given above resulted in very misleading values for the loads carried by the slabs supporting the falsework while a slab was being concreted. In fact, the props are not infinitely stiff and the props supporting the falsework are not necessarily in line with the backprops. The slabs also have different stiffnesses due to their different ages, though this has a relatively minor effect. These differences, for a case with a single level of backprops meant that, instead of the backprops carrying half the load from the slab being cast, only about 25% of the load was transmitted to the lower floor. As a result, the uppermost slab supporting the falsework attracted substantially greater loads than calculated by the normal means. It was also found that the amount of load transmitted by a second level of backprops, if these were used, was less than 10%. More than one level of backprops thus achieved very little. Overall, the most critical loading condition for a slab occurred during casting of the next slab above. During this operation, the slabs at Cardington were generally stressed to a level where F_{cr} was close to 1.0. F_{cr} would have exceeded 1.0 in most cases if the full design construction loading had actually been present at the time that sets of readings were taken. More recent studies have shown that, for slabs with a greater dead load relative to the imposed load than that for which the Cardington structure was designed, it is more or less impossible not to overload the slabs during construction.

The studies reported in [4] show that the deflections during service depend on the highest value of F_{cr} to which the slab has been subjected during its life up to the time considered. This is illustrated in Figure 1, which shows the deflection at the centre of edge panels at an age of 500 days as a function of the maximum value of F_{cr} to which they have been subjected. All panels were carrying the same load at the time for which Figure 1 is drawn. In all cases, the maximum value of F_{cr} occurred during construction. Vollum et al [3], who also studied the deflections in the Cardington structure and has also carried out extensive parameter studies, have developed a more sophisticated approach to deflection calculation but have reached the same basic conclusions about the influence of loading during construction on the deflection in later life. They use a different parameter to F_{cr} to characterise the state of loading but the effect is very similar.

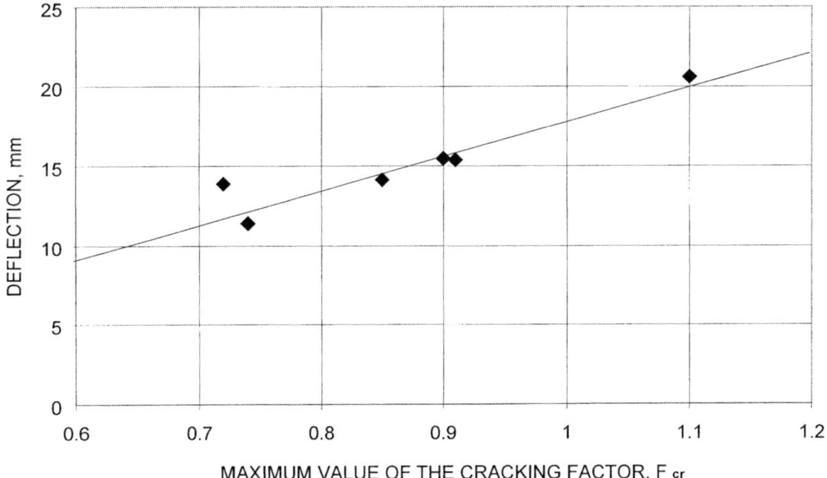

Figure 1 Measured deflections at the centre of edge panels at 260 days as a function of the maximum value of F_{cr} (from [4])

In summary, it can be seen that the construction process, which will in most cases be outside the control of the designer and outside his knowledge at the time he carries out the design, can have a very significant effect on the deflections in service.

ACCURACY OF DEFLECTION PREDICTION FORMULAE

The accuracy of deflection formulae depends to a great degree on their ability to predict the effects of tension stiffening. Tension stiffening is the contribution to the stiffness of a concrete member provided by the concrete in the tension zone after cracking has initiated. The tension stiffening effect is greatest just above the cracking load and tends to reduce as the load increases above this load. In members with high percentages of reinforcement, the tension stiffening effect tends to be relatively insignificant as the service load in such members tends to be well above the cracking load. This is not the case for slabs where the service loading is frequently close to the cracking load.

Figure 2 has been drawn to illustrate this. The curvatures under short term loads have been calculated firstly using the calculation method set out in Eurocode 2 (6), which allows for tension stiffening, and, secondly, on the assumption that the section is fully cracked and that the concrete in tension carries no stress. The service load moment has been calculated for each reinforcement percentage as 67% of the ultimate moment. It will be seen that the effect of the concrete in tension has a major influence on the calculated curvature at reinforcement percentages below about 0.5%.

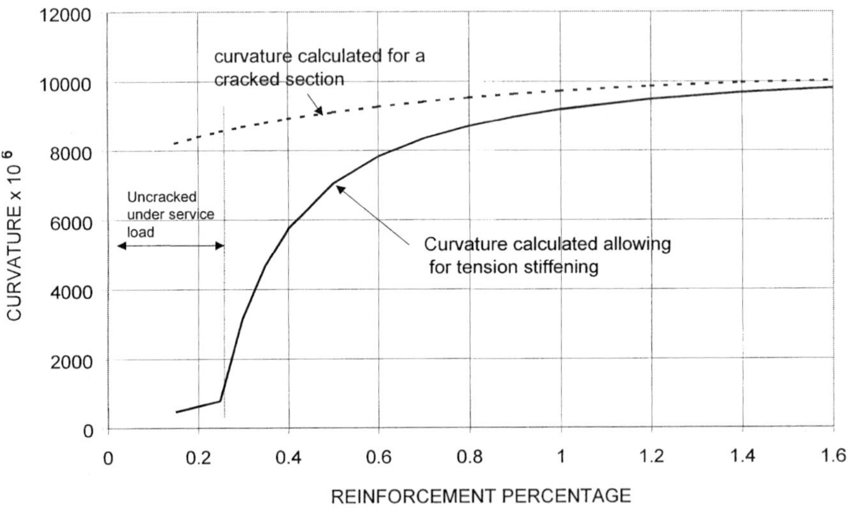

Figure 2 Comparison of curvatures calculated with and without tension stiffening

Figure 3 shows the curvatures calculated for a particular slab with 0.25% of tension reinforcement using three well established code methods: the methods given in Eurocode 2 [6], ACI 318 [7] and BS8110 [8]. The range of calculated deformations at the service load is very large, indicating that the constraints on the design of slabs imposed by different codes are very different. Which one is correct? All three methods give very similar results where the reinforcement percentage is in excess of 0.5%.

UNCERTAINTIES ABOUT THE TENSILE STRENGTH

Any study of the calculation of deflections will indicate the critical effect of the tensile strength assumed for situations where the service load is close to the cracking moment. In general, what is known about the concrete at the design stage is the specified characteristic compressive strength. This value is used to make an estimate of the tensile strength. Studies of the relationship between tensile and compressive strength show a high degree of scatter. Eurocode 2 gives values for an upper characteristic, a mean and a lower characteristic tensile
Figure 3. Deflections for a slab with 0.25% of tension reinforcement using various code provisions.

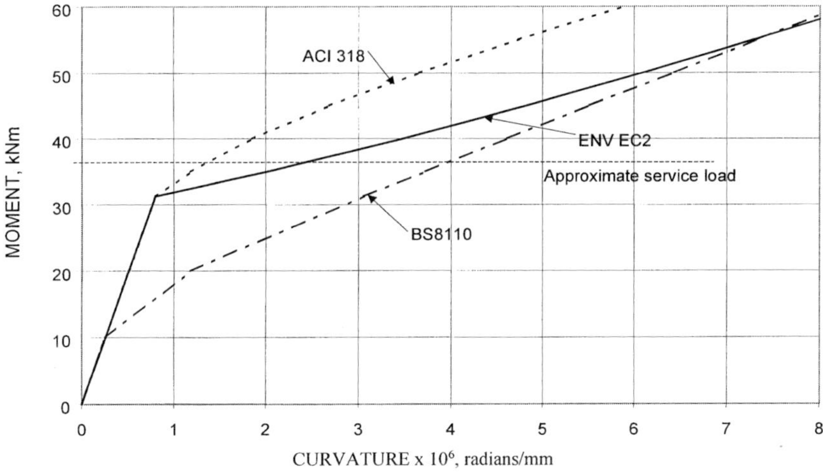

Figure 3 Deflections for a slab with 0.25% of tension reinforcement using various code provisions

strength for a given characteristic compressive strength. This range of values has been used to produce Figure 4 which aims to illustrate the variation in the calculated deflection which can result from this variation in tensile strength. The Eurocode 2 calculation method has been used to assess the deformations. It may be seen that, under the service load, the curvature calculated using the lower characteristic value of the tensile strength is more than 6 times the curvature calculated using the upper characteristic tensile strength. This variation does not include two other factors. Firstly, there is generally no limit on the maximum strength of concrete provided so it is perfectly possible for the contractor to use a much stronger concrete than that specified, possibly to assist with early striking of formwork. This means that the tensile strength may be substantially higher, and the curvature smaller than the minimum indicated in Figure 4. Secondly, few elements in structures are without some degree of restraint. For example, a slab cast monolithically with walls may be significantly restrained against shortening due to early thermal movements or shrinkage. This restraint will result in the development of tensile stresses which will result in cracking at a significantly lower load. This may result in deformations significantly greater than those corresponding to the lower characteristic concrete tensile strength.

CONCLUDING DISCUSSION

This paper has raised a number of issues relating to the calculation of the deformation of reinforced concrete slabs in design. It has been shown that the actual deflection will be significantly affected by the loading imposed during the construction and that these construction conditions may frequently be the most severe loading to which the structure is subjected. It has also been shown that the calculated deflection is highly dependent on the value assumed for the tensile strength of the concrete and that this is highly uncertain.

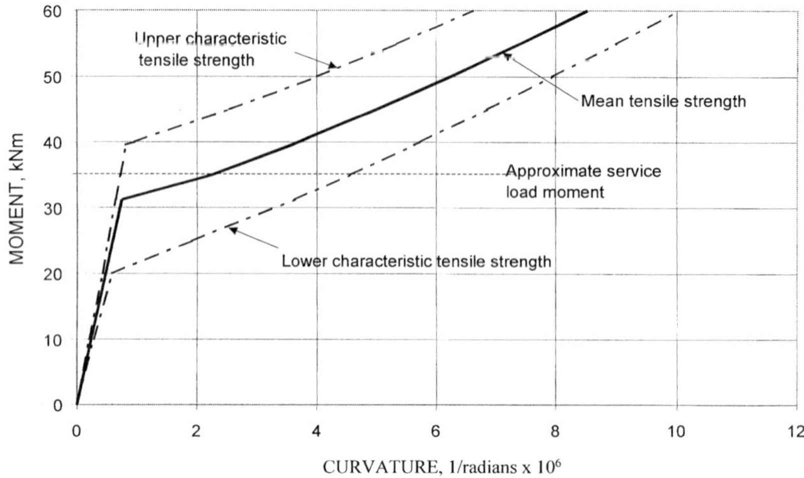

Figure 4 Effect on calculated deformations of a slab with 0.25% of tension reinforcement of variations in tensile strength

Finally, it has been pointed out that there can be large differences between the predictions of various codes. All these uncertainties are most critical in the design of slabs where the assessment of the deflection may be the critical factor defining the member design. More concerning is that the uncertainties in the tensile strength and loading applied during construction are possibly unavoidable at the design stage. Further research should permit more reliable formulae to be developed for the prediction of deflections where the conditions to which a slab has been subjected and the actual material properties in the slab are known but the issue of what should be predicted is a philosophical question. In design, the objective is to design a structure which can confidently be expected not to deflect more than the specified limits. If, as in research formulae, the prediction method aims to predict the most probable deflection, then 50% of structures would be expected to exceed the specified criteria. If, on the other hand, a method aims to predict something close to a worst case, then, in practice, deflections will be found to almost always be very much less than calculated. Since deflections can be measured, this tends to destroy confidence in the code and to result in pressure on the drafters to make the procedures less conservative. This is especially so if there are other codes which give smaller calculated deflections and designers complain that the code they are using puts them at a commercial disadvantage relative to designers using the other codes. This is currently the situation with BS8110. This can be seen from Figure 3 where the deflection calculated using the BS8110 method is about 4 times greater than that calculated using the ACI method but is BS8110 actually unnecessarily conservative or does it make rather more appropriate allowances for the possible uncertainties?

REFERENCES

1. BOYCE, W. H., Deflections of reinforced concrete slabs: case studies. International Congress on Challenges of Concrete Construction, Seminar 2: Concrete floors and slabs. Dundee Sept 2002.

2. GILBERT, R. I., Time-dependent deflection and cracking of reinforced concrete flat slabs. ACI Special Publication SP-203; Code provisions for deflection control in concrete structures. American Concrete Institute, Detroit, 2002. pp157 – 178.

3. VOLLUM, R. L., MOSS, R. M., HOSSAIN, T. R., Slab deflections in the Cardington in-situ concrete frame building. Magazine of Concrete Research, Volume 54 No. 1 February 2002. pp 23-34.

4. BEEBY, A. W., Criteria for the loading of slabs during construction. Proceedings of the Institution of Civil Engineers; Structures and Buildings 146 Issue 2, May 2001. pp 195 – 202.

5. BEEBY, A. W., The forces in backprops during construction of flat slab structures. Proceedings of the Institution of Civil Engineers; Structures and Buildings 146 Issue 3, August 2001. pp 307 – 317.

6. BRITISH STANDARDS INSTITUTION, DD ENV 1992-1-1: Eurocode 2: Design of concrete structures – Part 1: General rules and rules for buildings. BSI 1992 pp 254.

7. AMERICAN CONCRETE INSTITUTE. ACI318:2002: Building code requirements for structural concrete. American Concrete Institute, Detroit 2002.

8. BRITISH STANDARDS INSTITUTION. BS8110: Structural use of concrete Part 2: 1985: Code of practice for special circumstances. BSI, 1985 pp61.

CONGRESS CLOSING PAPER

CONCRETE: VADE MECUM

P C Hewlett

British Board of Agrément

United Kingdom

ABSTRACT. In putting together this Congress review from the many papers submitted, I have been looking for significant trends that can give direction to the way forward. Both the Seminars and Conferences have been taken into account but it has been written in advance. As a consequence the views expressed and conclusions drawn may well change as a result of the Congress itself and the exchanges that will occur during the event. There will be opportunity to cover such developments during the Closing Address Ceremony.

A number of ongoing challenges for concrete have been identified that suggest a way forward. The intent to change is serious but the consequences of not changing are even more so. Quo Vardis?

Keywords: Cement, Environment, Durability, Composite materials, Toughness, Pathology, Sustainability, Aesthetics, Waste, Deconstruction.

P C Hewlett is a chartered chemist and Chief Executive of the British Board of Agrément. He is visiting Industrial Professor in the Department of Civil Engineering at the University of Dundee and an active member of the Concrete Technology Unit. He is Chairman of the Technical Executive Committee of the UK Concrete Society and Chairman of the Editorial Board of the Magazine of Concrete Research.

Particular research interests cover durability, surface and bulk characteristics of concrete modified using chemical additions.

INTRODUCTION

A Congress is defined as a formal meeting of delegates with the purpose of discussing between those present issues such as special studies. In this regard, a Congress about concrete is both timely and needed and since concrete is a global material an International Congress seems very appropriate. This is now the 5th event held in Dundee on a triennial basis, the others being:

'Protection of Concrete' 1990
'Concrete 2000' 1993
'Concrete in the Service of Mankind' 1996
'Creating with Concrete' 1999

For those not involved with concrete it may seem strange after so many years of use, thousands of published papers and many books on the subject, all that could be known about concrete was known. So why does the debate and development continue?

Firstly, concrete is the most widely used construction material globally, it has become established in technically advanced countries and it has been estimated that some 1,200-2,400 kg per head of population are made per year [1]

Secondly, because of its adaptability it can be used for almost all construction situations, both structural and non-structural.

Thirdly, it is capable of being developed further in response to environmental concerns, energy considerations and new functional demands extending the material's performance in answer to engineering needs.

In other words, concrete and concreting are dynamic and will never reach a static position, unless of course our imaginations stagnate and we run out of ideas. For all of these reasons concrete is a material of opportunity and the debate will continue:

The theme of the present Congress is 'Challenges of Concrete Construction' embracing the weather, fire, seismic and marine situations and all that interlinks them. However, concrete faces other challenges from alternative materials such as steel, wood, glass, plastics and natural masonry. Alternatives do not readily have their justification in being straight material's replacements so much as finding their own niche in the design and functional requirements of buildings.

Other challenges covered at this Congress are,

1. Exploitation of the planet – Awakening of conscience in our own self interest
   – Environmental concerns/auditing/marking

2. Adoption of cleaner technologies

3. Sustainability – Design aspects
   – Brownfield development
   – Role of taxing and charging (landfill and aggregate tax)
   – Standardisation of components to assist reuse.

4. Elimination of waste
5. Whole life costing – life and death
6. Design for deconstruction
7. Alternatives to Portland cement
8. Wide/mandated use of cement replacement materials
9. High volume use of by-products
10. Energy conservation
11. Education for such involvements

Notwithstanding such challenges, concrete remains the most widely used construction material globally and that situation is likely to continue.

There is a constant probing for new developments reflecting drivers for change both direct and indirect and some suggested drivers are noted below.

- functional
- decorative
- competitiveness
- opportunity
- serviceability
- environmental issues
- safety
- fashion

These matters are dealt with in the Congress that comprises three Seminars and three Conferences.

CONGRESS REVIEW

Seminar 1 is concerned with 'Composite Materials used Internally and Externally', both organic and in-organic. These new materials offer better durability, lower weight and higher strength, ease of transportation, low thermal conductivity and reduced energy to make. For all of these reasons, adoption of composite materials technology should be welcomed but as with so many innovations in our industry, exploitation is guarded. It is clear that questions are being raised over slowness to adopt new ideas and innovations within concrete construction.

Conceptually fibre reinforced composites are not new and the ground rules for design and selecting them are established [2]. However, reinforced concrete is still conceptually large steel fibres (rebar) as reinforcement in an inorganic matrix rather than organic.

Proven performance based on unequivocal data are key to acceptance by specifiers and users alike to give confidence in the adoption of such new technologies.

Ironically, reinforced concrete itself was originally not backed up by a great deal of pre-use data as are fibre reinforced composites today and yet it was adopted with commitment. Why was this? Perhaps we are both set in our ways as well as worried about litigation. In this respect the composite beam example of Van Elp, Cattel and Heldt [3] is a good one that contradicts the apparent trend. The logical and prospective replacement of the established convention by optimum use of new materials resulting in a hybrid beam that outperforms the traditional reinforced concrete beam. The new types of beam have high load capacity, excellent fatigue resistance, outstanding durability, although detailed reaction to fire and comparable costs need to be addressed.

Are such radical developments welcomed and do we express the functional benefits well enough to persuade clients and specifiers and designers to adopt the new options? When considering new possibilities there is also the issue of appropriate technology. It is sometimes tacitly assumed that technologies rooted in advanced industrialised countries, may be used with equal effect in less developed countries where materials and skills may be different. Since concrete in one form or another is global, it should be possible to evolve appropriate applications for different locations. We talk readily of buildability but in the area of new ideas we should also consider adoptability.

There is merit in simplicity, both in concept, manufacture and use and a shift in Europe to prefabrication, giving better control over the process and a reduction in the need for established site skills.

Composite materials should respond in a ductile manner, eg be tough rather than simply strong. The issue of toughness is repeated throughout the Congress papers. The same trend applies to the development of lightweight structural materials, probably based on waste products, reflecting the emergence of conscience as well as functional and cost requirements. The phrase ‘priorities for change’ appears in Lowe’s keynote paper [4].

It is suggested that value should replace cost in selecting options. This is compatible with the concept of whole life costing, another strand repeated throughout the Congress.

Do regulations and standardisation stimulate or impede innovation? Regulation can stimulate by demanding new and more exacting performance levels. The means of showing compliance with the regulation, eg conforming to a standard or established technical specification may not assist change. Performance based not on prescriptive specifications are an answer but to implement such an approach requires a will to adopt and change.

Mention is made to high temperature and fire effects on organic binders and one paper by Ballaguru [5] introduced an inorganic adhesive for bonding carbon fibre sheets to concrete beams. The water-based adhesive is composed of an alumna-silicate and is stable up to 1000°C. Despite a very complicated application procedure it is indicative of what can be achieved.

One stimulant to development of fibre-reinforced plastics has been seismic performance. Typical fibres are carbon, glass and aramids maintaining cohesion beyond the point of failure.

The pursuit of lightweight and toughness is highly desirable and whilst carbon fibres have a role to play in achieving this objective it needs to be remembered that carbon can act as a noble metal in the galvanic series and whilst it will not corrode it may cause less noble metals relative to it to do so. The carbon itself is conductive and given the right combination of conditions may exacerbate the process.

Seminar 2 is concerned with 'Floors and Slabs' with an emphasis on flooring that seems to represent a typical case of 'we seem to know what to do but do not always do it'. Self imposed inadequacy!

Flooring is undoubtedly a substantial activity in the distribution depot and monolithic construction sectors that use some 1.5m^3 of concrete a year of concrete in the UK alone.

However, what in principle could be simpler than a slab? What could be simpler than a slab made of concrete? The physical principles of which are known – or are they? When you take regard of the conditions under which concrete for flooring is laid that assumption might be questioned.

There would appear to be too many homespun practices observes Harvey [7] but the problems with floors are global and commonplace. Notwithstanding all of this there would appear to be no substitute for concrete.

There is still much emphasis on Concrete Society Report TR34 [8]

Seidler [9] challenged that notwithstanding progress in concrete generally over the last 150-200 years, concrete flooring has not advanced significantly.

Incipient cheapness whilst a determining factor in this apparent lack of progress at some 27 Euros per square metre (approximately £17.50 per square metre) the equivalent of a medium quality fitted carpet! I cannot image the average fitted carpet functioning like an industrial floor. The requirements of industrial floors are clearly stated, namely,

- non shrink concrete
- monolithic construction
- non dusting
- adequate strength and surface resistance
- complex serviceability requirements

Chemical modifications can help overcome the known problems but the addition of polymers increases the cost substantially.

An industrial floor is a good example of a multi faceted performance requirement from something that is basically very simple. Why has this topic received so little attention? Value, longevity and robust performance rather than cost might be a better approach. In this sector cost seems to dominate.

Fast track construction leading to long strip and large bays are a trend to be acknowledged, together with joint-free slabs with speciality surface finishing and finishes. These are added demands to what superficially is a simple functional element.

Floor quality appears to have reached a plateau in terms of economically obtained quality. Again a plea for whole life costing.

The paper by Watanabe et al [10], suggests a form of flooring categorisation in the range 1-7, linked to the use to which the floor may be put and the development of a so-called U-scale as a measure of anti-static performance.

Seminar 3 deals with 'Repair, Rejuvenation and Enhancement of Concrete'.

Concrete repair and rehabilitation still dominate the concrete scene and yet concrete's failures are a relatively small proportion of concrete's use. Failure is also small, relative to new applications of concrete, eg, self compacting and ultra-high strength concrete and in relation to major projects such as the second Severn Crossing. However, in money and nuisance value terms, the profile of rehabilitation and repair is high and in that sense it attracts attention. It is the longer term inadequacies such as poor appearance, sulphate and chloride attack, sulphation and carbonation, that need to be addressed. Have these problems resulted from pursuing cost containment, lower cement contents, reduced cover rather than long term value?

The entire topic of concrete's pathology and degradation processes is worthy of our attention, how do we extend with confidence, the lifespan of concrete structures?

I refer to the paper of Tuutti [11] that quotes some telling figures relating capital values of buildings and structures to the value of all stocks as they apply to Sweden but may well reflect global trends in industrialised countries. Much of this stock has to be repaired and/or replaced on a 50 year cycle and that represents vast sums of money. If that is so, do we simply use concrete and design buildings, towns and cities as we have always done? Is such an approach sustainable?

Exploitation of new materials options that will have a longer service life and perhaps even be 'smart' will become the norm. Autogenous healing of cracked concrete is an example of a 'smart' material. It is difficult to predict where the initiative for change will come from. Will it be radical design, reflecting efficient function or will it be ad hoc picking up on personal preferences and available options? Will the drive be regulation or market opportunism? Did the development of self-compacting concrete start with an engineering need or more the availability of such a material's option, driven in turn by dispersant technology applied to cementitious suspensions rather than sound market research? In a capitalist economy the ultimate drive is financial well-being and planning and opportunism will live close together resulting in a somewhat volatile cocktail. So will concrete's future development be ad hoc and random or will sustainability, efficiency and environmental concerns dominate? Will such concerns only be responded to by wealthy economies with the less endowed creating their own appropriate technologies?

The existence of historical structures built from concrete is tangible evidence of the material's good latent durability. However, what remains is the best and a great deal has not lasted and for many reasons. Therefore the historical legacy has lessons to teach us and they are worthy of study.

In determining the effect of challenges to concrete, the diagnosis stage is very important. A comparison with forensic science is justified and the diagnostician has many techniques at his disposal. I am of course referring to the paper by Sims [12].

Threats, such as thaumasite sulphate attack (TSA), alkali-silica reaction (ASR), alkali-carbonate reaction (ACR) and delayed ettringite formation (DEF), all challenge concrete but we have to keep the potential problems in context. Notwithstanding this the avoidance of alkali-silica and alkali-carbonate reaction by selecting suitable aggregates remains a global issue.

Physico-chemical techniques have resulted in preventative and remedial measures such as cathodic protection, electro chemical realkalisation and chloride removal. All have a place and the last two may well have moved on from being something of a curiosity to full-scale practical application [13].

The themes developed in the Seminars enlarge and extend into the three Conferences. Conference 1 is concerned with 'Innovations and Developments in Concrete Materials and Construction' and Shah [14] sees the following targets for concrete in the 21st Century,

- more durable
- more constructable
- more predictable
- greener

I would add, more sustainable and more competitive.

By judicious use of investigative and monitoring techniques, all these objectives are attainable. However, since concrete is a global material, a case could be made for such developments to be globally supported with the results available to all. At the present time research cultures are very national and even regional, resulting in considerable duplication of effort on the one hand but also inventive variety on the other. The basic principle of such issues as rheology, hydration, corrosion and loading characteristics could be conducted in one or two locations whereas much effort is dispersed at numerous locations around the world, competing for funds and technical recognition. It is only when we draw people together in a Congress such as this that we identify the common features. It depends upon whether a global material such as concrete requires global development in a global economy with appropriate planning and prioritising or whether the more parochial approach is more beneficial if somewhat wasteful.

For instance, the use of electronic speckle pattern interferometry (ESPI) to study cracking in fibre reinforced concrete [14] compared with the K and F functions to describe fibre distributions resulting in stress intensity factors that govern crack propagation, assists in defining the capability limits of such reinforcements. Do such studies need to be duplicated? Is the study itself not definitive?

Shah et al's work[14] has shown that extrusion can improve fibre distribution and result in a stronger and tougher composite. Prefabricated components might well lend themselves to such techniques.

In the area of self compacting concrete, rheological studies assisted the optimisation of concrete mixes. A balance has to be struck between high flow and no segregation with yield values permitting deformability of the concrete to accommodate awkward shapes. The absolute rheological terms of yield value and viscosity do not relate directly to deformability, placeability and segregation that describe what the material has to do in practice. However, such an approach permits judgements and selection to be made.

Rossi [15] has developed ultra high strength performance fibre reinforced concrete (UHPFRC) to the point that concrete without conventional reinforcement might be a prospect. This concept is represented by the MSCC (multi scale concept concrete) that consists of short fibres (6mm) and long fibres (13mm) mixed together at 7% of the cement content.

Perhaps we have too many options e.g.

- MDF (macro defect free) concrete
- DSP (densified small particle concrete)
- CRC (compact reinforced composites) and its BPR variant (similar to CRC but with longer fibres)
- RPC (reactive powder concrete)
- MSCC (multi-scale cement composite)
- SIFCON (slurry infiltrated fibre concrete)
- ECC (engineered cementitous composite)

to mention but a few.

Having developed new materials, techniques of placing and finishing them have to be considered. These are not concretes as we know them. With approximately 1000 kg/m^3 of cement and cement:aggregate ratios of 4:1 and tensile strengths of 40-50 MPa being attainable!

Costs of these developments, at this stage, are not given but clearly it is cost effectiveness and value that matters – they are not like for like replacements of normal concretes. The importance of such extreme developments as this is that it shows what can be done and, like radical fashions, they set a pattern within which there is general advancement.

It is encouraging to note in a paper by Garshol and Constantiner [16] that at long last the concept of incorporating chemicals to control and modify the plastic and harden states of concretes and mortars has consolidated into normal practice. Indeed without certain admixtures and reactive additives these new concepts would remain only a wish.

Admixtures are now being tailored to the known chemistry of cements and the required physical and chemical characteristics of resulting mortars and concretes. Science is replacing intuitive flair and materials engineering replacing a 'try and see' approach. Integrating disciplines in this way will bring construction in line with other process engineered activities, eg, aircraft construction. Such trends are predicted in the Egan Report – 'Rethinking Construction' [17].

Having made and placed the concrete its maintained appearance does not receive the status it deserves. The visual impact of concrete is not popular with the community at large and Kronlof [18] is concerned with the aesthetics of concrete, as we all should be. The phrase 'concrete is beautiful when it makes the designer and user happy' can hardly be argued with.

Concrete's form and appearance should be predictable – no unpleasant surprises. Natural ageing should be taken into account and neatness should be a prime aim for concrete development but again a low cost outlook does not aid aesthetic development.

General expectations of society and designers may differ from those of the industry and users. Kukko [19] considers that designers and society may be concerned with image and public benefit, but industry is more concerned with economic and technical profits. Is this division true? If it is how can it ever be reconciled?

Production of high quality surfaces, edges and joints is a priority with high quality materials being required, resulting in ready for finishing details, eg, painting and wallpapering.

There are opportunities for higher strengths and toughness – thinner and slender structures, shells, lattices, profiled beams and columns, all are attainable. Kukko identifies some aims, namely,

- environmental friendliness
- improving the quality of life
- competitiveness
- improved employment prospects
- improved working conditions, in particular, safety

Concrete is 'a material for all reasons' and, if its permanency could be assured, we might invest more in some of these value added options.

Conference 2 deals with 'Sustainable Concrete Construction' and Nixon's paper [20] is both forthright, to the point and very relevant. He makes some telling criticisms of man's exploitation of the planet. He contends we should be concerned with,

- adaptable buildings
- minimum waste
- design for deconstruction
- low energy cements
- reduced energy in use by using concrete intelligently

In attaining these aims concrete is the premier construction material.

Firstly, some disturbing facts.

In the UK about 25% of the energy used in industry is accounted for by the manufacture and transportation of building materials.

- It is estimated that 8% of global CO_2 emissions result from concrete production.
- One ton of CO_2 is produced per ton of Portland cement.
- Cement production is growing, particularly in developing countries and what we gain by going in one direction to save the environment may be offset by trends in the opposite direction.
- In summary we need an alternative to Portland cement – such options are coming into play already.

Cements that require reduced energy for production (less by 16%) and in turn produce less carbon dioxide (less by 10%) are seemingly attainable. Cements based upon belite with properties comparable to Portland cements and with some evidence to indicate that durability of resulting concretes might even be improved have been produced on a commercial scale in China. There is also the recent TecEco development based on magnesium oxide.[21] What is the future for these alternatives?

Nixon also makes the point that we should use concrete innovatively, taking regard of its high thermal mass resulting in substantial economies in the running of buildings.

Jensen and Glavind [22] remind us that to make $1m^3$ of office space requires something in the region of 500 MJ whereas for the same office space it requires 15,000 MJ to heat and light. How can concrete assist in the use and running of buildings is an issue that needs to be addressed.

Jensen and Glavind continue a similar theme noting that 2-6% of worldwide CO_2 stems from cement production and cement manufacture is increasing at 5% per year, equivalent to an increase of 10 million tons of CO_2 per year.

Perhaps we will only take the environmental issues seriously when failure to achieve set aims is legislated for. The Eco Management and Audit Scheme (EMAS) coupled with a statutory instrument against which a company can be registered might be a way forward. Various tools are available to engendering an environmental culture but what creates the will to do so? The principles of life cycle assessment (LCA) and life cycle inventory (LCI) can have aims and set targets such as,

- CO_2 reduction by at least 30%
- 20% of all concrete should use residual products as aggregates
- use concrete industry's own residual products
- CO_2 neutral waste derived fuel being used at a rate of at least 10% of all fuel used in cement production

In the USA some 5 billion tons of non hazardous by-product materials are produced annually (NAIK [23]). Major inputs from agriculture (2.1 billion tons) and mineral sources (1.8 billion tons).The UK construction industry produces 20% of all UK waste [24].

The use of fly ash and bottom ash and clean coal ash in cement production with new energy generating technologies yielding different coal derived ashes assists the quest for energy containment. There are many new end uses for waste materials, for instance, sewage sludge for lightweight aggregates and for making clay bricks.

The task of carrying the environmental banner often falls to the lot of the manufacturer but contractors have a role to play as well (Goring [25]). Greater integration and co-operation emphasising quality and safety and less so cost, as has been the habit to date. These principles are set out in Egan's 'Rethinking Construction' [25]. How do these attitudes impinge upon concreting activities? Firstly, starting with design concepts – concrete can play a role by way of its thermal mass in providing better air quality and natural ventilation. To effect radical change we need an integrated approach involving concrete design and function and increasing the overlap between environmental concerns and how we build.

Torring and Lauritzen [26] estimate a potential of 400 million tons of reusable concrete, stone and brick from industrialised countries. We have to consider the means of deconstruction and reclaiming the materials used. – Joined up construction underwritten by a joined up sense of public conscience.

Pocklington and Glass [24] believe that the energy performance of buildings are a key to sustainability in which case there is good reason for using concrete. Phrases such as 'burn and bury', 'dilute and disperse' and 'end of pipe' are no longer acceptable. The term anthropogenic was used – greenhouse gas emissions caused by man. We need some form of fiscal encouragement to create a culture of change. We also need to plan for longer life and adaptability of buildings.

Further telling statistics supplied by Glavind and Munch-Petersen [27]. Some 5 km^3 of concrete are used per year globally and whilst some would contend that CO_2 produced per ton of cement is small in itself, it becomes large due to the amount of concrete produced. The prospect of quantified benchmarking of attainable objectives for CO_2 reduction say by 30% and recycled concrete used as aggregate, making an energy reduction of 20% is tantalising. The authors also consider that waste derived fuels should replace 10% of fossil fuels and their paper results in a specification for 'green' concrete types – some 14 in number. The authors were very conscious of solving one problem but creating a second order unwelcome legacy, eg, kiln dust containing zinc, vanadium, lead and copper as well as phosphorus pentoxide. We have to maintain a sense of proportion and perspective. The various phases of materials production and construction activity cannot be dealt with in isolation, one from the other.

Conference 3 deals with 'Concrete for Extreme Conditions' and the paper by De Vries [28] endeavours to put the problems of durability into perspective. On balance the performance of concrete is not as bad as many would contend. However, there are problems of poor workmanship and with new materials. Codes and specifications are not particular enough and matters of maintenance and repair not covered sufficiently well. De Vries would like to see performance and reliability based service life design and makes reference to the European Brite/Euram research project – 'Duracrete'. A plea is made to involve the client and give contractors a number of options.

Durability design should get as much attention as structural design. The Eastern Schedlt barrier has a service life-span of some 200 years! However, a period of 85 years was settled for the concrete when it was accepted that the cover will have to be replaced. An example of integrating maintenance with design life. We have to quantify the anticipated functional life-span. To do this a knowledge of the durability of materials is required and the effect of workmanship on achieving these properties needs to be addressed.

De Vries is also concerned with the interaction of structure and the environment and uses a probabilistic technique to determine the likelihood of failure and target service life. There are problems of defining the limit state requirements eg the onset of corrosion. Models exist for defining degradation conditions and the design can be modified to offset adverse predictions of service life based on such models – a preemptive approach.

Slater [29] concentrates on marine structures and these represent a severe exposure condition but relates the data so that it is relevant to all types of construction. The emphasis is on buildability and durability. Buildability covering such issues as safety and economy and durability fixed by design and exposure (environment). Reassuring to see a preference for low w:c ratios and a useful quoted rule of thumb 'a reduction in w:c ratio of .05 is the equivalent of an increase in a cover of 5mm'. Slater concludes with an excellent series of pragmatic recommendations.

Over recent times there has been much discussion on the existence of threshold chloride levels below which passivation is maintained and above which active corrosion commences. The paper by Paramasivam et al [30] is a good example of data obtained on an actual structure, in this case a 32 year old wharf from which 6 marine piles (driven) were recovered and analysed for chloride content and penetration. Comparisons were made of the actual with predicted ingress levels of chloride and the relationship between such levels and loss in mass of reinforcing bars. Reference is made to threshold chloride levels (Reference 2 of [30])again and the establishment of critical chloride levels in the range .03 - .1% weight/weight concrete, covering in the first instance the splash zone and the latter submerged (Reference 6 [30]). Alternatively, Browne (Reference 7 of [3]) again showed such values to be in the range .2 - .49% w/w cement. Therefore values of .034 - .068% w/w concrete were considered appropriate.

These figures are to be doubted where a plentiful supply of oxygen is available. There was broad agreement between predicted and actual values confirming the various models used. As an extension to the problems caused by chloride ingress and contamination, Masuda [31] considered salt damage to reinforced concrete buildings resulting from both seawater and airborne salt. Some 4,363 buildings were investigated, covering everything from domestic to industrial, schools, offices, hospitals, etc resulting in 60% or so being perfectly satisfactory in the age range 7-46 years old. The distance from the coast was a significant factor in determining the residual chloride levels. Insufficient cover was the cause of deterioration in most cases but the examination was primarily visual. The threshold levels in this study were broadly corroborated.

CONCLUSIONS

This Congress has brought together a great deal of data and experience that within itself has trends that indicate the way ahead and help to establish attitudes and create priorities. Some noted trends are given below.

1. Concrete is capable of considerable further performance-based development and should not posture as a low technology stereotype .

2. Sustainability will remain a motivator for regulators, designers and concrete material providers. Is there a sustainable alternative to Portland cement?

3. Adopted technology should be in proportion to prevailing local conditions.

4. Creating a concrete culture at operative level with recognition of skill status will help to exploit new developments and make the aim of best practice a reality.

5. Laboratory-based data must reconcile with what happens in practice — transfer of micro mechanisms to macro fact. Methods of diagnosis should be accurate and unambiguous, performed by those qualified to do so and interpretation should be subject to severe scrutiny.

6. The role of water needs more committed study. It is necessary for cement to transform into masonry but is also responsible for much of concrete's degradation.

7. Development of tough concrete rather than high strength but brittle concretes.

8. The visual appearance of concrete with time is of concern. Concrete should remain pristine and not take on a patina of industrial downgrading.

9. Coating and sealing may not be sufficient. We have to understand how concrete behaves to fluctuations in its surroundings at a micro mechanistic level.

10. Does a true threshold chloride level exist before steel corrosion occurs?

11. Can we consider structures that do not contain normal reinforcement but rely solely on metal fibres and a reconstituted matrix?

12. The use of recycled and waste materials should be encouraged using legislation and tax incentives for those that comply – a stick and carrot approach.

ACKNOWLEDGEMENTS

I am indebted to all the authors of those Congress papers to which I have made reference and indeed those that I have not, I also thank colleagues with whom I have endless discussions about concrete, without which opinions and viewpoints could not be formed and conclusions drawn.

REFERENCES

With the exception of References 1, 8, 17 and 21, all are drawn from papers given at the Dundee Congress 'Challenges of Concrete Construction' 5-11 September 2002. To simplify the listing only the relevant Seminar or Conference is identified.

1. GLASSER F. P., Private Communication

2. All Authors, Composite Materials in Concrete Construction, Seminar 1.

3. VAN ERP G., CATTEL C., HELDT T., Fibre Composites in Civil Engineering: An Opportunity for a Novel Approach to Traditional Reinforced Concrete Concepts. Proceedings of International Congress: Challenges of Concrete Construction, Seminar 1 – Composite Materials in Concrete Construction, Dundee, Scotland, 5-6 September, 2002, pp 1-16

4. LOWE P., Composite Materials in Concrete Construction, Proceedings of International Congress: Challenges of Concrete Construction, Seminar 1 – Composite Materials in Concrete Construction, Dundee, Scotland, 5-6 September, 2002, pp 17-30

5. BALAGURU P.N., Inorganic Polymer Composites in Concrete Construction: Properties, Opportunities and Challenges, Proceedings of International Congress: Challenges of Concrete Construction, Seminar 1 – Composite Materials in Concrete Construction, Dundee, Scotland, 5-6 September, 2002, pp 109-126

6. VAN GEMERT D., BROSENS K., Non-Metallic Reinforcements for Concrete Construction, Proceedings of International Congress: Challenges of Concrete Construction, Seminar 1 – Composite Materials in Concrete Construction, Dundee, Scotland, 5-6 September, 2002, pp 225-236

7. HARVEY D.J.J., Developing a Greater Understanding of the Nature and Usability of Concrete in Industrial Floor Applications, Proceedings of International Congress: Challenges of Concrete Construction, Seminar 2 – Concrete Floors and Slabs, Dundee, Scotland, 5-6 September, 2002, pp 183-194

8. CONCRETE SOCIETY, Concrete Industrial Ground Floors – A Guide to their Design and Construction, Technical Report No 34, 1994, pp 170.

9. SEIDLER P., How Polymers Improve Floors – Possibilities Today and Prospects for the Future, Proceedings of International Congress: Challenges of Concrete Construction, Seminar 2 – Concrete Floors and Slabs, Dundee, Scotland, 5-6 September, 2002, pp 1-14

10. WATANABE H., ONO H., KAIZU H., Performance Evaluation of Floors for Static Charge on Human Body – Proposal of a New Method, Proceedings of International Congress: Challenges of Concrete Construction, Seminar 2 –Concrete Floors and Slabs, Dundee, Scotland, 5-6 September, 2002, pp 233-244

11. TUUTTI K., Repair, Rejuvenation and Enhancement of Concrete – A Fast Growing Market, Proceedings of International Congress: Challenges of Concrete Construction, Seminar 3 – Repair, Rejuvenation and Enhancement of Concrete, Dundee, Scotland, 5-6 September, 2002, pp 1-10

12. SIMS I., Diagnosing and Avoiding the Causes of Concrete Degradation, Proceedings of International Congress: Challenges of Concrete Construction, Seminar 3 – Repair, Rejuvenation and Enhancement of Concrete, Dundee, Scotland, 5-6 September, 2002, pp 11-24

13. VENNESLAND O., Documentation of Electrochemical Maintenance Methods, Proceedings of International Congress: Challenges of Concrete Construction, Seminar 3 – Repair, Rejuvenation and Enhancement of Concrete, Dundee, Scotland, 5-6 September, 2002, pp 191-198

14. SHAH S.P., AKKAYA Y., BUI V.K., Innovations in Microstructure, Processing and Properties, Proceedings of International Congress: Challenges of Concrete Construction, Conference 1 - Innovations and Developments in Concrete Materials and Construction, Dundee, Scotland, 9-11 September, 2002, pp 1-16

15. ROSSI P., Developments of New Cement Composite Materials for Construction, Proceedings of International Congress: Challenges of Concrete Construction, Conference 1 - Innovations and Developments in Concrete Materials and Construction, Dundee, Scotland, 9-11 September, 2002, pp 17-30

16. GARSHOL K.F., CONSTANTINER D., Super-Concrete Examples: Complete Rheology Control and Passive Fire Protection, Proceedings of International Congress: Challenges of Concrete Construction, Conference 1 - Innovations and Developments in Concrete Materials and Construction, Dundee, Scotland, 9-11 September, 2002, pp 411-422

17. EGAN J., Rethinking Construction, July 1998, pp 39.

18. KRONLÖF A., Concrete Aesthetics: Flexible or Stiff – Humble or Arrogant, Proceedings of International Congress: Challenges of Concrete Construction, Conference 1 - Innovations and Developments in Concrete Materials and Construction, Dundee, Scotland, 9-11 September, 2002, pp 751-762

19. KUKKO H., Requirements for Advanced Concrete Materials, Proceedings of International Congress: Challenges of Concrete Construction, Conference 1 - Innovations and Developments in Concrete Materials and Construction, Dundee, Scotland, 9-11 September, 2002, pp 949-956

20. NIXON P.J., More Sustainable Construction: The Role of Concrete, Proceedings of International Congress: Challenges of Concrete Construction, Conference 2 – Sustainable Concrete Construction, Dundee, Scotland, 9-11 September, 2002, pp 1-12

21. GLASSER F.P., TecEco: Cements Based on Magnesium Oxide, A Private Communication, 2001, pp 9.

22. JENSEN B.L., GLAVIND M., Consider the Environment – Why and How, Proceedings of International Congress: Challenges of Concrete Construction, Conference 2 – Sustainable Concrete Construction, Dundee, Scotland, 9-11 September, 2002, pp 13-22

23. NAIK T.R., The Role of Combustion By-Products in Sustainable Construction Materials, Proceedings of International Congress: Challenges of Concrete Construction, Conference 2 – Sustainable Concrete Construction, Dundee, Scotland, 9-11 September, 2002, pp 117-130

24. POCKLINGTON D., GLASS J., Economics, Sustainability and Concrete, Proceedings of International Congress: Challenges of Concrete Construction, Conference 2 – Sustainable Concrete Construction, Dundee, Scotland, 9-11 September, 2002, pp 683-694

25. GORING P.G., Rethinking Sustainable Concrete Construction, Proceedings of International Congress: Challenges of Concrete Construction, Conference 2 – Sustainable Concrete Construction, Dundee, Scotland, 9-11 September, 2002, pp 439-456

26. TORRING M., LAURITZEN E., Total Recycling Opportunities – Tasting the Topics for the Conference Session, Proceedings of International Congress: Challenges of Concrete Construction, Conference 2 – Sustainable Concrete Construction, Dundee, Scotland, 9-11 September, 2002, pp 501-510

27. GLAVIND M., MUNCH-PETERSEN C., Green Concrete – A Life Cycle Approach, Proceedings of International Congress: Challenges of Concrete Construction, Conference 2 – Sustainable Concrete Construction, Dundee, Scotland, 9-11 September, 2002, pp 771-786

28. DE VRIES H., Durability of Concrete : A Major Concern to Owners of Reinforced Concrete Structures, Proceedings of International Congress: Challenges of Concrete Construction, Conference 3 – Concrete for Extreme Conditions, Dundee, Scotland, 9-11 September, 2002, pp 1-16

29. SLATER D., Marine and Underwater Concrete – Buildability and Durability, Proceedings of International Congress: Challenges of Concrete Construction, Conference 3 – Concrete for Extreme Conditions, Dundee, Scotland, 9-11 September, 2002, pp 189-204

30. PARAMASIVAM P., LIM C.T.E., ONG K.C.G., Performance of Reinforced Concrete Piles Exposed to Marine Environment, Proceedings of International Congress: Challenges of Concrete Construction, Conference 3 – Concrete for Extreme Conditions, Dundee, Scotland, 9-11 September, 2002, pp 525-536

31. MASUDA M.Y., Condition Survey of Salt Damage to Reinforced Concrete Buildings in Japan. Proceedings of International Congress: Challenges of Concrete Construction, Conference 3 – Concrete for Extreme Conditions, Dundee, Scotland, 9-11 September, 2002, pp 823-836

INDEX OF AUTHORS

SUBJECT INDEX

This index has been compiled from the keywords assigned to the papers, edited and extended as appropriate. The page references are to the first page of the relevant paper.